Lecture Notes in Computer Science

Commenced Publication in 1973
Founding and Former Series Editors:
Gerhard Goos, Juris Hartmanis, and Jan van Leeuwen

Germán Vidal (Ed.)

Logic-Based Program Synthesis and Transformation

21st International Symposium, LOPSTR 2011
Odense, Denmark, July 18-20, 2011
Revised Selected Papers

 Springer

Volume Editor

Germán Vidal
Universitat Politècnica de València
MiST, DSIC
Camino de Vera, S/N
46022 València, Spain
E-mail: gvidal@dsic.upv.es

ISSN 0302-9743   e-ISSN 1611-3349
ISBN 978-3-642-32210-5   e-ISBN 978-3-642-32211-2
DOI 10.1007/978-3-642-32211-2
Springer Heidelberg Dordrecht London New York

Library of Congress Control Number: 2012943721

CR Subject Classification (1998): D.1.6, D.2.4-5, F.4.1, G.2.2, F.3,
I.2.2-4, D.3.1, F.1.1

LNCS Sublibrary: SL 1 – Theoretical Computer Science and General Issues

Typesetting: Camera-ready by author, data conversion by Scientific Publishing Services, Chennai, India

Printed on acid-free paper

Springer is part of Springer Science+Business Media (www.springer.com)

Preface

This volume contains a selection of the papers presented at the 21st International Symposium on Logic-Based Program Synthesis and Transformation (LOPSTR 2011) held July 18–20, 2011, in Odense (Denmark). Information about the symposium can be found at: http://users.dsic.upv.es/~lopstr11/. Previous LOPSTR symposia were held in Hagenberg (2010), Coimbra (2009), Valencia (2008), Lyngby (2007), Venice (2006 and 1999), London (2005 and 2000), Verona (2004), Uppsala (2003), Madrid (2002), Paphos (2001), Manchester (1998, 1992, and 1991), Leuven (1997), Stockholm (1996), Arnhem (1995), Pisa (1994), and Louvain-la-Neuve (1993).

The aim of the LOPSTR series is to stimulate and promote international research and collaboration in logic-based program development. LOPSTR traditionally solicits contributions, in any language paradigm, in the areas of specification, synthesis, verification, analysis, optimization, specialization, security, certification, applications and tools, program/model manipulation, and transformational techniques. LOPSTR has a reputation for being a lively, friendly forum for presenting and discussing work in progress. Formal proceedings are produced only after the symposium so that authors can incorporate this feedback in the published papers.

In response to the call for papers, 28 contributions were submitted from 13 different countries. The Program Committee accepted six full papers for immediate inclusion in the formal proceedings, and eight more papers were accepted after a revision and another round of reviewing. Each paper was reviewed by at least three Program Committee members or external referees. In addition to the contributed papers, the program also included invited talks by John Gallagher (Roskilde University, Denmark), Fritz Henglein (DIKU, University of Copenhagen, Denmark) and Vitaly Lagoon (Cadence Design Systems, Boston, USA), two of them shared with PPDP 2011.

I am very grateful to all the authors of submitted papers and, especially, to the invited speakers, for their contributions to LOPSTR 2011. I want to thank the Program Committee members, who worked diligently to produce high-quality reviews for the submitted papers, as well as all the external reviewers involved in the paper selection. I would also like to thank Andrei Voronkov for his excellent EasyChair system that automates many of the tasks involved in chairing a conference.

LOPSTR 2011 was co-located with PPDP 2011, AAIP 2011 and WFLP 2011. Many thanks to the local organizers of these events, in particular to Peter Schneider-Kamp, the LOPSTR 2011 Symposium Chair. Finally, I gratefully acknowledge the institutions that sponsored this event: University of Southern Denmark and Danish Agency for Science, Technology and Innovation.

February 2012   German Vidal

Organization

Program Chair

Germán Vidal   Universitat Politècnica de València, Spain

Symposium Chair

Peter Schneider-Kamp   University of Southern Denmark, Denmark

Program Committee

Elvira Albert	Complutense University of Madrid, Spain
Malgorzata Biernacka	Institute of Computer Science, University of Wroclaw, Poland
Manuel Carro	Technical University of Madrid (UPM), Spain
Michael Codish	Ben-Gurion University, Israel
Danny De Schreye	Katholieke Universiteit Leuven, Belgium
Maribel Fernández	King's College London, UK
Raúl Gutiérrez	University of Illinois at Urbana-Champaign, USA
Mark Harman	University College London, UK
Frank Huch	Christian Albrechts University of Kiel, Germany
Michael Leuschel	University of Düsseldorf, Germany
Yanhong Annie Liu	State University of New York at Stony Brook, USA
Kazutaka Matsuda	Tohoku University, Japan
Fred Mesnard	Université de La Réunion, France
Ulrich Neumerkel	Technische Universität Wien, Austria
Alberto Pettorossi	DISP, University of Roma Tor Vergata, Italy
Carla Piazza	University of Udine, Italy
Peter Schneider-Kamp	University of Southern Denmark, Denmark
Hirohisa Seki	Nagoya Institute of Technology, Japan
Josep Silva	Universitat Politècnica de València, Spain
Germán Vidal	Universitat Politècnica de València, Spain
Jurgen Vinju	Centrum Wiskunde & Informatica, The Netherlands
Jianjun Zhao	Shanghai Jiao Tong University, China

Additional Reviewers

Arenas, Puri
Bolz, Carl Friedrich
Brandvein, Jonathan
Casagrande, Alberto
Cervesato, Iliano
Chang, Xi
Dal Palú, Alessandro
Dovier, Agostino
Escobar, Santiago
Fioravanti, Fabio
Fontaine, Marc
Formisano, Andrea
Genaim, Samir
Herranz, Ángel
Ivanovic, Dragan
Karna, Anil
Lagoon, Vitaly
Lakin, Matthew
Lawall, Julia

Lei, Dacheng
Lin, Bo
Montenegro, Manuel
Morales, José F.
Moreno, Ginés
Namet, Olivier
Nishida, Naoki
Nishimura, Susumu
Pilozzi, Paolo
Proietti, Maurizio
Senni, Valerio
Sneyers, Jon
Stroeder, Thomas
Tamarit, Salvador
Voets, Dean
Wielemaker, Jan
Yoo, Shin
Zaytsev, Vadim

Sponsors

University of Southern Denmark

Danish Agency for Science, Technology and Innovation

Table of Contents

Analysis of Logic Programs
Using Regular Tree Languages
(Extended Abstract)

John P. Gallagher*

Roskilde University, Denmark
IMDEA Software Institute, Madrid, Spain
jpg@ruc.dk

Abstract. The field of *finite tree automata* provides fundamental notations and tools for reasoning about sets of terms called regular or recognizable tree languages. We consider two kinds of analysis using regular tree languages, applied to logic programs. The first approach is to try to *discover* automatically a tree automaton from a logic program, approximating its minimal Herbrand model. In this case the input for the analysis is a program, and the output is a tree automaton. The second approach is to *expose or check properties* of the program that can be expressed by a given tree automaton. The input to the analysis is a program and a tree automaton, and the output is an abstract model of the program. These two contrasting abstract interpretations can be used in a wide range of analysis and verification problems.

Finite Tree Automata

A tree language is a set of trees, commonly represented as *terms*. Terms are ubiquitous in computing, representing entities as diverse as data structures, computation trees, and computation states. Here we consider only finite terms. Informally, a finite tree automaton (FTA) is a mathematical machine for recognising terms. FTAs provide a means of finitely specifying possibly infinite sets of ground terms, just as finite state automata specify sets of strings. Detailed definitions, algorithms and complexity results can be found in the literature [1]. An FTA includes a grammar for trees given by transition rules of the form $f(q_1, \ldots, q_n) \to q$, where f is a function symbol from a given signature Σ and $q, q_1, \ldots, q_n$ are states of the automaton. This rules states that a term $f(t_1, \ldots, t_n)$ is accepted at state q of the FTA if $t_1, \ldots, t_n$ are accepted at states $q_1, \ldots, q_n$ respectively. If q is a *final* state, then the term $f(t_1, \ldots, t_n)$ is recognised by the FTA.

A notational variant of FTAs is given by *alternating tree automata*, though the class of recognisable terms is the same as for FTAs. For the purposes of static analysis, a subset of alternating tree automata we call conjunctive FTA (CFTA) is useful. In a CFTA, transition rules have the form $f(C_1, \ldots, C_n) \to q$, which is like an FTA transition except that $C_1, \ldots, C_n$ are nonempty sets of

* Supported by Danish Research Council grants FNU 272-06-0574, FNU 10-084290.

G. Vidal (Ed.): LOPSTR 2011, LNCS 7225, pp. 1–3, 2012.

automata states. Such a rule states that a term $f(t_1, \ldots, t_n)$ is accepted at state q of the automaton if each t_i is accepted at all the states in C_i. A *bottom-up deterministic* finite tree automaton (DFTA) is one in which the set of transitions contains no two transitions with the same left-hand-side. For every FTA A there exists a DFTA A' that recognises the same set of terms.

An FTA is called *complete* iff for all n-ary functions f and states $q_1, \ldots, q_n$, there exists a state q and a transition $f(q_1, \ldots, q_n) \to q$. This implies that every term is accepted by some (not necessarily final) state of a complete FTA.

Deriving a CFTA from a Logic Program

The concrete semantics of a definite logic program P is defined by the usual T_P (immediate consequence) operator. We write it as follows, with subsidiary functions project, reduce, ground. Let I be a subset of the Herbrand base of the program.

$$T_P(I) = \{\mathsf{project}(H, \theta\phi)) \mid H \leftarrow B \in P, \theta \in \mathsf{reduce}(B, I),$$
$$\phi \in \mathsf{ground}(\mathsf{vars}(H) \setminus \mathsf{vars}(B))\}$$

The subsidiary functions have obvious meanings in the concrete semantics, yielding the familiar function such that $\mathsf{lfp}(T_P)$ is the least Herbrand model of P. This is the limit of the Kleene sequence $\emptyset, T_P(\emptyset), T_P^2(\emptyset), \ldots$.

We define a finite abstract domain of CFTAs F_P for a given program P, based on the symbols appearing in P, in the style described by Cousot and Cousot [3]. Consider the set of occurrences of non-variable subterms of the heads of clauses in P, including the heads themselves; call this set $\mathsf{headterms}(P)$. The aim is to analyse the instances of the elements of $\mathsf{headterms}(P)$. In particular, the set of instances of the clause heads in successful derivations is the least Herbrand model of P. Let P be a program, with function and predicate symbols Σ. Let Q be a finite set of identifiers labelling the elements of $\mathsf{headterms}(P)$ including a distinguished identifier τ. The base set of transitions Δ_P is the set of CFTA transitions that can be formed from Σ and states Q. The domain F_P is defined as $\mathsf{F}_P = \{\langle Q, \{\tau\}, \Sigma, \Delta \rangle \mid \Delta \subseteq \Delta_P\}$. In other words F_P is the set of CFTAs whose states are identifiers from the finite set Q, have a single accepting state τ and some set of transitions that is a subset of the base transitions of P. Q and Σ are finite, and hence F_P is finite.

Let γ be a concretisation function, where $\gamma(A)$ is the set of terms accepted by CFTA A. The abstract semantic function is a function $T_P^{\alpha_1} : \mathsf{F}_P \to \mathsf{F}_P$ and the safety condition is that $T \circ \gamma \subseteq \gamma \circ T_P^{\alpha_1}$, which is sufficient, together with the monotonicity of T_P, to prove that $\mathsf{lfp}(T_P) \subseteq \gamma(\mathsf{lfp}(T_P^{\alpha_1}))$ [2]. In short, $T_P^{\alpha_1}(A)$ returns a CFTA by solving each clause body wrt A resulting in a substitution of states of A for the body variables. This substitution is projected to the corresponding clause head variables, generating transitions for the returned CFTA.

The sequence $\emptyset, T_P^{\alpha_1}(\emptyset), T_P^{\alpha_1 2}(\emptyset), T_P^{\alpha_1 3}(\emptyset), \ldots$ is monotonically increasing wrt $\subseteq$. Convergence can be tested simply by checking the subset relation, which is a vital point for efficient implementations (since the language containment check would be very costly). Note that this abstract interpretation is not a Galois connection, in contrast to the second family of analyses discussed next.

Analysis of a Program with Respect to a Given DFTA

A complete DFTA induces a finite partition of the set of accepted terms, since every term is accepted at exactly one state. Given a complete DFTA R, let $\beta_R(t)$ be the partition (denoted by its DFTA state) to which term t belongs. The Herbrand base B_P for a program P is abstracted by mapping into a domain of abstract atoms A_P. More precisely, let $I \subseteq B_P$; define the abstraction function α as $\alpha(I) = \{p(\beta_R(t_1), \ldots, \beta_R(t_n)) \mid p(t_1, \ldots, t_n) \in I\}$. Given a set of abstract atoms $J \subseteq A_P$, the concretisation function γ is defined as $\gamma(J) = \{p(t_1, \ldots, t_n) \mid p(a_1, \ldots, a_n) \in J, t_i \text{ is accepted at state } a_i, 1 \leq i \leq n\}$. α and γ together define a Galois connection between the complete lattices $(2^{B_P}, \subseteq)$ and $(2^{A_P}, \subseteq)$. There is thus an optimal abstract semantic function $T_P^{\alpha_2} = \alpha \circ T_P \circ \gamma$ [2], and this can be implemented and computed by suitable instantiations of the functions reduce, project and ground mentioned earlier.

Regular Types and other Applications

Approximating a logic program by a CFTA has been called "descriptive type inference"; each inferred automaton state can be interpreted as a type. It has been used in debugging programs (highlighting inferred empty types) as well as checking whether inferred types correspond to intended types [5]. Analysis based on DFTAs has been used to perform mode analysis, binding time analysis and type checking for given types and modes expressed as DFTAs. Since it is possible to construct a DFTA from any given FTA, this kind of of analysis can be used to check any term property expressible as an FTA. Techniques for handling the potential blow-up caused by conversion to DFTA, as well as BDD representations of the DFTAs to improve scalability, are discussed in [4].

References

1. Comon, H., Dauchet, M., Gilleron, R., Löding, C., Jacquemard, F., Lugiez, D., Tison, S., Tommasi, M.: Tree automata techniques and applications (2007), http://www.grappa.univ-lille3.fr/tata (release October 12, 2007)
2. Cousot, P., Cousot, R.: Abstract interpretation frameworks. Journal of Logic and Computation 2(4), 511–547 (1992)
3. Cousot, P., Cousot, R.: Formal language, grammar and set-constraint-based program analysis by abstract interpretation. In: Proceedings of the Seventh ACM Conference on Functional Programming Languages and Computer Architecture, La Jolla, California, June 25-28, pp. 170–181. ACM Press, New York (1995)
4. Gallagher, J.P., Henriksen, K.S., Banda, G.: Techniques for Scaling Up Analyses Based on Pre-interpretations. In: Gabbrielli, M., Gupta, G. (eds.) ICLP 2005. LNCS, vol. 3668, pp. 280–296. Springer, Heidelberg (2005)
5. Gallagher, J.P., Puebla, G.: Abstract Interpretation over Non-deterministic Finite Tree Automata for Set-Based Analysis of Logic Programs. In: Adsul, B., Ramakrishnan, C.R. (eds.) PADL 2002. LNCS, vol. 2257, pp. 243–261. Springer, Heidelberg (2002)

Dynamic Symbolic Computation for Domain-Specific Language Implementation*

Fritz Henglein

Department of Computer Science, University of Copenhagen (DIKU)
henglein@diku.dk

Abstract. A domain-specific language (DSL) is a specification language designed to facilitate programming in a certain application domain. A well-designed DSL reflects the natural structure of the modeled domain, enforces abstraction, and its implementation exploits domain-specific properties for safety and performance. Expounding on the latter, in this paper we describe a simple, but powerful method for gradually enhancing a base implementation with computational performance improvements. It consists of adding constructors for expensive operations if this facilitates asymptotic performance improvements in some contexts, with at most constant-factor overhead. The resulting implementation can be thought of as executing "standard" computation steps from the base implementation interleaved with symbolic computation steps on the newly introduced constructors. Evaluation of constructor expressions is lazy in a strong sense: It does not only delay evaluation by thunkifying its standard evaluation, but performs run-time inspection to evaluate constructor expressions in a nonstandard way.

Extending earlier work by Henglein and Larsen [1] on generic multiset programming we arrive at novel multiset representations with constructors for unions, Cartesian products, scalar multiplication, and mapping binary functions over Cartesian products. We show how the resulting implementation can be used to implement finite probability distributions with exact (rational) probabilities. It supports computing both the expected value and, what we believe to be novel, the variance of random variables over n-ary product distributions efficiently by avoiding enumeration of the product sample space.

1 Introduction

A common problem in bulk data processing is to gather, correlate and aggregate data. Consider for example, the problem of producing a list of depositors' names and their account balances from separate lists containing depositor information and balance information, respectively. How to program such a function?

---

* This research has been partially supported by the Danish National Advanced Technology Foundation under Projects *3d generation Enterprise Resource Planning Systems (3gERP)* and by the Danish Strategic Research Council for the *HIPERFIT* Research Center (hiperfit.dk).

G. Vidal (Ed.): LOPSTR 2011, LNCS 7225, pp. 4–24, 2012.

An attractive declarative solution is *functional programming* using list comprehension [2,3,4] corresponding to relational calculus, or an algebraic formulation using combinators corresponding to relational algebra [1].

Following the latter approach, let us define higher-order functions for applying functions componentwise, for defining *(equi)join conditions*, and for constructing the Cartesian list product:

```
(f *** g) (x, y) = (f x, g y)
(f .==. g) (x, y) = (f x == g y)
prod s t = [ (x, y) | x ← s, y ← t ]
```

For example,

```
((+2) *** not)(5, True) = (7, False)
((+2) .==. (+8))(9, 3) = True
prod [4, 8] [3, 5] = [(4, 3), (4, 5), (8, 3), (8, 5) ]
```

Example 1. We can define the function listing depositors' names and their balances as follows.

```
depAcctBal depositors accounts =
  map (depName *** acctBalance)
      (filter (depId .==. acctId)
              (depositors 'prod' accounts))
```

where `map` and `filter` are standard list combinators generalizing projection and selection, respectively, in relational algebra.[1]   □

Coding relational algebra operations as straightforward list processing functions makes for an *a priori* attractive library for language-integrated querying. In practice, however, this is not viable for performance reasons. The culprit is the product function `prod`. In the example above, `depositors 'prod' accounts` *multiplies* the Cartesian product *out* by nested iteration through the two lists. This requires $\Theta(n^2)$ time for input lists of size n. Lazy evaluation, as in Haskell or relational database query engines, avoids a space blow-up, but not the exhaustive iteration through the Cartesian product.

In this paper we describe and illustrate a method for developing the implementation step by step into one that is asymptotically more efficient. The query will eventually look like this:

```
depAcctBal depositors accounts =
  perform (depName :***: acctBalance)
          (select ((depId, acctId) 'Is' eqInt)
                  (depositors 'X' accounts))
```

and can, with simple definitions, be made to look identical to the first formulation. The difference is that, in its final version, it executes in $O(n)$ time in the worst case, without requiring its designer to do anything different.

The method consists of identifying expensive operations, reifying them as constructors and exploiting algebraic properties to achieve asymptotic performance

---

[1] `depositors 'prod' accounts` is infix notation for `prod depositors accounts`.

improvements vis a vis standard evaluation in the original implementation. In this fashion it mixes "symbolic" computation steps with "standard" steps at run time. For this reason we call the method *dynamic symbolic computation*.

1.1 Contributions

We claim the following contributions:

- We present dynamic symbolic computation as an iterative *design method* for adding performance improvements to a base implementation.
- We show how dynamic symbolic computation naturally leads to the addition of symbolic operators for Cartesian product, scalar multiplication and formal maps on Cartesian products in multiset programming.
- As an application, we show how dynamic symbolic computation facilitates implementation of finite probability distributions, with efficient computations of not only exact expectation, but also variance on product distributions.

The introduction of symbolic Cartesian products, with the complementary method of efficient join computation based on generic discrimination, has been presented before [1]. The extensions with scalar multiplication and maps over products, and their application to probability distributions are new, however.

The paper is written in an informal style, deliberately leaving out details and missing opportunities for generalizations or theoretical analysis. This is to emphasize the design considerations. The reader hopefully experiences this as a feature rather than a bug.

1.2 Required Background, Notation and Terminology

We use the functional core notation of Haskell [5] as our programming language, extended with Generalized Algebraic Data Types (GADTs) [6,7], as implemented in Glasgow Haskell.[2] GADTs provide a convenient *type-safe* framework for shallow embedding of *little languages* [8]. Hudak, Peterson and Fasel [9] provide a brief and gentle introduction to Haskell. Apart from type safety, all other aspects of our library can be easily coded up in other general-purpose programming languages, both eager and lazy. Since we deliberately do not use monads, type classes or any other Haskell-specific language constructs except for GADTs, we believe basic knowledge of functional programming is sufficient for understanding the code we provide.

For consistency we use the term *multiset*, not bag, for sets with positive integral multiplicities associated with their elements. We use *Cartesian product* or simply *product* to refer to the collection of all pairs constructed from two collections. We do not use the term "cross-product", used in the database literature, because of its common, but different meaning as the Gibbs vector product in mathematics. We call the number of elements in a multiset its *cardinality* or its

---

[2] See www.haskell.org/ghc

count, which is common in database systems. We reserve the word *size* for a measure of storage space. For example, the size of a multiset representation is the number of nodes and edges when viewing it as an acyclic graph, which may be larger or smaller than its cardinality.

1.3   Outline

In Section 2 we demonstrate dynamic symbolic computation by adding symbolic unions and Cartesian products to multiset representations. Much of that section is taken from Henglein and Larsen [1, Section 2]. In Sections 3 and 4 we continue where Henglein and Larsen leave off: We add constructors for scalar multiplication and for mapping over Cartesian products. This facilitates representing finite probability distributions, with efficient computation of expectation and variance for random variables over product distributions, shown in Section 5. We discuss related and possible future work (Sections 6 and 7) and briefly offer conclusions in Section 8.

2   Symbolic Union and Product

The key idea in this paper is to introduce data type constructors for expensive operations and exploit algebraic properties involving them to improve performance. Henglein and Larsen [10,11,1] have developed this for generic multiset programming. They add constructors U and X for unions and Cartesian products, respectively, to a base implementation of multisets as lists. This gives us the following data type definition for *multiset representations*:

```
data MSet a where
    O    :: MSet a                          -- the empty multiset
    S    :: a → MSet a                       -- singleton multiset
    U    :: MSet a → MSet a → MSet a         -- multiset union
    X    :: MSet a → MSet b → MSet (a, b)    -- Cartesian product
```

A practical, but not essentially different alternative to O and S is using a single MSet constructor with a list argument.

The purpose of introducing U and X instead of just using lists is that algebraic properties of unions and Cartesian products let us avoid having to flatten unions and multiply out products in certain application contexts. In Example 1 there are two application contexts where algebraic properties of Cartesian products can be exploited:

- The predicate depId .==. acctId in the selection is an equijoin condition. When applied to depositors 'X' accounts, we can employ an efficient join algorithm without multiplying the product out.
- The function depName *** acctBalance works componentwise on its input pairs. Semantically, the result of mapping it over a product s 'X' t is the product of mapping depName over s and acctBalance and t, which allows us to avoid multiplying out s 'X' t altogether.

To recognize opportunities for symbolic processing of products, we shall also need symbolic representations of functions and predicates used in projections and selections, respectively. We illustrate the design process of developing them, driven by our example query.

2.1 Selection

Consider the subexpression

```
filter (depId .==. acctId) (depositors 'prod' accounts)
```

in our query. We would like to replace `filter` by a function `select` that operates on multisets instead of lists. We need to represent those Boolean-valued functions symbolically, however, for which we can take advantage of symbolic products in second argument position. To that effect we introduce a data type of *predicates*, symbolic Boolean-valued functions, starting with a single constructor for representing arbitrary Boolean-valued functions:

```
data Pred a where
     Pred :: (a → Bool) → Pred a
     ...
```

The dots represent additional predicate constructors, which we subsequently add to achieve computational advantage over predicates represented as "black-box" Boolean-valued functions only.

One class of predicates we can exploit are *equivalence join conditions*, just called *join conditions* henceforth. They generalize equality predicates on attributes in relational algebra. A join condition is a predicate on pairs. It has three components: Two functions f, g and an equivalence relation E. A pair (x, y) satisfies the join condition if and only if $f(x) \equiv_E g(y)$, that is $f(x)$ and $g(y)$ are E-equivalent. To represent a join condition symbolically we introduce a constructor `Is` and add it to `Pred a`:

```
data Pred a where
     Pred :: (a → Bool) → Pred a
     Is   :: (a → k, b → k) → Equiv k → Pred (a, b)
```

The data type `Equiv k` of *equivalence representations* contains symbolic representations of equivalence relations. They are introduced by Henglein [12]. For the present purpose, all we need to know is that there exists an efficient *generic discriminator*

```
disc :: Equiv k → [(k, v)] → [[v]]
```

which partitions key-value pairs for an equivalence relation E represented by e in the following sense: `disc` e partitions the input into groups of values associated with E-equivalent keys.

Example 2. For equality on the number type `Int` denoted by

```
eqInt32 :: Equiv Int
```

the expression `disc eqInt32 [(5,10), (8,20), (5,30)]` evaluates to `[[10, 30], [20]]`. □

Let us now define the function `select` corresponding to `filter` on lists. First we note that `select` takes predicates as inputs, not Boolean-valued functions:

```
select :: Pred a → MSet a → MSet a
```

Its *default implementation* processes 0, S and U-constructed multisets straightforwardly, and symbolic products only after *normalizing* them:

```
select p 0                = 0
select p (S x)            = if sat p x then S x else 0
select p (s ‘U‘ t)        = select p s ‘U‘ select p t
select p s                = select p (norm s) -- default clause
```

where

```
sat :: Pred a → a → Bool
```

converts a predicate into the Boolean-valued function it represents. The function `norm` produces a *normalized* multiset representation equivalent to its input. A multiset is normalized if contains no occurrence of X.

```
norm :: MSet a → MSet a
norm 0 = 0
norm (S x) = S x
norm (s ‘U‘ t) = norm s ‘U‘ norm t
norm (0 ‘X‘ t) = 0
norm (s ‘X‘ 0) = 0
norm (S x ‘X‘ S y) = S (x, y)
norm (s ‘X‘ (t1 ‘U‘ t2)) = norm (s ‘X‘ t1) ‘U‘ norm (s ‘X‘ t2)
norm ((s1 ‘U‘ s2) ‘X‘ t) = norm (s1 ‘X‘ t) ‘U‘ norm (s2 ‘X‘ t)
norm (s ‘X‘ (t1 ‘X‘ t2)) = norm (s ‘X‘ norm (t1 ‘X‘ t2))
norm ((s1 ‘X‘ s2) ‘X‘ t) = norm (norm (s1 ‘X‘ s2) ‘X‘ t)
```

Note that `norm (s ‘X‘ t)` where s and t are normalized is tantamount to multiplying the product out; in particular, it takes $\Theta(n^2)$ time if the cardinalities of s and t are n.

If the default clause for `select` were the only clause for products, there would be no point in having a symbolic product constructor: We might as well keep multisets normalized throughout. The point is that we can add clauses that exploit properties of X. For example, the clause for a join condition applied to a product is

```
select ((f, g) ‘Is‘ e) (s ‘X‘ t) = join (f, g) e (s, t)
```

where

```
join :: (a → k, b → k) → Equiv k → (MSet a, MSet b) → MSet (a, b)
join (f, g) e (s, t) = ... disc e ...
```

is an efficient generic join algorithm based on `disc` [10]. The important point of this clause is not the particular choice of join algorithm invoked, but that `select` discovers *dynamically* when it is advantageous to branch off into an efficient join algorithm. Furthermore, it is important that `join` returns its output as symbolic unions of symbolic products. This ensures that the output size[3] is linearly bounded by the sum of the sizes of the input multisets.

The selection in our example query can then be formulated as

```
select ((depId, acctId) 'Is' eqInt32) (depositors 'X' accounts)
```

It returns its result as a multiset of the form (s1 'X' t1) 'U' ... 'U' (sn 'X' tn). This is an important difference to query evaluation in conventional RDBMSs, which follow the System R [13] architecture. There, intermediate results are exclusively represented as streams of records, corresponding to lists in Haskell. To avoid unnecessarily multiplying out the component products of joins consumed by other parts of a query, aggressive query optimization is employed prior to issuing a query plan for execution.

There are more opportunities for exploiting properties of predicates and/or multisets [1]; e.g. we can represent the constant-true function by predicate `TT` and add the clause

```
select TT s = s
```

which returns the second argument without iterating through it.

2.2   Projection

Consider now the outermost projection applied in the query from Example 1:

```
map (depName *** acctBalance) ...
```

We want to replace the list combinator `map` with an efficient function `perform` operating on multisets. We call it `perform` instead of `project` since it maps arbitrary functions over multisets, not only projections from records to their attributes as in relational algebra.

To represent functions symbolically we introduce, analogous to predicates, a data type we call *performable functions*

```
data Func a b where
    Func : (a → b) → Func a b
    ...
```

with an *extension* function

```
ext : Func a b → (a → b)
```

returning the ordinary function denoted by a performable function.

Performable functions are basically symbolic arrows [14]. The projection functions on pairs are represented by `Fst` and `Snd`:

---

[3] Recall that this is not the *cardinality* of the output, but, up to a constant factor, the storage space required for representing an element of `MSet a`.

```
data Func a b where
    Func : (a → b) → Func a b
    Fst :  Func (a, b) a
    Snd :  Func (a, b) b
    ...
```

To represent functions operating componentwise, we add to Func a b a *parallel composition* constructor:

```
data Func a b where
    Func : (a → b) → Func a b
    Fst :  Func (a, b) a
    Snd :  Func (a, b) b
    (:***:) : Func (a, c) → Func (b, d) → Func (a, b) (c, d)
    ...
```

Example 3. The projection depName *** acctBalance in our query from Example 1 is symbolically represented by the constructor expression

```
depName :***: acctBalance
```

where

```
depName = Snd
acctBalance = Snd
```

are the projections to the depositor name and account balance fields, assuming that both depositor and account records are represented by pairs with the identifiers in the first components and the associated information in the second components.   □

We now define perform, which maps a performable function over all elements of a multiset. Analogous to select, we note that the first argument is symbolic, a performable function:

```
perform :: Func a b → MSet a → MSet b
```

Its default implementation normalizes symbolic products:

```
perform :: Func a b → MSet a → MSet b
perform f 0          = 0
perform f (S x)      = S (ext f x)
perform f (s 'U' t) = perform f s 'U' perform f t
perform f s          = perform f (norm s) -- default clause
```

As for select we add special clauses based on algebraic identities for products:

```
...
perform Fst (s 'X' t) = times (count t) s
perform Snd (s 'X' t) = times (count s) t
perform (f :***: g) (s 'X' t) = perform f s 'X' perform g t
perform f s          = perform f (norm s) -- default clause
```

Here `count` computes the cardinality of a multiset. It is the prime example for exploitation of symbolic unions and products, since no normalization is required at all:

```
count :: MSet a → Integer
count 0          = 0
count (S x)      = 1
count (s 'U' t)  = count s + count t
count (s 'X' t)  = count s * count t
```

The function `times` implements *scalar multiplication* on multisets under union:

```
times :: Integer → MSet a → MSet a
times 0 s = 0
times 1 s = s
times n s = s 'U' times (n-1) s
```

Recall that `MSets` represent multisets, not sets, which accounts for the need to scale the result of a projection to one component of a product by the cardinality of the other component.

2.3   Query Example with Dynamic Symbolic Computation

Using multiset representations with symbolic unions and products, with the dynamic symbolic computation clauses in the functions processing them, our query example looks remarkably similar to the formulation based on a naïve list implementation in Example 1:

```
perform (depName :***: acctBalance)
        (select (depId .==. acctId)
                (depositors 'X' accounts))
```

where `.==.` defines the join condition

```
f .==. g = (f, g) 'Is' eqInt32
```

and `depId`, `depName`, `acctId` and `acctBalance` project the desired components from the records:

```
depId = Fst
depName = Snd
acctId = Fst
acctBalance = Snd
```

The important difference is that this query executes in linear, not quadratic, time in the cardinalities of `depositors` and `accounts`. The `select` subexpression invokes a discrimination-based join, which returns its result in the form (`s1` `'X'` `t1`) `'U'` ... `'U'` (`sn` `'X'` `tn`). The two special clauses for `perform` above ensure that `perform (depName :***: acctBalance)` is executed by mapping the component functions `depName` and `acctBalance` without multiplying the products out. The output is of linear *size* and is computed in linear time in the sum of the *size* of the inputs `depositors` and `accounts`. This is the case even though, in general, the *cardinality* of the output of the query may be as large as quadratic in the cardinalities of the input.

3   Symbolic Scalar Multiplication

The clauses for performing a projection on a Cartesian product

```
perform Fst (s 'X' t) = times (count t) s
perform Snd (s 'X' t) = times (count s) t
```

in Section 2.2 require the auxiliary function `times`. It generates a repeated union of the argument multiset; e.g.

```
times 5 s = s 'U' (s 'U' (s 'U' (s 'U' s)))
```

Note the repeated occurrences of `s`. We can think of `times` as producing a unary representation of a scalar product from a binary one, its input. Clearly this is wasteful. We may as well keep the binary representation. Employing the method of dynamic symbolic computation, we thus add a new multiset constructor for scalar multiplication:

```
data MSet a where
    0     :: MSet a                            -- the empty multiset
    S     :: a → MSet a                        -- singleton multiset
    U     :: MSet a → MSet a → MSet a          -- multiset union
    X     :: MSet a → MSet b → MSet (a, b)     -- Cartesian product
    (:.)  :: Integer → MSet a → MSet a         -- scalar multiplication
```

The clauses for performing a projection become

```
perform Fst (s 'X' t) = count t :. s
perform Snd (s 'X' t) = count s :. t
```

We only attain a benefit if the conversion of formal scalar products into the corresponding unary representation can be elided in some, and preferably many, application contexts of multisets. This is indeed the case since `count`, `perform` and `select` distribute over scalar multiplication and we can add the corresponding clauses to their implementations:

```
count (r :. s)      = r * count s
perform f (k :. s)  = k :. perform f s
select p (k :. s)   = k :. select p s
```

4   Efficient Reduction on Scalar and Cartesian Products

So far we have focused on relational algebra operations. What about *aggregation* functions over multisets?

Reduction implements the unique homomorphism from multisets over a set S, viewed as the free commutative monoid generated by S, to another commutative monoid over S. In other words, it replaces U, 0 in multisets by other operations u, n, where the binary operation u is required to be associative and commutative, with n as neutral element. This gives us the following default implementation for `reduce`:

```
reduce :: (a → a → a, a) → MSet a → a
reduce (u, n) 0 = n
reduce (u, n) (S x) = x
reduce (u, n) (s 'U' t) = reduce (u, n) s 'u' reduce (u, n) t
reduce (u, n) s = reduce (u, n) (norm s) -- default clause
```

4.1   Reduction of Scalar Products

We can often implement reduction by passing an additional argument m for
interpreting scalar multiplication:

```
mreduce :: (a → a → a, Integer → a → a, a) → MSet a → a
mreduce (u, m, n) 0 = n
mreduce (u, m, n) (S x) = x
mreduce (u, m, n) (s 'U' t) = reduce (a, m, z) s 'u' reduce (a, m, z) t
mreduce (u, m, n) (k :. s) = k 'm' reduce (u, m, n) s
mreduce (u, m, n) s = mreduce (u, m, n) (red s) -- default clause
```

For this to be correct, $(u, m, 0)$ must be a left module over a, that is the following
equations must be satisfied:

$$r \text{ 'm' } (s \text{ 'u' } t) = (r \text{ 'm' } s) \text{ 'u' } (r \text{ 'm' } t)$$
$$(r + r') \text{ 'm' } s = (r \text{ 'm' } s) \text{ 'u' } (r' \text{ 'm' } s)$$
$$(r * r') \text{ 'm' } s = r \text{ 'm' } (r' \text{ 'm' } s)$$
$$1 \text{ 'm' } 0 = 0$$

For example, we can compute the sum of a multiset of integers by

```
msum :: MSet Integer → Integer
msum = mreduce ((+), (*), 0)
```

4.2   Reduction of Maps over Products

The definition of msum in Section 4.1 avoids multiplying out scalar multiplication,
but it still multiplies out all Cartesian products involved in the construction of
its argument.

Example 4. Consider the computation of the sum of all rolls of two dice, one
m-sided and the other n-sided:

```
die1 = mset [1..m]
die2 = mset [1..n]
res = msum (perform (Func (+)) (die1 'X' die2))
```

where mset converts a list into a multiset. In the implementation we have developed so far, perform multiplies the Cartesian product out, and msum iterates
over the resulting $m \cdot n$ numbers.   □

The sum in the example can be computed more efficiently, however, using the
algebraic identity

```
msum (perform (Func (+) (s 'X' t) =
   count t * msum s + count s * msum t
```

In other words, there is a reduction context of s 'X' t that gives rise to an asymptotic performance improvement vis a vis the present implementation of msum. To be able to detect this context at run time we add a constructor for mapping a function over a Cartesian product[4]

```
data MSet a where
   ...
   (:$)  :: Func (a, b) c → (MSet a, MSet b) → MSet c
                          -- symbolic map over product
```

and we add a symbolic representation of addition to the performable functions:

```
data Func a b where
   ...
   (:+:)  :: Func (Integer, Integer) Integer    -- addition
```

The definition of perform is changed to produce symbolic maps over Cartesian products:

```
perform :: Func a b → MSet a → MSet b
...
perform (:+:) (s 'X' t)   = (:+:) :$ (s, t)
perform f s               = perform f (norm s) -- default clause
```

Let us consider msum = mreduce ((+), (*), 0) after inlining the definition of mreduce:

```
msum :: MSet Integer → Integer
msum 0        = 0
msum (S x)    = x
msum (s 'U' t) = msum s + msum t
msum (k :. s) = k * msum s
msum s        = msum (norm s)    -- default clause
```

Now we add a clause to msum that dispatches to a sum computation without multiplying out the product:

```
...
msum ((:+:) :$ (s, t)) = count t * msum s + count s * msum t
msum s        = msum (norm s)    -- default clause
```

Example 5. Computing the sum of all dice rolls from Example 4 by

```
res = msum (perform (:+:) (die1 'X' die2))
```

takes $O(m + n)$ time.   □

The added clause for :+: in msum also averts quadratic time computation in cases where the Cartesian products are generated dynamically.

---

[4] This does *not* correspond to the function zipWith in Haskell!

Example 6. Computing the sum of only those dice rolls where both dice are either even or both are odd is naturally expressed as:

```
msum (perform (:+:) (select ((id, id) 'Is' parityEq)
                            (die1 'X' die2)))
```

where `parityEq` denotes integer equivalence modulo 2. The computation of `select` produces unions of products, and the sum computation computes the sums of the Cartesian products without multiplying out the products.   □

We have added the clause for summing over a Cartesian product to a separate definition of `msum`. An alternative is to do so in the definition of `mreduce` and defining `msum` as an instantiation as before. This requires changing the type of `mreduce` to be passed performable functions as the first argument instead of standard functions, something we have seen before in connection with filtering (`select`) and mapping (`perform`).

There are other algebraic identities that can be exploited to avoid multiplying out Cartesian products. For example, the identities

```
msum (perform (Func (*)) (s 'U' t) = msum s * msum t
msum (perform (Func ((^2) . (+)) (s 'U' t) =
     count t * msum (perform (Func (^2) s)
   + count s * msum (perform (^2) t)
   + 2 * msum s * msum t
```

are useful when computing the variance of a probability distribution (see below). Indeed, there are many more useful identities, which can be added one by one or by including a full-blown symbolic computation processor. Note that dynamic symbolic computation as a method lets us do this as part of an iterative development process: We always have a running implementation.

5   Application: Finite Probability Distributions

Multisets provide a straightforward representation for finite probability distributions[5] with rational event probabilities:

```
type Dist a = MSet a
```

The probability of an element of `a` in a probability distribution represented by a multiset is its multiplicity—how often it occurs in the multiset—divided by the multiset's cardinality.

Distribution combinators can be defined as multiset operations. A distribution can be given by enumerating its elements:

```
dist :: [a] → Dist a
dist = mset
```

---

[5] Strictly speaking probability spaces, but we shall follow common informal usage in referring to them as distributions.

A list without repetitions results in the *uniform* distribution over its elements. Repeating elements in the input provides a way of equipping elements with individual probabilities. For example, a biased coin that yields heads with probability $2/3$ can be modeled by

```
data Coin = Heads | Tails
biasedCoin = dist [Heads, Heads, Tails]
```

The (independent) product distribution is simply Cartesian product:

```
dprod :: Dist a → Dist b → Dist (a, b)
dprod = X
```

It equips each element (x, y) with the probability $p_x \cdot p_y$ where p_x is the probability of x drawn from one distribution and p_y the probability of y drawn from the other distribution.

The choice combinator, which chooses from one distribution with rational probability p and from the other with probability $1 - p$, can be encoded as follows:

```
choice :: Rational → Dist a → Dist a → Dist a
choice p s t =
  let (n, d) = (numerator p, denominator p)
      (v, w) = (n * count t, (d - n) * count s)
  in (v :. s) 'U' (w :. t)
```

The v and w factors ensure that each multiset is weighted correctly before being unioned: The count-factors adjust the cardinalities of the two multisets so they are the same before entering into a weighted unioned determined by the input probability.

Random variables are modeled by performable functions

```
type RandVar a b = Func a b
```

Pushing a random variable forward over a distribution is done by mapping the random variable over the multiset representing the distribution:

```
dpush :: RandVar a b → Dist a → Dist b
dpush = perform
```

The expected value of a random variable over a given distribution can be computed as the sum of its push-forward, divided by its cardinality. Similarly, the variance can be computed from summing the square of the random variable[6], the sum and the cardinality of the multiset representing the distribution. These functions can be coded straightforwardly as follows:

```
dsum :: Dist a → RV a Integer → Integer
dsum p rv = msum (dpush rv p)

linsum :: Dist Integer → Integer
linsum p = dsum p Id
```

---

[6] Strictly speaking: The squaring function composed with the random variable.

```
sqsum :: Dist Integer → Integer
sqsum p = dsum p (Sq :.: Id)


mean :: Dist Integer → Rational
mean p = linsum p % count p


variance :: Dist Integer → Rational
variance p =
  let n = count p
      s = linsum p
      s2 = sqsum p
  in n * s2 - s^2 % n^2
```

The performable functions Id, Sq and the combinator : . : denote the identity, the squaring function and function composition, respectively. The binary operator % returns the fraction of two integers as a rational number. Together with the last clause added in Section 4, both linsum and sqsum are computed in time $O(m \cdot n)$ for the Bernoulli distribution of rolling n m-sided dice independently. This is in contrast to a naïve implementation, which enumerates all m^n elements of the sample space, or an implementation using normalized representations, which takes $\Theta(m^2 \cdot n^2)$ time.

Example 7. Consider the sum of rolling n 6-sided dice.

```
die :: Dist Integer
die = uniform [1..6]

dice :: Int → Dist Integer
dice 1 = die
dice n = dpush (:+:)  (die 'dprod' dice (n-1))
```

The evaluations of

```
mean (dice 100)
variance (dice 100)
```

quickly return 350 % 1 and 875 % 3, respectively. These computations can be further improved by adding symbolic multiset constructors for uniform distributions over intervals.   □

6   Related Work

Most efficiency-minded DSL-implementations follow a static symbolic computation approach, in which symbolic processing and subsequent (standard) execution are staged. This often follows the natural phase distinction where compilation, without access to run-time data, is followed by execution once the data are available. Dynamic symbolic computation is intuitively more like interpretation: Nothing happens before the run-time data are ready, but during execution some symbolic processing takes place. The principal trade-offs are that dynamic symbolic processing can take run-time data into account and no support for full-fledged language processing (parser, analysis, rewriting system, code generator) for program transformation or compilation is required, but it incurs run time overhead for matching.

Below we review some select work close to the topics of this paper to convey a sense of the methodological differences between static and dynamic symbolic computation as DSL implementation methodologies.

6.1 Append Lists versus Difference List

Append lists are list representations with a constant-time constructor for appending (concatenating) two lists. Elements of MSet a can be thought of as append lists where U denotes concatenation. Append lists are also known as join lists, ropes [15], Boom lists [16,17,18,19], or catenable lists [20]. Matching against the append-constructor is semantically dangerous since it may result in incoherent results: different append lists denoting the same list may give different results. If converting to an ordinary cons lists, lists only nil and cons-constructors, is the only permissible matching against the constructors of append lists then we may as well *shortcut* the constructor applications of an append-list with this conversion. This results in the well-known *difference lists* [21]. The transformation of a program by *freezing* [22] the function append into a constructor and then transforming it away prior to program execution is, methodologically, a form of static symbolic computation.

Retaining the Append-constructor can be thought of as a dynamic symbolic computation approach. Append lists incur run-time overhead from implementing a data type with 3 constructors as opposed to 2, but they also allow for exploiting algebraic properties; e.g.

```
rev (Append xs ys) = Append (rev ys) (rev xs)
```

in the implementation of list reversal.

Voigtländer [23] generalizes the difference list approach and demonstrates how append, reverse and map "vanish" [24]. There are numerous other incarnations of static symbolic computing where *fusing* isomorphisms to eliminate intermediate computations and data structures play a central role; e.g. short-cut fusion [25] (build/fold), stream fusion [26] (stream/unstream) and vector fusion [27,28]. Interestingly, the addition of a delayed array constructor to manifest arrays in Data Parallel Haskell [29] facilitates a form of dynamic symbolic computation: Applicability of vertical loop fusion is detected and performed at run time.

Closest to our focus on avoiding multiplying out multiset products is the paper by Emoto, Fischer and Hu on semiring fusion [30]. Their work can be thought of as following Voigtländer's general methodology, but with semiring fusion on multisets instead of list fusion and with a clever filter embedding technique to facilitate MapReduce-style *generate-test-aggregate* programming with asymptotic performance improvements vis a vis standard evaluation. Our dynamic symbolic computation approach performs optimizations that are analogous to semiring fusion, but neither subsume it nor are subsumed by it. For example, the clause

```
msum ((Sq :.: (:+:)) :$ (s, t)) =
  count t * msum (perform Sq s) + count s * msum (perform Sq t) +
  2 * msum s * msum t
```

exploits a property on products that averts having to multiply out the product. This cannot be captured as a semiring homomorphism clause, which in our notation has the form

```
hom ((:++:) :$ (s, t)) = hom s * hom t
```

where (:++:) is a constructor denoting list concatenation, and (*) is the multiplication in the target semiring.

6.2 Probabilistic DSLs

Ramsey and Pfeffer [31] present a stochastic λ-calculus and provide multiple interpretations, one for generating sampling functions and another for efficient computation of expected values for independent product distributions by *variable elimination* [32]. Efficient computation of variance does not seem to be supported due to the absence of special treatment of arithmetic functions. Erwig and Kollmansberger [33] provide a Haskell library for experimenting with finite probability distributions using monadic programming, which supports do-notation for intuitively specifying stochastic processes. Their implementation is expressive, but its representation of distributions in *normalized* form as lists of pairwise distinct elements with associated probabilities requires explicit enumeration of the sample space for computing expectation and variance, even for independent product distributions. Larsen [34] demonstrates how symbolic representation of Erwig and Kollmansberger's probability distributions in monadic form avoids the space blow-up of a naïve list-based implementation. Kiselyov and Shan [35] show how delimited continuations can be used to transparently mix deterministic with probabilistic computation steps: A probabilistic computation step is implemented by performing its continuation on each element in the support of a probability distribution. The results are collected lazily and associated with the probabilities of each element. Using shift and reset in OCaml, they turn *all* of OCaml into a probabilistic programming language without the run-time overhead of an embedded domain-specific language. Delimited continuations are black-box functions that cannot be analyzed as we do by introducing symbolic functions (arrows) Func a b. Consequently, independent product distributions are multiplied out, and distributions with a large support result in (potentially prohibitive) sample space enumeration. Like Erwig and Kollmansberger they represent distributions in normalized form. Our DSL also employs normalization (not shown here), but its symbolic processing power facilitates even more efficient computation of expectation and variance without explicitly having to produce the whole distribution. Park, Pfenning and Thrun [36] present an expressive calculus whose expressions denote *sampling functions*, functions from $(0, 1]$ to some probability domain. These may denote discrete or continuous distributions, or distributions that are neither. They do not support exact computation of expectation and variance of probability distributions, but provide the basis for approximation using Monte-Carlo sampling. Mogensen [37] presents a DSL for rolling dice in role-playing games. He provides both a random sampling interpretation for "rolling the dice" and another interpretation, which computes the explicit distribution in normalized form. Interestingly, he employs an efficient symbolic representation for independent product distributions and handles some discrete, but nonfinite distributions. Bhat, Agarwal, Vuduc and Gray [38] design a language for specifying discrete-continuous probability distributions in nonrecursive monadic style and symbolically generating closed-form probability density functions (PDF). Their type system guarantees that the calculated PDFs are well-defined and correct.

7   Future Work

Our DSL for probability distributions is of very limited expressive power, but is intended to be developed into a full-fledged DSL for modeling stochastic processes used in financial modeling for pricing and risk analysis. Such a DSL must support dependent probability distributions such as Markov chains, implemented by both efficient matrix operations and (symbolic) linear algebra optimizations for highest performance, which is a necessity in finance. In particular, the DSL should recognize that rolling y in

```
do x ← die
   y ← die
   return x + y
```

is independent of rolling x to bring the techniques presented here for computing expectation and variance of independent product distributions to bear. *Loop lifting* techniques [39,40,41] in combination with *compositional data types* [42,43,44] look like promising techniques for performing this efficiently in a type-safe fashion at run time.

The introduction of scalar multiplication suggests generalizing multisets to *fuzzy sets* by allowing nonintegral numbers such as rational or floating-point numbers in scalar multiplication. Restricting multiplicities to the range $[0 \dots 1]$ provides a direct representation of (finite) probability distributions, without the need for dividing weights by the cardinality of a multiset, as in Section 5. This simplifies the code. For example,

```
choice :: Probability → Dist a → Dist a → Dist a
choice p s t = (p :. s) 'U' ((1 - p) :. t)
```

Using single- or double-precision floating point multiplicities instead of rationals we obtain performance advantages at the cost of numerical approximation.

Other work could be investigating dynamic symbolic computation as a methodology for adding rewrite rules in rewriting logic [45] for (asymptotic) performance improvement to orthogonal rewrite systems that start out with normalizing to canonical terms.

8   Conclusions

Dynamic symbolic computation is a light-weight iterative implementation method. It consists of adding lazy (symbolic) operators to a base implementation if that gives an *asymptotic* performance improvement by employing symbolic computation steps in *some* (ideally frequently occurring) contexts at a constant-time (ideally minimal) run-time cost. The starting point is a canonical implementation ("standard evaluation"), which serves as a reference implementation. In particular, the method guarantees that there is a complete implementation at any given point during development. We have employed it to develop an efficient library for multiset programming and used it for finite probability distributions. The idea of enriching a base implementation with symbolic processing at run-time should have wider practical applicability, notably to DSLs with rich mathematical equational theories.

Acknowledgements. I would like to thank the organizers of LOPSTR 2011 and PPDP 2011 for inviting me to give a talk, which provided the impetus to develop the idea of dynamic symbolic computation as a methodology. Thanks also to José Meseguer for discussions on the connections to rewriting logic. Torben Grust has provided helpful explanations on how comprehension-style expressions can be transformed into product form, which facilitate algebraic optimizations based on Cartesian products. Thanks to Zhenjiang Hu for explaining and discussing his work on semiring fusion. Last, but not least, many thanks to Ken Friis Larsen for ongoing cooperation and discussions on how to develop efficient domain-specific languages for querying and probability distributions.

References

1. Henglein, F., Larsen, K.: Generic multiset programming with discrimination-based joins and symbolic Cartesian products. Higher-Order and Symbolic Computation (HOSC) 23, 337–370 (2010); (Publication date: November 24, 2011)
2. Trinder, P., Wadler, P.: List comprehensions and the relational calculus. In: Proc. 1988 Glasgow Workshop on Functional Programming, Rothesay, Scotland, pp. 115–123 (August 1988)
3. Peyton Jones, S., Wadler, P.: Comprehensive comprehensions. In: Proc. 2007 Haskell Workshop, Freiburg, Germany (2007)
4. Giorgidze, G., Grust, T., Schreiber, T., Weijers, J.: Haskell Boards the Ferry. In: Hage, J., Morazán, M.T. (eds.) IFL 2010. LNCS, vol. 6647, pp. 1–18. Springer, Heidelberg (2011)
5. Peyton Jones, S.: The Haskell 98 language. J. Functional Programming (JFP) 13(1), 0–146 (2003)
6. Cheney, J., Hinze, R.: First-class phantom types. CUCIS TR2003-1901, Cornell University (2003)
7. Xi, H., Chen, C., Chen, G.: Guarded recursive datatype constructors. In: Proc. 30th ACM SIGPLAN-SIGACT Symposium on Principles of Programming Languages, pp. 224–235. ACM (2003)

8. Bentley, J.: Programming pearls: Little languages. Commun. ACM 29(8), 711–721 (1986)
9. Hudak, P., Peterson, J., Fasel, J.H.: A gentle introduction to Haskell Version 98. Online Tutorial (May 1999)
10. Henglein, F.: Optimizing relational algebra operations using discrimination-based joins and lazy products. In: Proc. ACM SIGPLAN 2010 Workshop on Partial Evaluation and Program Manipulation, January 18-19, pp. 73–82. ACM, New York (2010); Also DIKU TOPPS D-report no. 611
11. Henglein, F., Larsen, K.: Generic multiset programming for language-integrated querying. In: Proceedings of the 6th ACM SIGPLAN Workshop on Generic Programming (WGP), pp. 49–60. ACM (2010)
12. Henglein, F.: Generic top-down discrimination for sorting and partitioning in linear time. Invited Submission to Journal of Functional Programming (JFP) (December 2010)
13. Selinger, P.G., Astrahan, M.M., Chamberlin, D.D., Lorie, R.A., Price, T.G.: Access path selection in a relational database management system. In: Proc. 1979 ACM SIGMOD Int'l. Conf. on Management of Data, SIGMOD 1979, pp. 23–34. ACM, New York (1979)
14. Hughes, J.: Generalising monads to arrows. Science of Computer Programming 37(1-3), 67–111 (2000)
15. Boehm, H.J., Atkinson, R., Plass, M.: Ropes: An alternative to strings. Software: Practice and Experience 25(12), 1315–1330 (1995)
16. Meertens, L.: Algorithmics–towards programming as a mathematical activity. In: Proc. CWI Symp. on Mathematics and Computer Science, pp. 289–334. North-Holland (1986)
17. Bird, R.: An introduction to the theory of lists. Technical Report PRG-56, Oxford University (October 1986)
18. Backhouse, R.: An exploration of the Bird-Meertens formalism. In: STOP Summer School on Constructive Algorithmics (1989)
19. Meijer, E., Fokkinga, M., Paterson, R.: Functional Programming with Bananas, Lenses, Envelopes and Barbed Wire. In: Hughes, J. (ed.) FPCA 1991. LNCS, vol. 523, pp. 124–144. Springer, Heidelberg (1991)
20. Okasaki, C.: Purely Functional Data Structures. Cambridge University Press, Cambridge (1998)
21. Hughes, J.: A novel representation of lists and its application to the function "reverse". Information Processing Letters 22, 141–144 (1986)
22. Kühnemann, A., Glück, R., Kakehi, K.: Relating Accumulative and Non-accumulative Functional Programs. In: Middeldorp, A. (ed.) RTA 2001. LNCS, vol. 2051, pp. 154–168. Springer, Heidelberg (2001)
23. Voigtländer, J.: Concatenate, reverse and map vanish for free. In: Proc. Int'l. Conf. on Functional Programming (ICFP), Pittsburgh, PA, pp. 14–25 (2002)
24. Wadler, P.: The concatenate vanishes. Unpublished manuscript (December 1987) (revised November 1989)
25. Gill, A., Launchbury, J., Jones, S.P.: A short cut to deforestation. In: Proceedings of Functional Programming Languages and Computer Architecture (FPCA), pp. 223–232. ACM Press (1993)

26. Coutts, D., Leshchinskiy, R., Stewart, D.: Stream fusion: from lists to streams to nothing at all. In: Hinze, R., Ramsey, N. (eds.) ICFP, pp. 315–326. ACM (2007)
27. Axelsson, E., Claessen, K., Dévai, G., Horváth, Z., Keijzer, K., Lyckegard, B., Persson, A., Sheeran, M., Svenningsson, J., Vajdax, A.: Feldspar: A domain specific language for digital signal processing algorithms. In: Proc. 8th IEEE/ACM Int'l. Conf. on Formal Methods and Models for Codesign (MEMOCODE), pp. 169–178. IEEE (2010)
28. Chakravarty, M., Keller, G., Lee, S., McDonell, T., Grover, V.: Accelerating Haskell array codes with multicore GPUs. In: Proc. 6th Workshop on Declarative Aspects of Multicore Programming (DAMP), pp. 3–14. ACM (2011)
29. Keller, G., Chakravarty, M., Leshchinskiy, R., Peyton Jones, S., Lippmeier, B.: Regular, shape-polymorphic, parallel arrays in Haskell. ACM SIGPLAN Notices 45(9), 261–272 (2010)
30. Emoto, K., Fischer, S., Hu, Z.: Generate, Test, and Aggregate —a Calculation-Based Framework for Systematic Parallel Programming with MapReduce. In: Seidl, H. (ed.) ESOP 2012. LNCS, vol. 7211, pp. 254–273. Springer, Heidelberg (2012)
31. Ramsey, N., Pfeffer, A.: Stochastic lambda calculus and monads of probability distributions. ACM SIGPLAN Notices 37(1), 154–165 (2002)
32. Dechter, R.: Bucket elimination: A unifying framework for probabilistic inference. NATO ASI Series D, Behavioural and Social Sciences 89, 75–104 (1998)
33. Erwig, M., Kollmansberger, S.: Probabilistic functional programming in Haskell (functional pearl). J. Functional Programming 16(01), 21–34 (2006)
34. Larsen, K.F.: Memory efficient implementation of probability monads. Unpublished manuscript (August 2011)
35. Kiselyov, O., Shan, C.-C.: Embedded Probabilistic Programming. In: Taha, W.M. (ed.) DSL 2009. LNCS, vol. 5658, pp. 360–384. Springer, Heidelberg (2009)
36. Park, S., Pfenning, F., Thrun, S.: A probabilistic language based on sampling functions. ACM TOPLAS 31(1), 4 (2008)
37. Mogensen, T.: Troll, a language for specifying dice-rolls. In: Proc. 2009 ACM Symp. on Applied Computing (SAC), pp. 1910–1915. ACM (2009)
38. Bhat, S., Agarwal, A., Vuduc, R., Gray, A.: A type theory for probability density functions. In: Proc. 39th SIGACT-SIGPLAN Symp. on Principles of Programming Languages (POPL). ACM press (January 2012)
39. Grust, T., Sakr, S., Teubner, J.: XQuery on SQL hosts. In: Proc. 30th Int'l. Conf. on Very Large Bata Bases, vol. 30, p. 263 (2004)
40. Grust, T.: Purely relational FLWORs. In: Proc. XIME-P (2005)
41. Grust, T., Rittinger, J., Schreiber, T.: Avalanche-safe LINQ compilation. Proc. VLDB Endow 3, 162–172 (2010)
42. Swierstra, W.: Data types à la carte. J. Functional Programming 18(4), 423–436 (2008)
43. Bahr, P., Hvitved, T.: Compositional data types. In: Proc. 7th ACM SIGPLAN Workshop on Generic Programming (WGP), pp. 83–94. ACM (2011)
44. Bahr, P., Hvitved, T.: Parametric compositional data types. In: Proc. Mathematically Structured Functional Programming, MSFP (2012)
45. Clavel, M., Durán, F., Eker, S., Lincoln, P., Martí-Oliet, N., Meseguer, J., Quesada, J.: Maude: Specification and programming in rewriting logic. Theoretical Computer Science 285(2), 187–243 (2002)

Resource-Driven CLP-Based
Test Case Generation

Elvira Albert[1], Miguel Gómez-Zamalloa[1], and José Miguel Rojas[2]

[1] DSIC, Complutense University of Madrid, E-28040 Madrid, Spain
[2] Technical University of Madrid, E-28660 Boadilla del Monte, Madrid, Spain

Abstract. Test Data Generation (TDG) aims at automatically obtaining test inputs which can then be used by a software testing tool to validate the functional behaviour of the program. In this paper, we propose *resource-aware* TDG, whose purpose is to generate test cases (from which the test inputs are obtained) with associated *resource consumptions*. The framework is parametric w.r.t. the notion of resource (it can measure memory, steps, etc.) and allows using software testing to detect bugs related to non-functional aspects of the program. As a further step, we introduce *resource-driven* TDG whose purpose is to guide the TDG process by taking resource consumption into account. Interestingly, given a *resource policy*, TDG is guided to generate test cases that adhere to the policy and avoid the generation of test cases which violate it.

1 Introduction

Test data generation (TDG) is the process of automatically generating *test inputs* for interesting test *coverage criteria*. Examples of coverage criteria are: *statement coverage* which requires that each line of the code is executed; *loop-k* which limits to a threshold k the number of times we iterate on loops. The standard approach to generate test cases statically is to perform a *symbolic execution* of the program [7, 8, 12, 14, 17, 18, 20], where the contents of variables are expressions rather than concrete values. Symbolic execution produces a system of constraints over the input variables consisting of the conditions to execute the different paths. The conjunction of these constraints represents the equivalence class of inputs that would take this path. In what follows, we use the term *test cases* to refer to such constraints. Concrete instantiations of the test cases that satisfy the constraints are generated to obtain test inputs for the program. Testing tools can later test the *functionality* of an application by executing such test inputs and checking that the output is as expected. The CLP-based approach to TDG of *imperative*[1] programs [4, 11] consists of two phases: (1) first, an imperative program is translated into an equivalent CLP program and (2) symbolic execution is performed on the CLP program by relying on the standard CLP's evaluation mechanisms (extended with a special treatment for heap-allocated data [11]) which provide symbolic execution for free.

---

[1] The application of this approach to TDG of logic programs must consider failure [20] and, to functional programs, should consider laziness, higher-order, etc. [9].

G. Vidal (Ed.): LOPSTR 2011, LNCS 7225, pp. 25–41, 2012.

Non-functional aspects of an application, like its resource consumption, are often more difficult to understand than functional properties. Profiling tools execute a program for concrete inputs to assess the associated resource consumption, a non-functional aspect of the program. Profilers can be parametric w.r.t. the notion of resource which often includes cost models like time, number of instructions, memory consumed, number of invocations to methods, etc. Usually, the purpose of profiling is to find out which parts of a device or software contribute most to its poor performance and find bugs related to the resource consumption.

In this paper, we propose *resource-aware TDG* which strives to build performance into test cases by additionally generating their resource consumption, thus enriching standard TDG with non-functional properties. The main idea is that, during the TDG process, we keep track of the exercised instructions to obtain the test case. Then, in a simple post-process we map each instruction into a corresponding cost, we obtain for each class of inputs a detailed information of its resource consumption (including the resources above). Our approach is not reproducible by first applying TDG, then instantiating the test cases to obtain concrete inputs and, finally, performing profiling on the concrete data. This is because, for some cost criteria, resource-aware TDG is able to generate symbolic (i.e., non-constant) costs. E.g., when measuring memory usage, the amount of memory might depend on an input parameter (e.g., the length of an array to be created is an input argument). The resource consumption of the test case will be a symbolic expression that profilers cannot compute.

A well-known problem of TDG is that it produces a large number of test cases even for medium size programs. This introduces scalability problems as well as complicates human reasoning on them. An interesting aspect of resource-aware TDG is that resources can be taken into account in order to filter out test cases which do not consume more (or less) than a given amount of resources, i.e, one can consider a *resource policy*. This leads to the idea of resource-*driven* TDG, i.e., a new heuristics which aims at guiding the TDG process to generate test cases that adhere to the resource policy. The potential interest is that we can prune the symbolic execution tree and produce, more efficiently, test cases for inputs which otherwise would be very expensive (and even impossible) to obtain.

Our approach to resource-driven CLP-based TDG consists of two phases. First, in a pre-process, we obtain (an over-approximation of) the set of *traces* in the program which lead to test cases that adhere to the resource policy. We sketch several ways of automatically inferring such traces, starting from the simplest one that relies on the call graph of the program to more sophisticated ones that enrich the abstraction to reduce the number of unfeasible paths. An advantage of formalizing our approach in a CLP-based setting is that traces can be partially defined and the TDG engine then completes them. Second, executing standard CLP-based TDG with a (partially) instantiated trace generates a test case that satisfies the resource policy (or it fails if the trace is unfeasible). An interesting aspect is that, if the trace is fully instantiated, TDG becomes deterministic and solutions can be found very efficiently. Also, since there is no need to backtrack, test cases for the different traces can be computed in parallel.

2 CLP-Based Test Case Generation

This section summarizes the CLP-based approach to TDG of [11] and extends it to incorporate *traces* in the CLP programs that will be instrumental later to define the resource-aware framework. CLP-based TDG consists of two main steps: (1) imperative programs are translated into an extended form of equivalent CLP-programs which incorporate built-in operations to handle dynamic data, and, (2) symbolic execution is performed on the CLP-translated programs by relying on the standard evaluation mechanisms of CLP with special operations to treat such built-ins. The next two sections overview these steps.

2.1 CLP-Translation with Traces

The translation of imperative (object-oriented) programs into equivalent CLP program has been subject of previous work (see, e.g.,[1, 10]). Therefore, we will not go into details of how the transformation is done, but rather simply recap the features of the translated programs in the next definition.

Definition 1 (CLP-translated program). *Given a method m with input arguments $\bar{x}$ and output arguments $\bar{y}$. Its CLP-translation consists of a set of predicates $m, m_1, \ldots, m_n$ such that each of them is defined by a set of rules of the form "$m(\mathtt{I},\mathtt{O},\mathtt{H_{in}},\mathtt{H_{out}}):- g,b_1,\ldots,b_n$." where:*

1. *m is the entry predicate (named as the method) and its arguments $\mathtt{I}$ and $\mathtt{O}$ are lists of variables that correspond to $\bar{x}$ and $\bar{y}$.*
2. *For the remaining predicates $m_1,\ldots,m_n$, $\mathtt{I}$ and $\mathtt{O}$ are, resp., the list of input and output arguments of this predicate.*
3. *$\mathtt{H_{in}}$ and $\mathtt{H_{out}}$ are, resp., the input and output heaps to each predicate.*
4. *If a predicate m_i is defined by multiple rules, the guards in each rule contain mutually exclusive conditions. We denote by m_i^k the $k-th$ rule defining m_i.*
5. *$g,b_1,\ldots,b_n$ are CLP-representations of equivalent instructions in the imperative language (as usual, a SSA transformation is performed on the variables), method invocations are replaced by calls to corresponding predicates, and operations that handle data in the heap are translated into built-in predicates (e.g., new_object(H,Class,Ref,H'), get_field(H,Ref,Fld,Val), etc.).*

Given a rule m_i^k, we denote by $instr(m_i^k)$ the sequence of instructions in the original program that have been translated into rule m_i^k.

As the imperative program is deterministic, the CLP translation is deterministic as well (point 4 in Def. 1). Observe that the global memory (or heap) is explicitly represented in the CLP program by means of logic variables. When a rule is invoked, the input heap $\mathtt{H_{in}}$ is received and, after executing its body, the heap might be modified, resulting in $\mathtt{H_{out}}$. The operations that modify the heap will be shown in the example. Note that the above definition proposes a translation to CLP as opposed to a translation to pure logic (e.g. to predicate logic or even to propositional logic, i.e., a logic that is not meant for "programming"). This is because we then want to execute the resulting translated

```
class Vector {
    int [] elems; int size, cap;
    Vector(int iCap) throws Exception{
        if (iCap > 0){
            elems = new int[iCap];
            cap = iCap; size = 0;
        } else
            throw new Exception();
    }
    void add(int x){
        if (size >= cap)
            realloc();
        elems[size++] = x;
    }
    void realloc(){
        int nCap = cap*2;
        int [] nElems = new int[nCap];
        for (int i=0; i<cap; i++) {
            nElems[i] = elems[i];
        }
        cap = nCap; elems = nElems;
    }
}
```

Fig. 1. Java source code example (1)

programs to perform TDG and this requires, among other things, handling a constraint store and then generating actual data from such constraints. CLP is a natural paradigm to perform this task.

In the next definition, we add a *trace term* as an additional argument to each rule of Def. 1 to keep track of the sequence of rules that are executed.

Definition 2 (CLP-translated program with trace). *Given the rule of Def. 1, its CLP-translation with trace is:* "$m(\mathrm{I},\mathrm{O},\mathrm{H_{in}},\mathrm{H_{out}},\mathrm{T}):- g,b'_1,\ldots,b'_n.$", *where* T *is the trace term for* m *of the form* $\mathsf{m}(\mathsf{k},\mathsf{P},\langle T_{c_i},\ldots,T_{c_m}\rangle)$. *Here,* P *is the list of trace parameters, i.e., the subset of the variables in rule* m^k *on which the resource consumption depends;* $c_i,\ldots,c_m$ *is the (possibly empty) subsequence of method calls in* $b_1,\ldots,b_n$. T_{c_j} *is a free logic variable representing the trace term associated to the call* c_j. *Calls in the body of the rule are extended with their corresponding trace terms, i.e., for all* $1 \le j \le n$, *if* $b_j \equiv p(I_p,O_p,H_{in_p},H_{out_p})$, *then* $b'_j \equiv p(I_p,O_p,H_{in_p},H_{out_p},T_{c_j})$; *otherwise* $b'_j \equiv b_j$.

Example 1. Our example in Fig. 1 shows class Vector, that contains a reference to an array of integers elems and two integer fields to keep track of its size and capacity (size and cap). The initial capacity of the array is set by the constructor (method init). The interesting aspect of class Vector is that, when

adding an element using method add, if the size has already reached the maximum capacity determined by field cap the size of the array is duplicated (by
method realloc) before actually adding the new element. Fig. 2 shows the (simplified and pretty-printed) CLP translation obtained by the PET system [4]
from the bytecode associated to class Vector. For brevity, we have omitted the
predicates that model the exceptional behavior. Observe that each method is
transformed into a set of predicates, some of them defined by several (mutually
exclusive) guarded rules. In particular, method add is transformed into predicates add, if and addc. Variable names in the decompiled program correspond
to the original names in the Java source. The operations that handle the heap
remain as built-in predicates. Heap references are written as terms of the form
r(Ref). Function *instr* in Def. 1 keeps the mapping between rules and bytecode
instructions. For instance, $instr(\text{init}^1)=\langle$iload icap, ifgt, aload this, iload
icap, newarray int, putfield elems, aload this, aload icap, putfield cap,
aload this, iconst 0, putfield size, return$\rangle$ is the sequence of bytecode instructions that have been translated into rule init. A trace term is made up
by the predicate name and number, the set of input arguments on which the
cost depends (e.g., rule realloc and its trace parameter NCap) and it recursively
includes the trace terms for the predicates it calls.

2.2   Symbolic Execution

When the imperative language does not use dynamic memory, CLP-translated
programs can be executed by using the standard CLP's execution mechanism
with all arguments being free variables. However, in order to generate arbitrary
heap-allocated data structures, it is required to define heap-related operations
which build the heap associated with a given path by using only the constraints
induced by the visited code. We rely in the CLP-implementation presented in
[11], where operations to create, read and modify heap-allocated data structures
are presented in detail. Briefly, at symbolic execution-time, the heap is represented as a list of locations which are pairs formed by a unique reference and a
cell. Each cell can be an object or an array. An object contains its type and its
list of fields, each one represented as a pair of the form *(signature, content)*. An
array contains its type, its length and its list of elements. Note that our CLP
translated programs manipulate the heap as a black-box through its associated
operations. For instance, a new object is created through a call to predicate
new_object(H_{In},Class,Ref,H_{Out}), where H_{In} is the current heap, Class is the new
object's type, Ref is a unique reference in the heap for accessing the new object
and H_{Out} is the new heap after allocating the object. Read-only operations do
not produce any output heap (e.g. get_field(H,Ref,Field,Value)). The remaining
operations are implemented likewise.

It is well-known that the execution tree to be traversed in symbolic execution is
in general infinite. This is because iterative constructs such as loops and recursion
whose number of iterations depends on the input values usually induce an infinite
number of execution paths when executed with unknown input values. It is
therefore essential to establish a *termination criterion*. In the context of TDG,

```
add([r(This),X],[],H,H₁,add(1,[],[T])) :- get_field(H,This,size,Size),
    get_field(H,This,cap,Cap), if([Size,Cap,r(This),X],[],H,H₁,T).
if¹([Size,Cap,r(This),X],[],H,H₁,if(1,[],[T])) :- Size #< Cap,
    addc([r(This),X],[],H,H₁,T).
if²([Size,Cap,r(This),X],[],H,H₂,if(2,[],[T₁,T₂])) :- Size #>= Cap,
    realloc([r(This)],[],H,H₁,T₁), addc([r(This),X],[],H₁,H₂,T₂).
addc([r(This),X],[],H,H₂,addc(1,[],[])) :- get_field(H,This,elems,r(Es)),
    get_field(H,This,size,Size), set_array(H,Es,Size,X,H₁),
    NSize #= Size+1, set_field(H₁,This,size,NSize,H₂).
realloc([r(This)],[],H,H₂,realloc(1,[NCap],[T])) :-
    get_field(H,This,cap,Cap), NCap #= Cap*2, new_array(H,int,NCap,NEs,H₁),
    loop([r(This),NCap,r(NEs),0],[],H₁,H₂,T).
loop([r(This),NCap,r(NEs),I],[],H,H₁,loop(1,[],[T])) :-
    get_field(H,This,cap,Cap), cond([Cap,I,r(This),NCap,r(NEs)],[],H,H₁,T).
cond¹([Cap,I,r(This),NCap,r(NEs)],[],H,H₂,cond(1,[],[])) :- I #>= Cap,
    set_field(H,This,cap,NCap,H₁), set_field(H₁,This,elems,r(NEs),H₂).
cond²([Cap,I,r(This),NCap,r(NEs)],[],H,H₂,cond(2,[],[T])) :- I #< Cap,
    get_field(H,This,elems,r(Es)), get_array(H,Es,I,E), set_array(H,NEs,I,E,H₁),
    NI #= I+1, loop([r(This),NCap,r(NEs),NI],[],H₁,H₂,T).
init¹([r(This),ICap],[],H,H₄,init(1,[ICap],[])) :- ICap #> 0,
    new_array(H,int,ICap,E,H₁),set_field(H₁,This,elems,r(E),H₂),
    set_field(H₂,This,cap,ICap,H₃),set_field(H₃,This,size,0,H₄).
init²([r(This),ICap],[],H,H₁,init(2,[ICap],[])) :- ICap #=< 0,
    new_object(H,'Exception',E,H₁).
```

Fig. 2. CLP translation

termination is usually ensured by the *coverage criterion* which guarantees that the set of paths generated produces test cases which meet certain degree of code coverage and the process terminates. In what follows, we denote by $\mathcal{T}_m^C$ the finite symbolic execution tree of method m obtained using coverage criterion C.

Definition 3 (test case with trace and TDG). *Given a method m, a coverage criterion C and a successful branch b in $\mathcal{T}_m^C$ with root $m(Args_{in}, Args_{out}, H_{in}, H_{out}, T)$, a test case with trace for m w.r.t. C is a 6-tuple of the form:* $\langle \sigma(Args_{in}), \sigma(Args_{out}), \sigma(H_{in}), \sigma(H_{out}), \sigma(T), \theta \rangle$, *where σ and θ are the set of bindings and constraint store, resp., associated to b. TDG generates the set of test cases with traces obtained for all successful branches in $\mathcal{T}_m^C$.*

The root of the execution tree is a call to method m with its associated arguments. Calls to methods are inlined in this scheme (one could use compositional TDG [5] as well). Each test case represents a class of inputs that will follow the same execution path, and its trace the sequence of rules applied along such path.

Example 2. Let us consider loop-1 as coverage criterion. In our example, loop-1 forces the array in the input vector to be at most of length 1. Note that we include

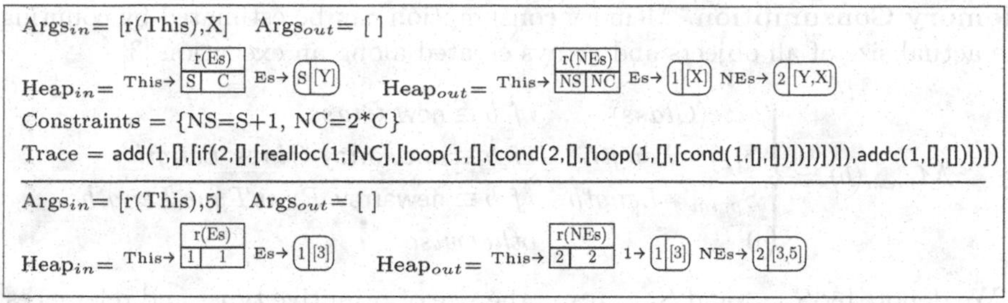

Args$_{in}$ = [r(This),X]   Args$_{out}$ = [ ]

Heap$_{in}$ = This→ [S | C] (r(Es)) Es→ (S [Y])   Heap$_{out}$ = This→ [NS | NC] (r(NEs)) Es→ (1 [X]) NEs→ (2 [Y,X])

Constraints = {NS=S+1, NC=2*C}

Trace = add(1,[],[if(2,[],[realloc(1,[NC],[loop(1,[],[cond(2,[],[loop(1,[],[cond(1,[],[])])])])])]),addc(1,[],[])])])

Args$_{in}$ = [r(This),5]   Args$_{out}$ = [ ]

Heap$_{in}$ = This→ [1 | 1] (r(Es)) Es→ (1 [3])   Heap$_{out}$ = This→ [2 | 2] (r(NEs)) 1→ (1 [3]) NEs→ (2 [3,5])

Fig. 3. Example of test case (up) and test input (down) for add with loop-1

the reference to the This object as an explicit input argument in the CLP translated program. The symbolic execution tree of add(Args$_{in}$,Args$_{out}$,H$_{in}$,H$_{out}$,T) will contain the following two successful derivations (ignoring exceptions):

1. If the size of the Vector object is less than its capacity, then the argument X is directly inserted in elems.
2. If the size of the Vector object is greater than or equal to its capacity, then method realloc is invoked before inserting X.

Fig. 3 shows in detail the second test case. Heaps are graphically represented by using rounded boxes for arrays (the array length appears to the left and the array itself to the right) and square boxes for Vector objects (field elems appears at the top, fields size and cap to the left and right bottom of the square, resp.). The trace-term T contains the rules that were executed along the derivation. At the bottom of the figure, an (executable) instantiation of this test case is shown.

3   Resource-Aware Test Case Generation

In this section, we present the extension of the TDG framework of Sec. 2 to build resource consumption into the test cases. First, in Sec. 3.1 we describe the cost models that we will consider in the present work. Then, Sec. 3.2 presents our approach to resource-aware TDG.

3.1   Cost Models

A cost model defines how much the execution of an instruction costs. Hence, the resource consumption of a test case can be measured by applying the selected cost model to each of the instructions exercised to obtain it.

Number of Instructions. The most traditional model, denoted $\mathcal{M}_{ins}$, is used to estimate the number of instructions executed. In our examples, since the input to our system is the bytecode of the Java program, we count the number of bytecode instructions. All instructions are assigned cost 1.

Memory Consumption. Memory consumption can be estimated by counting the actual size of all objects and arrays created along an execution [3].

$$\mathcal{M}_{mem}(b) = \begin{cases} size(Class) & if\ b \equiv \mathsf{new}\ Class \\ S_{ref} * Length & if\ b \equiv \mathsf{anewarray}\ Class\ Length \\ S_{prim} * Length & if\ b \equiv \mathsf{newarray}\ PrimType\ Length \\ 0 & otherwise \end{cases}$$

We denote by S_{prim} and S_{ref}, resp., the size of primitive types and references. In the examples, by assuming a standard JVM implementation, we set both values to 4 bytes. The size of a class is the sum of the sizes of the fields it defines. Note that, if one wants to consider garbage collection when assessing memory consumption, then the behaviour of the garbage collection should be simulated during the generation of test cases. In this paper, we assume that no garbage collection is performed.

Number of calls. This cost model, $\mathcal{M}_{call}$, counts the number of invocations to methods. It can be specialized to $\mathcal{M}_{call}^{m}$ to count calls to a specific method m which, for instance, can be one that triggers a billable event (e.g. send SMS).

Example 3. The application of the cost models $\mathcal{M}_{ins}$, $\mathcal{M}_{mem}$ and $\mathcal{M}_{call}$ to the sequence of instructions in rule init (i.e., $instr(init)$ of Ex. 1) results in, resp., 14 bytecode instructions, $4 * ICap$ bytes and 0 calls.

3.2   Resource-Aware TDG

Given the test cases with trace obtained in Def. 3, the associated cost can be obtained as a simple post-process in which we apply the selected cost models to all instructions associated to the rules in its trace.

Definition 4 (test case with cost). *Consider a test case with trace* $Test \equiv \langle Args_{in}, Args_{out}, H_{in}, H_{out}, \text{Trace}, \theta \rangle$, *obtained in Def. 3 for method m w.r.t. C. Given a cost model $\mathcal{M}$, the cost of Test w.r.t. $\mathcal{M}$, is defined as:*

$$C(Test, \mathcal{M}) = \mathsf{cost}(\text{Trace}, \mathcal{M})$$

where function cost *is recursively defined as:*

$$\mathsf{cost}(\mathtt{m(k,P,L)}, \mathcal{M}) = \begin{cases} \sum_{\forall i \in instr(m^k)} \mathcal{M}(i) & if\ \mathtt{L} = [\ ] \\ \sum_{\forall i \in instr(m^k)} \mathcal{M}(i) + \sum_{\forall l \in \mathtt{L}} \mathsf{cost}(l, \mathcal{M}) & otherwise \end{cases}$$

For the cost models in Sec. 3.1, we define the test case with cost *as a tuple of the form* $\langle Test, C(Test, \mathcal{M}_{ins}), C(Test, \mathcal{M}_{mem}), C(Test, \mathcal{M}_{call}) \rangle$.

This could also be done by profiling the resource consumption of the execution of the test-case. However, observe that our approach goes beyond the capabilities of TDG + profiling, as it can also obtain *symbolic* (non-constant) resource usage estimations while profilers cannot. Besides, it saves us from the non trivial implementation effort of developing a profiler for the language.

```
static Vector[] multiples(int[] ns, int div, int icap){
    Vector v = new Vector(icap);
    for (int i=0; i<ns.length; i++)
        if (ns[i]%div == 0)
            v.add(ns[i]);
    return r;
}
```

Fig. 4. Java source code example (2)

Args$_{in}$ = [r(Ns),Div,1] Args$_{out}$ = r(0)

Heap$_{in}$ = Ns→ 4 [E$_1$,E$_2$,E$_3$,E$_4$] Heap$_{out}$ = Ns→ 4 [E$_1$,E$_2$,E$_3$,E$_4$] 1→ r(4) 4 4 2→ 1 [E$_1$] 3→ 2 [E$_1$,E$_2$] 4→ 4 [E$_1$,E$_2$,E$_3$,E$_4$]

Constraints = {Div≠0, E$_1$ mod Div=0, E$_2$ mod Div=0, E$_3$ mod Div=0, E$_4$ mod Div=0}

$\mathcal{M}_{mem}$=2*$S_{PrimType}$+S_{Ref}+$S_{PrimType}$*1+$S_{PrimType}$*2+$S_{PrimType}$*4=40 bytes

$\mathcal{M}_{ins}$ = 270 bytecode instructions $\mathcal{M}_{call}^{realloc}$ = 2

Args$_{in}$ = [r(Ns),Div,ICap] Args$_{out}$ = r(0)

Heap$_{in}$ = Ns→ 2 [E$_1$,E$_2$] Heap$_{out}$ = Ns→ 4 [E$_1$,E$_2$] 1→ r(2) 4 ICap 2→ ICap [E$_1$,E$_2$,...]

Constraints = {ICap≥4, Div≠0, E$_1$ mod Div=0, E$_2$ mod Div≠0}

Trace = multiples(1,[],[init(1,[ICap],[]),mloop(1,[],[mcond(2,[],[mif(2,[],[add(1,[],[if(1,[],[addc(1,[],[])])]),
                    mloop(1,[],[mcond(2,[],[mif(1,[],[mloop(1,[],[mcond(1,[],[])])])])])])])])])])

$\mathcal{M}_{mem}$ = 2*$S_{PrimType}$+S_{Ref}+$S_{PrimType}$*ICap = 12+4*ICap

$\mathcal{M}_{ins}$ = 86 bytecode instructions $\mathcal{M}_{call}^{realloc}$ = 0

Args$_{in}$=[r(Ns),Div,3] Args$_{out}$=r(0)

Heap$_{in}$= Ns→ 4 [E$_1$,E$_2$,E$_3$,E$_4$] Heap$_{out}$= Ns→ 4 [E$_1$,E$_2$,E$_3$,E$_4$] 0→ r(2) 4 6 1→ 3 [E$_1$,E$_2$,E$_3$] 2→ 6 [E$_1$,E$_2$,E$_3$,E$_4$,0,0]

Constraints = {Div≠0, E$_1$ mod Div=0, E$_2$ mod Div=0, E$_3$ mod Div=0, E$_4$ mod Div=0}

$\mathcal{M}_{mem}$ = (2*$S_{PrimType}$+S_{Ref}+$S_{PrimType}$*3+$S_{PrimType}$*6) = 48 bytes

$\mathcal{M}_{ins}$ = 247 bytecode instructions $\mathcal{M}_{call}^{realloc}$ = 1

Fig. 5. Selected test cases with cost for method multiples with loop-4

Example 4. We use a slightly more complex example from now on. Fig. 4 shows method multiples, which receives an input array of integers ns and outputs an object of type Vector, created with initial capacity icap, containing all the elements of ns that are multiples of the second input argument div. Let us consider the TDG of the CLP translation of method multiples with loop-4 as coverage criterion. We get 54 test cases, which correspond to all possible executions for input arrays of length not greater than 4, i.e., at most 4 iterations of the *for* loop. Fig. 5 shows three test cases. The upper one corresponds to the test case that executes the highest number of instructions, in which method realloc is executed 2 times (worst case for $\mathcal{M}_{call}^{realloc}$ as well). The one in the middle corresponds to one of the paths with the highest parametric memory consumption (for brevity, only the trace for this case is shown), and the one at the bottom corresponds to that with the highest constant memory consumption. In the middle one, the input array Ns is of length 2, both elements in the array are multiples of Div, and the initial capacity is constrained by ICap $\geq$ 4. With such input configuration, the array is fully traversed and its two elements are inserted in the resulting Vector object. By applying the cost model $\mathcal{M}_{mem}$ to the trace in the figure, we obtain a symbolic heap consumption which is parametric on ICap (observe that ICap is a parameter of the second and third calls in the trace). Importantly, this parameter remains as a variable because method realloc is not executed. Symbolic execution of realloc would give a concrete value to ICap when determining the number of iterations of its loop. The test case at the bottom in contrast executes realloc once, as the vector runs out of capacity at the fourth iteration of the loop. Hence, its capacity is duplicated.

Resource-aware TDG has interesting applications. It can clearly be useful to detect, early within the software development process, bugs related to an excessive consumption of resources. Additionally, one of the well-known problems of TDG is that, even for small programs, it produces a large set of test cases which complicate the software testing process which, among other things, requires reasoning on the correctness of the program by verifying that the obtained test cases lead to the expected result. Resource-aware TDG can be used in combination with a *resource policy* in order to filter out test cases which do not adhere to the policy. The resource policy can state that the resource consumption of the test cases must be larger (or smaller) than a given threshold so that one can focus on the (potentially problematic) test cases which consume a certain amount of resources.

Example 5. Let us recall that in Ex. 4 we had obtained 54 test cases. By using a resource policy to focus on those cases that consume more than 48 bytes, we filter out 23 test cases. In a realistic scenario, the user must provide the testing framework with resource consumption parameters. For instance, by setting the amount of memory available in the resource policy, TDG could help us detect (potentially buggy) behaviours of the program under test which exceed the memory limit.

Furthermore, one can display to the user the test cases ordered according to the amount of resources they consume. For instance, for the cost model $\mathcal{M}_{mem}$, the test cases in Ex. 4 would be shown first. It is easy to infer the condition ICap > 9, which determines when the parametric test case is the most expensive one. Besides, one can implement a *worst-case* resource policy which shows to the user only the test case that consumes more resources among those obtained by the TDG process (e.g., the one at the top together with the previous condition for $\mathcal{M}_{mem}$), or display the n test cases with highest resource consumption (e.g., the two cases in Fig. 5 for $n = 2$).

4   Resource-Driven TDG

This section introduces *resource-driven TDG*, a novel heuristics to guide the symbolic execution process which improves, in terms of scalability, over the resource-aware approach, especially in those cases where restrictive resource policies are supplied. The main idea is to try to avoid, as much as possible, the generation of paths during symbolic execution that do not satisfy the policy. If the resource policy imposes a maximum threshold, then symbolic execution can stop an execution path as soon as the resource consumption exceeds it. However, it is often more useful to establish resource policies that impose a minimum threshold. In such case, it cannot be decided if a test case adheres to the policy until it is completely generated. Our heuristics to avoid the unnecessary generation of test cases that violate the resource policy is based on this idea: 1) in a pre-process, we look for traces corresponding to potential paths (or sub-paths) that adhere to the policy, and 2) we use such traces to guide the symbolic execution.

An advantage of relying on a CLP-based TDG approach is that the trace argument of our CLP-transformed programs can be used, not only as an output, but also as an input argument. Let us observe also that, we could either supply fully or partially instantiated traces, the latter ones represented by including free logic variables within the trace terms. This allows guiding, completely or partially, the symbolic execution towards specific paths.

Definition 5 (guided TDG). *Given a method m, a coverage criterion C, and a (possibly partial) trace π, guided TDG generates the set of test cases with traces, denoted $gTDG(m, C, \pi)$, obtained for all successful branches in $\mathcal{T}_m^C$.*

Observe that the symbolic execution guided by one trace (a) generates exactly one test case if the trace is complete and corresponds to a feasible path, (b) none if it is unfeasible, or (c) can also generate several test cases in case it is partial. In this case the traces of all test cases are instantiations of the partial trace.

Example 6. Let us consider the partial trace multiples(1, [ ], [init(1, [ICap], [ ]), mloop(1,[ ], [mcond(2, [ ], [mif(2, [ ], [A1, mloop(1, [ ], [mcond(2,[ ], [mif(2,[ ], [A2, mloop(1,[ ], [mcond(2, [ ], [mif(2, [ ], [A3, mloop(1, [ ], [mcond(2,[ ], [mif(2,[ ], [A4, mloop(1,[ ], [mcond(1, [ ], [ ])...]), which represents the paths that iterate four times in the *for* loop of method multiples (rules mloop[1] and mcond[2] in the CLP translated program), always following the *then* branch of the *if* statement (rule mif[2]), i.e. invoking method add. The trace is partial since it does not specify where the execution goes after method add is called (in other words, whether method realloc is executed or not). This is expressed by the free variables (A1, A2, A3 and A4) in the trace-term arguments. The symbolic execution guided by such trace produces four test cases which differ on the constraint on ICap, which is resp. ICap=1, ICap=2, ICap=3 and ICap$\geq$4. The first and the third test cases are the ones shown at the top and at the bottom resp. of Fig 5. All the executions represented by this partial trace finish with the evaluation to false of the loop condition (rule mcond[1]).

By relying on an oracle O that provides the traces, we now define resource-driven TDG as follows.

Definition 6 (resource-driven TDG). *Given a method m, a coverage criterion C and a resource-policy R, resource-driven TDG generates the set of test cases with traces defined by*

$$\bigcup_{i=1}^{n} gTDG(m, C, \pi_i)i$$

where $\{\pi_1, \ldots, \pi_n\}$ is the set of traces computed by an oracle O w.r.t R and C.

In the context of symbolic execution, there is an inherent need of carrying out a constraint store over the input variables of the program. When the constraint store becomes unsatisfiable, symbolic execution must discard the current execution path and backtrack to the last branching point in the execution tree. Therefore, in general it is not possible to parallelize the process. This is precisely what we gain with deterministic resource-guided TDG. Because the test cases are computed as the union of independent executions, they can be parallelized. Experimenting on a parallel infrastructure remains as future work.

This definition relies on a generic oracle. We will now sketch different techniques for defining specific oracles. Ideally, an oracle should be *sound, complete* and *effective*. An oracle is sound if every trace it generates satisfies the resource policy. It is complete if it generates all traces that satisfy the policy. Effectiveness is related to the number of unfeasible traces it generates. The larger the number, the less effective the oracle and the less efficient the TDG process. For instance, assuming a worst-case resource policy, one can think of an oracle that relies on the results of a static cost analyzer [1] to detect the methods with highest cost. It can then generate partial traces that force the execution go through such costly methods (combined with a terminating criterion). Such oracle can produce a trace as the one in Ex. 6 with the aim of trying to maximize the number of times method add (the potentially most costly one) is called. This kind of oracle can be quite effective though it will be in general unsound and incomplete.

4.1   On Soundness and Completeness of Oracles

In the following we develop a concrete scheme of an oracle which is sound, complete, and parametric w.r.t. both the cost model and the resource policy. Intuitively, an oracle is complete if, given a resource policy and a coverage criterion, it produces an over-approximation of the set of traces (obtained as in Def. 3) satisfying the resource policy and coverage criterion. We first propose a naive way of generating such an over-approximation which is later improved.

Definition 7 (trace-abstraction program). *Given a CLP-translated program with traces P, its trace-abstraction is obtained as follows: for every rule of P, (1) remove all atoms in the body of the rule except those corresponding to rule calls, and (2) remove all arguments from the head and from the surviving atoms of (1) except the last one (i.e., the trace term).*

The trace-abstraction of a program corresponds to its control-flow graph, and can be directly used as a trace-generator that produces a superset of the (usually infinite) set of traces of the program. The coverage criterion is applied in order to obtain a concrete and finite set of traces. Note that this is possible as long as the coverage criterion is structural, i.e., it only depends in the program structure (like loop-k). The resource policy can then be applied over the finite set: (1) in a post-processing where the traces that do not satisfy the policy are filtered out or (2) depending on the policy, by using a specialized search method.

As regards soundness, the intuition is that an oracle is sound if the resource consumption for the selected cost model is *observable* from the traces, i.e, it can be computed and it is equal to the one computed after the guided TDG.

Definition 8 (resource observability). *Given a method m, a coverage criterion C and a cost-model $\mathcal{M}$, we say that $\mathcal{M}$ is observable in the trace-abstraction for m, if for every feasible trace π generated from the trace-abstraction using C, we have that $\mathsf{cost}(\pi, \mathcal{M}) = \mathsf{cost}(\pi', \mathcal{M})$, where π' is a corresponding trace obtained for $gTDG(m, C, \pi)$.*

Observe that π can only have variables in trace parameters (second argument of a trace-term). This means that the only difference between π and π' can be made by means of instantiations (or associated constraints) performed during the symbolic execution on those variables. Trivially, $\mathcal{M}_{ins}$ and $\mathcal{M}_{call}$ are observable since they do not depend on such trace parameters. Instead, $\mathcal{M}_{mem}$ can depend on trace parameters and is therefore non-observable in principle on this trace-abstraction, as we will discuss later in more detail.

Enhancing trace-abstractions. Unfortunately the oracle proposed so far is in general very far from being effective since trace-abstractions can produce a huge amount of unfeasible traces. To solve this problem, we propose to enhance the trace-abstraction with information (constraints and arguments) taken from the original program. This can be done at many degrees of precision, from the empty enhancement (the one we have seen) to the full one, where we have the original program (hence the original resource-aware TDG). The more information we include, the less unfeasible traces we get, but the more costly the process is. The goal is thus to find heuristics that enrich sufficiently the abstraction so that many unfeasible traces are avoided and with the minimum possible information.

A quite effective heuristic is based on the idea of adding to the abstraction those program variables (input arguments, local variables or object fields) which get instantiated during symbolic execution (e.g., field size in our example). The idea is to enhance the trace-abstraction as follows. Let us start with a set of variables V initialized with those variables (this can be soundly approximated by means of static analysis). For every $v \in V$, we add to the program all occurrences of v and the guards and arithmetic operations in which v is involved. The remaining variables involved in those guards are added to V and the process is repeated until a fixpoint is reached. Fig. 6 shows the trace-abstraction with the proposed enhancement for our working example, in which variables Size and Cap (fields), ICap (input argument) and I (local variable) are added.

Resource Observability for $\mathcal{M}_{mem}$. As already mentioned, $\mathcal{M}_{mem}$ is in general non-observable in trace-abstractions. The problem is that the memory consumed by the creation of arrays depends on dynamic values which might be

```
multiples(ICap,multiples(1,[],[T₁,T₂])):- init(ICap,Size,Cap,T₁),
                              mloop(Size,Cap,T₂).
mloop(Size,Cap,mloop(1,[],[T])) :- mcond(Size,Cap,T).
mcond¹(_,_,mcond(1,[],[])).
mcond²(Size,Cap,mcond(2,[],[T])) :- mif(Size,Cap,T).
mif¹(Size,Cap,mif(1,[],[T])) :- mloop(Size,Cap,T).
mif²(Size,Cap,mif(2,[],[T₁,T₂])) :- add(Size,Cap,NSize,NCap,T₁),
                              mloop(NSize,NCap,T₂).
add(Size,Cap,NSize,NCap,add(1,[],[T])) :- if(Size,Cap,NSize,NCap,T).
if¹(Size,Cap,NSize,Cap,if(1,[],[T])) :- Size #\= Cap, addc(Size,NSize,T).
if²(Size,Cap,NSize,NCap,if(2,[],[T₁,T₂])) :- Size #= Cap,
                       realloc(Cap,NCap,T₁), addc(Size,NSize,T₂).
addc(Size,NSize,addc(1,[],[])) :- NSize #= Size+1.
realloc(Cap,NCap,realloc(1,[NCap],[T])) :- NCap #= Cap*2, loop(Cap,0,T).
loop(Cap,I,loop(1,[],[T])) :- cond(Cap,I,T).
cond¹(Cap,I,cond(1,[],[])) :- I #>= Cap.
cond²(Cap,I,cond(2,[],[T])) :- I #< Cap, NI #= I+1, loop(Cap,NI,T).
init¹(ICap,0,ICap,init(1,[ICap],[])).
init²(ICap,0,ICap,init(2,[ICap],[])).
```

Fig. 6. Enhanced trace-abstraction program

not present in the trace-abstraction. Again, this problem can be solved by enhancing the trace-abstraction with the appropriate information. In particular, the enhancement must ensure that the variables involved in the creation of new arrays (and those on which they depend) are added to the abstraction. This information can be statically approximated [2, 15, 16].

Instances of Resource-driven TDG. The resource-driven scheme has been deliberately defined as generic as possible and hence it could be instantiated in different ways for particular resource policies and cost-models producing more effective versions of it. For instance, for a worst-case resource policy, the oracle must generate all traces in order to know which is the one with maximal cost. Instead of starting a guided symbolic execution for all of them, we can try them one by one (or k by k in parallel) ordered from higher to lower cost, so that as soon as a trace is feasible the process stops. By correctness of the oracle, the trace will necessarily correspond to the feasible path with highest cost.

Theorem 1 (correctness of trace-driven TDG). *Given a cost model $\mathcal{M}$, a method m, a coverage criterion C and a sound oracle O on which $\mathcal{M}$ is observable, resource-driven TDG for m w.r.t. C using O generates the same test cases as resource-aware TDG w.r.t. C for the cost model $\mathcal{M}$.*

Soundness is trivially entailed by the features of the oracle.

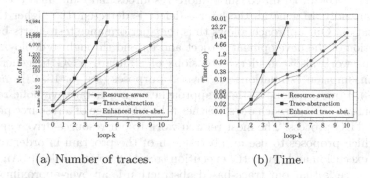

(a) Number of traces.   (b) Time.

Fig. 7. Preliminary experimental results

4.2 Performance of Trace-Driven TDG

We have performed some preliminary experiments on our running example using different values for k for the loop-k coverage criterion (X axis) and using a worst-case resource policy for the $\mathcal{M}_{ins}$ cost model. Our aim is to compare resource-aware TDG with the two instances of resource-driven TDG, the one that uses the naive trace-abstraction and the enhanced one. Fig. 7a depicts the number of traces which will be explored in each case. It can be observed that the naive trace-abstraction generates a huge number of unfeasible traces and the growth is larger as k increases. Indeed, from $k = 6$ on, the system runs out of memory when computing them. The enhanced trace-abstraction reduces drastically the number of unfeasible traces and besides the difference w.r.t. this number in resource-aware is a (small) constant. Fig. 7b shows the time to obtain the worst-case test case in each case. The important point to note is that resource-driven TDG outperforms resource-aware TDG in all cases, taking in average half the time w.r.t. the latter. We believe our results are promising and suggest that the larger the symbolic execution tree is (i.e., the more exhaustive TDG aims to be), the larger the efficiency gains of resource-driven TDG are. Furthermore, in a real system, the different test cases for resource-driven TDG could be computed in parallel and hence the benefits would be potentially larger.

5 Conclusions and Related Work

In this paper, we have proposed resource-aware TDG, an extension of standard TDG with resources, whose purpose is to build resource consumption into the test cases. Resource-aware TDG can be lined up in the scope of performance engineering, an emerging software engineering practice that strives to build performance into the design and architecture of systems. Resource-aware TDG can serve different purposes. It can be used to test that a program meets performance criteria up to a certain degree of code coverage. It can compare two systems to find which one performs better in each test case. It could even help finding out what parts of the program consume more resources and can cause the system to perform badly. In general, the later a defect is detected, the higher the cost of remediation. Our approach allows thus that performance test efforts begin at the inception of the development project and extend through to deployment.

Previous work also considers extensions of standard TDG to generate resource consumption estimations for several purposes (see [6, 13, 21] and their references). However, none of those approaches can generate symbolic resource estimations, as our approach does, neither take advantage of a resource policy to guide the TDG process. The most related work to our resource-driven approach is [19], which proposes to use an abstraction of the program in order to guide symbolic execution and prune the execution tree as a way to scale up. An important difference is that our trace-based abstraction is an over-approximation of the actual paths which allows us to select the most expensive paths. In contrast, their abstraction is an under-approximation which tries to reduce the number of test cases that are generated in the context of concurrent programming, where the state explosion can be problematic. Besides, our extension to infer the resources from the trace-abstraction and the idea to use it as a heuristics to guide the symbolic execution is new.

Acknowledgements. This work was funded in part by the Information & Communication Technologies program of the European Commission, Future and Emerging Technologies (FET), under the ICT-231620 *HATS* project, by the Spanish Ministry of Science and Innovation (MICINN) under the TIN-2008-05624 *DOVES* project, the UCM-BSCH-GR35/10-A-910502 *GPD* Research Group and by the Madrid Regional Government under the S2009TIC-1465 *PROMETIDOS-CM* project.

References

1. Albert, E., Arenas, P., Genaim, S., Puebla, G., Zanardini, D.: Cost Analysis of Java Bytecode. In: De Nicola, R. (ed.) ESOP 2007. LNCS, vol. 4421, pp. 157–172. Springer, Heidelberg (2007)
2. Albert, E., Arenas, P., Genaim, S., Puebla, G., Zanardini, D.: Removing useless variables in cost analysis of java bytecode. In: SAC 2008. ACM (2008)
3. Albert, E., Genaim, S., Gómez-Zamalloa, M.: Heap Space Analysis for Java Bytecode. In: Proc. of ISMM 2007. ACM Press (2007)
4. Albert, E., Gómez-Zamalloa, M., Puebla, G.: PET: A Partial Evaluation-based Test Case Generation Tool for Java Bytecode. In: Proc. of. PEPM 2010. ACM Press (2010)
5. Albert, E., Gómez-Zamalloa, M., Rojas, J.M., Puebla, G.: Compositional CLP-Based Test Data Generation for Imperative Languages. In: Alpuente, M. (ed.) LOPSTR 2010. LNCS, vol. 6564, pp. 99–116. Springer, Heidelberg (2011)
6. Antunes, J., Neves, N.F., Veríssimo, P.: Detection and prediction of resource-exhaustion vulnerabilities. In: Proc. of ISSRE 2008. IEEE Computer Society (2008)
7. Clarke, L.A.: A System to Generate Test Data and Symbolically Execute Programs. IEEE Transactions on Software Engineering 2(3), 215–222 (1976)

8. Engel, C., Hähnle, R.: Generating Unit Tests from Formal Proofs. In: Gurevich, Y., Meyer, B. (eds.) TAP 2007. LNCS, vol. 4454, pp. 169–188. Springer, Heidelberg (2007)
9. Fischer, S., Kuchen, H.: Systematic generation of glass-box test cases for functional logic programs. In: Proc. of PPDP 2007. ACM (2007)
10. Gómez-Zamalloa, M., Albert, E., Puebla, G.: Decompilation of Java Bytecode to Prolog by Partial Evaluation. JIST 51, 1409–1427 (2009)
11. Gómez-Zamalloa, M., Albert, E., Puebla, G.: Test Case Generation for Object-Oriented Imperative Languages in CLP. In: TPLP, ICLP 2010 Special Issue (2010)
12. Gotlieb, A., Botella, B., Rueher, M.: A CLP Framework for Computing Structural Test Data. In: Palamidessi, C., Moniz Pereira, L., Lloyd, J.W., Dahl, V., Furbach, U., Kerber, M., Lau, K.-K., Sagiv, Y., Stuckey, P.J. (eds.) CL 2000. LNCS (LNAI), vol. 1861, pp. 399–413. Springer, Heidelberg (2000)
13. Holzer, A., Januzaj, V., Kugele, S.: Towards resource consumption-aware programming. In: Proc. of ICSEA 2009. IEEE Computer Society (2009)
14. King, J.C.: Symbolic Execution and Program Testing. Communications of the ACM 19(7), 385–394 (1976)
15. Leuschel, M., Sørensen, M.H.: Redundant Argument Filtering of Logic Programs. In: Gallagher, J.P. (ed.) LOPSTR 1996. LNCS, vol. 1207, pp. 83–103. Springer, Heidelberg (1997)
16. Leuschel, M., Vidal, G.: Forward Slicing by Conjunctive Partial Deduction and Argument Filtering. In: Sagiv, M. (ed.) ESOP 2005. LNCS, vol. 3444, pp. 61–76. Springer, Heidelberg (2005)
17. Meudec, C.: Atgen: Automatic test data generation using constraint logic programming and symbolic execution. Softw. Test., Verif. Reliab. 11(2), 81–96 (2001)
18. Müller, R.A., Lembeck, C., Kuchen, H.: A symbolic java virtual machine for test case generation. In: IASTED Conf. on Software Engineering (2004)
19. Rungta, N., Mercer, E.G., Visser, W.: Efficient Testing of Concurrent Programs with Abstraction-Guided Symbolic Execution. In: Păsăreanu, C.S. (ed.) SPIN 2009. LNCS, vol. 5578, pp. 174–191. Springer, Heidelberg (2009)
20. Degrave, F., Schrijvers, T., Vanhoof, W.: Towards a Framework for Constraint-Based Test Case Generation. In: De Schreye, D. (ed.) LOPSTR 2009. LNCS, vol. 6037, pp. 128–142. Springer, Heidelberg (2010)
21. Zhang, J., Cheung, S.C.: Automated test case generation for the stress testing of multimedia systems. Softw., Pract. Exper. 32(15), 1411–1435 (2002)

A Declarative Embedding of XQuery in a Functional-Logic Language[*]

Jesús M. Almendros-Jiménez[1], Rafael Caballero[2],
Yolanda García-Ruiz[2], and Fernando Sáenz-Pérez[3]

[1] Dpto. Lenguajes y Computación, Universidad de Almería, Spain
[2] Dpto. de Sistemas Informáticos y Computación, UCM, Spain
[3] Dpto. de Ingeniería del Software e Inteligencia Artificial, UCM, Spain
jalmen@ual.es, {rafa,fernan}@sip.ucm.es , ygarciar@fdi.ucm.es

Abstract. This paper addresses the problem of integrating a fragment of XQuery, a language for querying XML documents, into the functional-logic language $\mathcal{TOY}$. The queries are evaluated by an interpreter, and the declarative nature of the proposal allows us to prove correctness and completeness with respect to the semantics of the subset of XQuery considered. The different fragments of XML that can be produced by XQuery expressions are obtained using the non-deterministic features of functional-logic languages. As an application of this proposal we show how the typical *generate and test* techniques of logic languages can be used for generating test-cases for XQuery expressions.

1 Introduction

XQuery has been defined as a query language for finding and extracting information from XML [15] documents. Originally designed to meet the challenges of large-scale electronic publishing, XML also plays an important role in the exchange of a wide variety of data on the Web and elsewhere. For this reason many modern languages include libraries or encodings of XQuery, including logic programming [1] and functional programming [6]. In this paper we consider the introduction of a simple subset of XQuery [18,20] into the functional-logic language $\mathcal{TOY}$ [11].

One of the key aspects of declarative languages is the emphasis they pose on the logic semantics underpinning declarative computations. This is important for reasoning about computations, proving properties of the programs or applying declarative techniques such as abstract interpretation, partial evaluation or algorithmic debugging [14]. There are two different declarative alternatives that can be chosen for incorporating XML into a (declarative) language:

1. Use a domain-specific language and take advantage of the specific features of the host language. This is the approach taken in [9], where a rule-based

---

[*] Work partially supported by the Spanish projects STAMP TIN2008-06622-C03-01, DECLARAWEB TIN2008-06622-C03-03, Prometidos-CM S2009TIC-1465 and GPD UCM-BSCH-GR58/08-910502.

G. Vidal (Ed.): LOPSTR 2011, LNCS 7225, pp. 42–56, 2012.

language for processing semi-structured data that is implemented and embedded into the functional-logic language Curry, and also in [13] for the case of logic programming.

2. Consider an existing query language such as XQuery, and embed a fragment of the language in the host language, in this case $\mathcal{TOY}$. This is the approach considered in this paper.

Thus, our goal is to include XQuery using the purely declarative features of the host languages. This allows us to prove that the semantics of the fragment of XQuery has been correctly included in $\mathcal{TOY}$. To the best of our knowledge, it is the first time a fragment of XQuery has been encoded in a functional-logic language. A first step in this direction was proposed in [5], where XPath [16] expressions were introduced in $\mathcal{TOY}$. XPath is a subset of XQuery that allows navigating and returning fragments of documents in a similar way as the path expressions used in the *chdir* command of many operating systems. The contributions of this paper with respect to [5] are:
- The setting has been extended to deal with a simple fragment of XQuery, including *for* statements for traversing XML sequences, *if/where* conditions, and the possibility of returning XML elements as results. Some basic XQuery constructions such as *let* statements are not considered, but we think that the proposal is powerful enough for representing many interesting queries.
- The soundness of the approach is formally proved, checking that the semantics of the fragment of XQuery is correctly represented in $\mathcal{TOY}$.

Next section introduces the fragment of XQuery considered and a suitable operational semantics for evaluating queries. The language $\mathcal{TOY}$ and its semantics are presented in Section 3. Section 4 includes the interpreter that performs the evaluation of simple XQuery expressions in $\mathcal{TOY}$. The theoretical results establishing the soundness of the approach with respect to the operational semantics of Section 2 are presented in Section 4.1. Section 5 explains the automatic generation of test cases for simple XQuery expressions. Finally, Section 6 concludes summarizing the results and proposing future work.

An extended version of the paper including proofs of the theoretical results can be found at [2].

2   XQuery and Its Operational Semantics

XQuery allows the user to query several documents, applying join conditions, generating new XML fragments, and using many other features [18,20]. The syntax and semantics of the language are quite complex [19], and thus only a small subset of the language is usually considered. The next subsection introduces the fragment of XQuery considered in this paper.

2.1   The Subset SXQ

In [4] a declarative subset of XQuery, called XQ, is presented. This subset is a core language for XQuery expressions consisting of *for*, *let* and *where/if* statements.

$$query ::= (\ ) \mid query\ query \mid tag$$
$$\mid doc(File) \mid doc(File)/axis :: \nu \mid var \mid var/axis :: \nu$$
$$\mid \textbf{for } var \textbf{ in } query \textbf{ return } query$$
$$\mid \textbf{if } cond \textbf{ then } query$$
$$cond ::= var{=}var \mid query$$
$$tag ::= \langle a \rangle\, var \ldots var\, \langle /a \rangle \mid \langle a \rangle\, tag\, \langle /a \rangle$$

Fig. 1. Syntax of SXQ, a simplified version of XQ

In this paper we consider a simplified version of XQ which we call SXQ and whose syntax can be found in Figure 1. where *axis* can be one of *child, self, descendant* or *dos* (i.e. descendant or self), and ν is a node test. The differences of SXQ with respect to XQ are:

1. XQ includes the possibility of using variables as tag names using a constructor *lab($x)*.
2. XQ permits enclosing any query Q between tag labels $\langle a \rangle Q \langle /a \rangle$. SXQ only admits either variables or other tags inside a tag.

Our setting can be easily extended to support the *lab($x)* feature, but we omit this case for the sake of simplicity in this presentation. The second restriction is more severe: although *let*s are not part of XQ, they could be simulated using *for* statements inside tags. In our case, forbidding other queries different from variables inside tag structures imply that our core language cannot represent *let* expressions. This limitation is due to the non-deterministic essence of our embedding, since a *let* expression means collecting all the results of a query instead of producing them separately using non-determinism. In spite of these limitations, the language SXQ is still useful for solving many common queries as the following example shows.

Example 1. Consider an XML file *"bib.xml"* containing data about books, and another file *"reviews.xml"* containing reviews for some of these books (see [17], sample data 1.1.2 and 1.1.4 to check the structure of these documents and an example). Then we can list the reviews corresponding to books in *"bib.xml"* as follows:

```
for $b in doc("bib.xml")/bib/book,
    $r in doc("reviews.xml")/reviews/entry
where $b/title = $r/title
for $booktitle in $r/title,
    $revtext in $r/review
return <rev> $booktitle  $revtext </rev>
```

The variable $b takes the value of the different books, and $r the different reviews. The *where* condition ensures that only reviews corresponding to the book are considered. Finally, the last two variables are only employed to obtain the book title and the text of the review, the two values that are returned as output of the query by the *return* statement.

It can be argued that the code of this example does not follow the syntax of Figure 1. While this is true, it is very easy to define an algorithm that converts a query formed by *for, where* and *return* statements into a SXQ query (as long as it only includes variables inside tags, as stated above). The idea is simply to convert the *where* into *ifs*, following each *for* by a *return*, and decomposing XPath expressions including several steps into several *for* expressions by introducing a new auxiliary variable and each one consisting of a single step.

Example 2. The query of Example 1 using SXQ syntax:

```
for $x1 in doc("bib.xml")/child::bib return
for $x2 in $x1/child::book  return
for $x3 in doc("reviews.xml")/child::reviews return
for $x4 in $x3/entry  return
if ($x2/title = $x4/title) then
    for $x5 in $x4/title return
      for $x6 in $x4/review return   <rev> $x5  $x6 </rev>
```

We end this subsection with a few definitions that are useful for the rest of the paper. The set of variables in a query Q is represented as $Var(Q)$. Given a query Q, we use the notation $Q_{|p}$ for representing the subquery Q' that can be found in Q at position p. Positions are defined as usual in syntax trees:

Definition 1. *Given a query Q and a position p, $Q_{|p}$ is defined as follows:*

$$
\begin{aligned}
Q_{|\varepsilon} &= Q \\
(Q_1\ Q_2)_{|(i \cdot p)} &= (Q_i)_{|p}\ i \in \{1,2\} \\
(\text{for var in } Q_1 \text{ return } Q_2)_{|(i \cdot p)} &= (Q_i)_{|p}\ i \in \{1,2\} \\
(\text{if } Q_1 \text{ then } Q_2)_{|(i \cdot p)} &= (Q_i)_{|p}\ i \in \{1,2\} \\
(\text{if var=var then } Q_1)_{|(1 \cdot p)} &= (Q_1)_{|p}
\end{aligned}
$$

Hence the position of a subquery is the path in the syntax tree represented as the concatenation of children positions $p_1 \cdot p_2 \ldots \cdot p_n$. For every position p, $\varepsilon \cdot p = p \cdot \varepsilon = p$. In general $Q_{|p}$ is not a proper SXQ query, since it can contain *free variables*, which are variables defined previously in *for* statements in Q. The set of variables of Q that are *relevant* for $Q_{|p}$ is the subset of $Var(Q)$ that can appear free in any subquery at position p. This set, denoted as $Rel(Q,p)$ is defined recursively as follows:

Definition 2. *Given a query Q, and a position p, $Rel(Q,p)$ is defined as:*

1. $\emptyset$,   *if $p = \varepsilon$.*
2. $Rel(Q_1, p')$,   *if $Q \equiv Q_1\ Q_2$, $p = 1 \cdot p'$.*
3. $Rel(Q_2, p')$,   *if $Q \equiv Q_1\ Q_2$, $p = 2 \cdot p'$.*
4. $Rel(Q_1, p')$,   *if $Q \equiv \text{for var in } Q_1 \text{ return } Q_2$, $p = 1 \cdot p'$.*
5. $\{var\} \cup Rel(Q_2, p')$,   *if $Q \equiv \text{for var in } Q_1 \text{ return } Q_2$, $p = 2 \cdot p'$.*
6. $Rel(Q_1, p')$,   *if $Q \equiv \text{if } Q_1 \text{ then } Q_2$, $p = 1 \cdot p'$.*

7. $Rel(Q_2, p')$, $if\ Q \equiv$ if Q_1 then Q_2, $p = 2 \cdot p'$.

Observe that cases $Q \equiv ()$, $Q \equiv tag$, $Q \equiv var$, $Q \equiv var/\chi :: \nu$, and var = var correspond to $p \equiv \varepsilon$.

Without loss of generality we assume that all the relevant variables for a given position are indexed starting from 1 at the outer level. We also assume that every for statement introduces a new variable. A query like for X in ((for Y in ...) (for Y in ...)) ... is then renamed to an equivalent query of the form for X_1 in ((for X_2 in ...) (for X_3 in ...)) ... (notice that the two Y variables occurred in different scopes).

2.2 XQ Operational Semantics

Figure 2 introduces the operational semantics of XQ that can be found in [4]. The only difference with respect to the semantics of this paper is that there is no rule for the constructor *lab*, for the sake of simplicity.

As explained in [4], the previous semantics defines the denotation of an XQ expression Q with k relevant variables, under a graph-like representation of a data forest $\mathcal{F}$, and a list of indexes $\bar{e}$ in $\mathcal{F}$, denoted by $[\![Q]\!]_k(\mathcal{F}, \bar{e})$. In particular, each relevant variable $\$x_i$ of Q has as value the tree of $\mathcal{F}$ indexed at position e_i. $\chi^{\mathcal{F}}(e_i, v)$ is a boolean function that returns *true* whenever v is the subtree of $\mathcal{F}$ indexed at position e_i. The operator $construct(a, (\mathcal{F}, [w_1...w_n]))$, denotes the construction of a new tree, where a is a label, $\mathcal{F}$ is a data forest, and $[w_1 ... w_n]$ is a list of nodes in $\mathcal{F}$. When applied, *construct* returns an indexed forest $(\mathcal{F} \cup T', [root(T')])$, where T' is a tree with domain a new set of nodes, whose root is labeled with a, and with the subtree rooted at the i-th (in sibling order) child of $root(T')$ being an isomorphic copy of the subtree rooted by w_i in $\mathcal{F}$. The symbol $\uplus$ used in the rules takes two indexed forests $(\mathcal{F}_1, l_1), (\mathcal{F}_2, l_2)$ and returns an indexed forest $(\mathcal{F}_1 \cup \mathcal{F}_2, l)$, where $l = l_1 \cdot l_2$. Finally, $tree(e_i)$ denotes the maximal tree within the input forest that contains the node e_i, hence $<_{doc}^{tree(e_i)}$ is the document order on the tree containing e_i.

$$[\![(\ )]\!]_k(\mathcal{F}, \bar{e}) = (\mathcal{F}, [\ ])$$
$$[\![Q_1\ Q_2]\!]_k(\mathcal{F}, \bar{e}) = [\![Q_1]\!]_k(\mathcal{F}, \bar{e}) \uplus [\![Q_2]\!]_k(\mathcal{F}, \bar{e})$$
$$[\![\text{for } \$x_{k+1} \text{in } Q_1 \text{ return } Q_2]\!]_k(\mathcal{F}, \bar{e}) = \text{let } (\mathcal{F}', \bar{l}) = [\![Q_1]\!]_k(\mathcal{F}, \bar{e}) \text{ in }$$
$$\uplus_{1 \le i \le |\bar{l}|} [\![Q_2]\!]_{k+1}(\mathcal{F}', \bar{e} \cdot l_i)$$
$$[\![\$x_i]\!]_k(\mathcal{F}, [e_1, \ldots, e_k]) = (\mathcal{F}, [e_i])$$
$$[\![\$x_i/\chi :: \nu]\!]_k(\mathcal{F}, [e_1, \ldots, e_k]) = (\mathcal{F}, \text{list of nodes } v \text{ such that } \chi^{\mathcal{F}}(e_i, v) \text{ and}$$
$$\text{label name of } v = \nu \text{ in order } <_{doc}^{tree(e_i)})$$
$$[\![\text{if } C \text{ then } Q_1]\!]_k(\mathcal{F}, \bar{e}) = \text{if } \pi_2([\![C]\!]_k(\mathcal{F}, \bar{e})) \ne [\ ] \text{ then } [\![Q_1]\!]_k(\mathcal{F}, \bar{e})$$
$$\text{else } (\mathcal{F}, [\ ])$$
$$[\![\$x_i = \$x_j]\!]_k(\mathcal{F}, [e_1, \ldots, e_k]) = \text{if } e_i = e_j \text{ then } construct(yes, (\mathcal{F}, [\ ]))$$
$$\text{else } (\mathcal{F}, [\ ])$$

Fig. 2. Semantics of Core XQuery

Without loss of generality this semantics assumes that all the variables relevant for a subquery are numbered consecutively starting by 1 as in Example 2. It also assumes that the documents appear explicitly in the query. That is, in Example 2 we must suppose that instead of *doc("bib.xml")* we have the XML corresponding to this document. Of course this is not feasible in practice, but simplifies the theoretical setting and it is assumed in the rest of the paper.

These semantic rules constitute a term rewriting system (TRS in short, see [3]), with each rule defining a single reduction step. The symbol $:=^*$ represents the reflexive and transitive closure of $:=$ as usual. The TRS is terminating and confluent (the rules are not overlapping). Normal forms have the shape $(\mathcal{F}, e_1, \dots, e_n)$ where $\mathcal{F}$ is a forest of XML fragments, and e_i are nodes in $\mathcal{F}$, meaning that the query returns the XML fragments (**indexed by**) $e_1, \dots, e_n$. The semantics evaluates a query starting with the expression $[\![Q]\!]_0(\emptyset, ())$. Along intermediate steps, expressions of the form $[\![Q']\!]_k(\mathcal{F}, \bar{e}_k)$ are obtained. The idea is that Q' is a subquery of Q with k relevant variables (which can occur free in Q'), that must take the values $\bar{e}_k$. The next result formalizes these ideas.

Proposition 1. *Let Q be a SXQ query. Suppose that*

$$[\![Q]\!]_0(\emptyset, ()) :=^* [\![Q']\!]_n(\mathcal{F}, \bar{e}_n)$$

Then:

- *Q' is a subquery of Q, that is, $Q' = Q_p$ for some p.*
- *$Rel(Q, p) = \{X_1, \dots, X_n\}$.*
- *Let S be the set of free variables in Q'. Then $S \subset Rel(Q, p)$.*
- *$[\![Q']\!]_n(\mathcal{F}, \bar{e}_n) = [\![Q'\theta]\!]_0(\emptyset, ())$, with $\theta = \{X_1 \mapsto e_1, \dots, X_n \mapsto e_n\}$*

Proof. Straightforward from Definition 2, and from the XQ semantic rules of Figure 2.

A more detailed discussion about this semantics and its properties can be found in [4].

3 $\mathcal{TOY}$ and Its Semantics

A $\mathcal{TOY}$ [11] program is composed of data type declarations, type alias, infix operators, function type declarations and defining rules for functions symbols. The syntax of *partial expressions* in $\mathcal{TOY}$ $e \in Exp_\perp$ is $e ::= \perp \mid X \mid h \mid (e\ e')$ where X is a variable and h either a function symbol or a data constructor. Expressions of the form $(e\ e')$ stand for the application of expression e (acting as a function) to expression e' (acting as an argument). Similarly, the syntax of *partial patterns* $t \in Pat_\perp \subset Exp_\perp$ can be defined as $t ::= \perp \mid X \mid c\ t_1 \dots t_m \mid f\ t_1 \dots t_m$ where X represents a variable, c a data constructor of arity greater or equal to m, and f a function symbol of arity greater than m, being t_i partial patterns for all $1 \le i \le m$. Each rule for a function f in $\mathcal{TOY}$ has the form:

$$\underbrace{f\ t_1 \dots t_n}_{\text{left-hand side}} \rightarrow \underbrace{r}_{\text{right-hand side}} \Leftarrow \underbrace{C_1, \dots, C_k}_{\text{condition}} \text{ where } \underbrace{s_1 = u_1, \dots, s_m = u_m}_{\text{local definitions}}$$

where u_i and r are expressions (that can contain new extra variables), C_j are strict equalities, and t_i, s_i are patterns. In $\mathcal{TOY}$, variable names must start with either an uppercase letter or an underscore (for anonymous variables), whereas other identifiers start with lowercase.

Data type declarations and type alias are useful for representing XML documents in $\mathcal{TOY}$:

```
data node       = txt     string
                | comment string
                | tag     string [attribute] [node]
data attribute = att     string string
type xml        = node
```

The data type **node** represents nodes in a simple XML document. It distinguishes three types of nodes: texts, tags (element nodes), and comments, each one represented by a suitable data constructor and with arguments representing the information about the node. For instance, constructor **tag** includes the tag name (an argument of type **string**) followed by a list of attributes, and finally a list of child nodes. The data type **attribute** contains the name of the attribute and its value (both of type **string**). The last type alias, **xml**, renames the data type **node**. Of course, this list is not exhaustive, since it misses several types of XML nodes, but it is enough for this presentation.

$\mathcal{TOY}$ includes two primitives for loading and saving XML documents, called **load_xml_file** and **write_xml_file** respectively. For convenience all the documents are started with a dummy node **root**. This is useful for grouping several XML fragments. If the file contains only one node N at the outer level, the **root** node is unnecessary, and can be removed using this simple function:

```
load_doc F = N <== load_xml_file F == xmlTag "root" [] [N]
```

where F is the name of the file containing the document. Observe that the strict equality == in the condition forces the evaluation of **load_xml_file** F and succeeds if the result has the form **xmlTag "root" [] [N]** for some N. If this is the case, N is returned.

The constructor-based ReWriting Logic (CRWL) [7] has been proposed as a suitable declarative semantics for functional-logic programming with lazy non-deterministic functions. The calculus is defined by five inference rules (see Figure 3): (BT) that indicates that any expression can be approximated by bottom, (RR) that establishes the reflexivity over variables, the decomposition rule (DC), the (JN) (join) rule that indicates how to prove strict equalities, and the function application rule (FA). In every inference rule, $r, e_i, a_j \in Exp_\perp$ are partial expressions and $t, t_k \in Pat_\perp$ are partial patterns. The notation $[P]_\perp$ of the inference rule FA represents the set $\{(l \to r \Leftarrow C)\theta \mid (l \to r \Leftarrow C) \in P, \ \theta \in Subst_\perp\}$ of partial instances of the rules in program P ($Subst_\perp$ represents the set of partial

BT $\qquad\qquad\qquad\qquad e \rightarrow \bot$

RR $\qquad\qquad\qquad\qquad X \rightarrow X \qquad\qquad\qquad$ with $X \in Var$

DC $\qquad\qquad \dfrac{e_1 \rightarrow t_1 \ \dots \ e_m \rightarrow t_m}{h \ \overline{e}_m \rightarrow h \ \overline{t}_m} \qquad\qquad h \ \overline{t}_m \in Pat_\bot$

JN $\qquad\qquad\qquad \dfrac{e \rightarrow t \quad e' \rightarrow t}{e \ == \ e'} \qquad\qquad t \in Pat$ (total pattern)

FA $\qquad \dfrac{e_1 \rightarrow t_1 \ \dots \ e_n \rightarrow t_n \quad C \quad r \ \overline{a}_k \rightarrow t}{f \ \overline{e}_n \ \overline{a}_k \ \rightarrow t}$
$\qquad\qquad$ if $(f \ \overline{t}_n \rightarrow r \Leftarrow C) \in [P]_\bot , t \neq \bot$

Fig. 3. CRWL Semantic Calculus

substitutions that replace variables by partial terms). The most complex inference rule is FA (Function Application), which formalizes the steps for computing a *partial pattern* t as approximation of a function call $f \ \overline{e}_n$:

1. Obtain partial patterns t_i as suitable approximations of the arguments e_i.
2. Apply a program rule $(f \ \overline{t}_n \rightarrow r \Leftarrow C) \in [P]_\bot$, verify the condition C, and check that t approximates the right-hand side r.

In this semantic notation, local declarations $a = b$ introduced in $\mathcal{TOY}$ syntax by the reserved word **where** are part of the condition C as approximation statements of the form $b \rightarrow a$.

The semantics in $\mathcal{TOY}$ allows introducing non-deterministic functions, such as the following function **member** that returns all the elements in a list:

```
member:: [A]  -> A
member  [X | Xs] =  X
member  [X | Xs] =  member Xs
```

Another example of $\mathcal{TOY}$ function is the definition of the infix operator `.::.` for XPath expressions (the operator `::` in XPath syntax):

```
(.::.) :: (A -> B) -> (B -> C) -> (A -> C)
(F .::. G) X = G (F X)
```

As the examples show, $\mathcal{TOY}$ is a typed language. However the type declaration is optional and in the rest of the paper they are omitted for the sake of simplicity. *Goals* in $\mathcal{TOY}$ are sequences of strict equalities. A strict equality $e_1 == e_2$ holds (inference JN) if both e_1 and e_2 can be reduced to the same total pattern t. For instance, the goal **member [1,2,3,4] == R** yields four answers, the four values for R that make the equality true: $\{R \mapsto 1\}, \dots, \{R \mapsto 4\}$.

4   Transforming SXQ into $\mathcal{TOY}$

In order to represent SXQ queries in $\mathcal{TOY}$ we use some auxiliary datatypes:

```
type xPath = xml-> xml

data sxq  = xfor xml sxq sxq | xif cond sxq | xmlExp xml |
                    xp path | comp sxq sxq
data cond = xml := xml | cond sxq
data path = var xml | xml :/ xPath | doc string xPath
```

The structure of the datatype `sxq` allows representing any SXQ query (see SXQ syntax in Figure 1). It is worth noticing that a variable introduced by a *for* statement has type `xml`, indicating that the variable always contains a value of this type. $\mathcal{TOY}$ includes a primitive `parse_xquery` that translates any SXQ expression into its corresponding representation as a term of this datatype, as the next example shows:

Example 3. The translation of the SXQ query of Example 2 into the datatype `sxq` produces the following $\mathcal{TOY}$ data term:

```
Toy> parse_xquery  "for $x1 in doc(\"bib.xml\")/child::bib return
          for $x2 in .....    <rev> $x5  $x6 </rev>" == R
yes
{R --> xfor X1 (xp (doc "bib.xml" (child .::. (nameT "bib"))))
    (xfor X2 (xp ( X1 :/ (child .::.(nameT "book"))))
    (xfor X3 (xp (doc "reviews.xml" (child   .::. (nameT "reviews"))))
    (xfor X4 (xp ( X3 :/ (child .::.(nameT "entry"))))
    (xif  ((xp(X2 :/ (child .::.(nameT "title")))) :=
              (xp(X4 :/ (child .::.(nameT "title")))))
    (xfor X5 (xp ( X4 :/ (child .::.(nameT "title"))))
    (xfor X6 (xp ( X4 :/ (child .::.(nameT "review"))))
      (xmlExp (xmlTag "rev" [] [X5,X6]))))))))))
}
```

The interpreter assumes the existence of the infix operator `.::.` that connects axes and tests to build steps, defined as the sequence of applications in Section 3.

The rules of the $\mathcal{TOY}$ interpreter that processes SXQ queries can be found in Figure 4. The main function is `sxq`, which distinguishes cases depending of the form of the query. If it is an XPath expression then the auxiliary function `sxqPath` is used. If the query is an XML expression, the expression is just returned (this is safe thanks to our constraint of allowing only variables inside XML expressions). If we have two queries (`comp` construct), the result of evaluating any of them is returned using non-determinism. The **for** statement (`xfor` construct) forces the evaluation of the query `Q1` and binds the variable `X` to the result. Then the result query `Q2` is evaluated. The case of the **if** statement is analogous. The XPath subset considered includes tests for attributes (`attr`), label names (`nameT`), general elements (`nodeT`) and text nodes (`textT`). It also includes the axes `self`, `child`, `descendant` and `dos`. Observe that we do not

```
sxq (xp E)          = sxqPath E
sxq (xmlExp X)      = X
sxq (comp Q1 Q2)    = sxq Q1
sxq (comp Q1 Q2)    = sxq Q2
sxq (xfor X Q1 Q2)  = sxq Q2 <== X== sxq Q1
sxq (xif (Q1:=Q2) Q3) = sxq Q3 <== sxq Q1 == sxq Q2
sxq (xif (cond Q1) Q2) = sxq Q2 <== sxq Q1 == _


sxqPath (var X) = X
sxqPath (X :/ S) = S X
sxqPath (doc F S) = S (load_xml_file F)


%%% XPATH %%%%
attr A (xmlTag S Attr L ) = xmlText T <== member Attr == xmlAtt A T
nameT S (xmlTag S Attr L ) = xmlTag S Attr L
nodeT X = X
textT (xmlText S) = xmlText S
commentT S (xmlComment S) = xmlComment S


self X = X
child (xmlTag _Name _Attr L) = member L
descendant X = child X
descendant X = descendant Y <== child X == Y
dos = self
dos = descendant
```

Fig. 4. $\mathcal{TOY}$ transformation rules for SXQ

include reverse axes like `ancestor` because they can be replaced by expressions including forward axes, as shown in [12]. Other constructions such as filters can be easily included (see [5]). The next example uses the interpreter to obtain the answers for the query of our running example.

Example 4. The goal `sxq (parse_xquery "for....")) == R` applies the interpreter of Figure 4 to the code of Example 2 (assuming that the string after `parse_xquery` is the query in Example 2), and returns the $\mathcal{TOY}$ representation of the expected results:

```
<rev>
 <title>TCP/IP Illustrated</title>
 <review>  One of the best books on TCP/IP.  </review>
</rev>
...
```

4.1 Soundness of the Transformation

One of the goals of this paper is to ensure that the embedding is semantically correct and complete. This section introduces the theoretical results establishing these properties. If V is a set of indexed variables of the form $\{X_1, \ldots, X_n\}$ we use the notation $\theta(V)$ to indicate the sequence $\theta(X_1), \ldots, \theta(X_n)$. In these results it is implicitly assumed that there is a bijective mapping f from XML format to the datatype xml in $\mathcal{TOY}$. Also, variables in XQuery $\$x_i$ are assumed to be represented in $\mathcal{TOY}$ as X_i and conversely. However, in order to simplify the presentation, we omit the explicit mention to f and to f^{-1}.

Lemma 1. *Let P be a $\mathcal{TOY}$ program, Q' an SXQ query, and Q, p such that $Q \equiv Q'_{|p}$. Define $V = Rel(Q', p)$ (see Definition 2), and $k = |V|$. Let θ be a substitution such that $\mathcal{P} \vdash (\text{sxq } Q\theta == t)$ for some pattern t.*
Then $[\![Q]\!]_k(\mathcal{F}, [\theta(V)]) :=^* (\mathcal{F}', L)$, for some forests $\mathcal{F}, \mathcal{F}'$ and with L verifying $t \in L$.

The theorem that establishes the correctness of the approach is an easy consequence of the Lemma.

Theorem 1. *Let P be the $\mathcal{TOY}$ program of Figure 4, Q an SXQ query, t a $\mathcal{TOY}$ pattern, and θ a substitution such that $\mathcal{P} \vdash (\text{sxq } Q\theta == t)$ for some θ. Then $[\![Q]\!]_0(\emptyset, []) :=^* (\mathcal{F}, L)$, for some forest $\mathcal{F}$, and L verifying $t \in L$.*

Proof. In Lemma 1 consider the position $p \equiv \varepsilon$. Then $Q' \equiv Q$, $V = \emptyset$ and $k = 0$. Without loss of generality we can restrict in the conclusion to $\mathcal{F} = \emptyset$, because $\theta(V) = \emptyset$ and therefore $\mathcal{F}$ is not used during the rewriting process. Then the conclusion of the theorem is the conclusion of the lemma.

Thus, our approach is correct. The next Lemma allows us to prove that it is also complete, in the sense that the $\mathcal{TOY}$ program can produce every answer obtained by the XQ operational semantics.

Lemma 2. *Let $\mathcal{P}$ be the $\mathcal{TOY}$ program of Figure 4. Let Q' be a SXQ query and Q, p such that $Q \equiv Q'_{|p}$. Define $V = Rel(Q', p)$ (see Definition 2) and $k = |V|$. Suppose that $[\![Q]\!]_k(\mathcal{F}, \bar{e}_k) :=^* (\mathcal{F}', \bar{a}_n)$ for some $\mathcal{F}, \mathcal{F}', \bar{e}_k, \bar{a}_n$.*
Then, for every a_j, $1 \le j \le n$, there is a substitution θ such that $\theta(X_i) = e_i$ for $X_i \in V$ and a CRWL-proof proving $\mathcal{P} \vdash \text{sxq } Q\theta == a_j$.

As in the case of correctness, the completeness theorem is just a particular case of the Lemma:

Theorem 2. *Let $\mathcal{P}$ be the $\mathcal{TOY}$ program of Figure 4. Let Q be a SXQ query and suppose that $[\![Q]\!]_k(\emptyset, []) :=^* (\mathcal{F}, \bar{a}_n)$ for some $\mathcal{F}, \bar{a}_n$. Then for every a_j, $1 \le j \le n$, there is $\mathcal{P} \vdash (\text{sxq } Q)\theta == a_j$ for some substitution θ.*

Proof. As in Theorem 1, suppose $p \equiv \varepsilon$ and thus $Q' \equiv Q$. Then $V = \emptyset$ and $k = 0$. Then, if $[\![Q]\!]_0(\emptyset, \emptyset) :=^* (\mathcal{F}, \bar{a}_n)$ it is easy to check that $[\![Q]\!]_0(\mathcal{F}', \emptyset) :=^* (\mathcal{F}, \bar{a}_n)$ for any $\mathcal{F}'$. Then the conclusion of the lemma is the same as the conclusion of the Theorem.

The proofs of Lemmata 1 and 2 can be found in [2].

5 Application: Test Case Generation

In this section we show how an embedding of SXQ into $\mathcal{TOY}$ can be used for obtaining test-cases for the queries. For instance, consider the erroneous query of the next example.

Example 5. Suppose that the user also wants to include the publisher of the book among the data obtained in Example 1. The following query tries to obtain this information:

```
Q = for $b in doc("bib.xml")/bib/book,
        $r in doc("reviews.xml")/reviews/entry,
        where $b/title = $r/title
        for $booktitle in $r/title,
            $revtext in $r/review,
            $publisher in $r/publisher
        return   <rev> $booktitle $publisher $revtext </rev>
```

However, there is an error in this query, because in `$r/publisher` the variable `$r` should be `$b`, since the publisher is in the document *"bib.xml"*, not in *"reviews.xml"*. The user does not notice that there is an error, tries the query (in $\mathcal{TOY}$ or in any XQuery interpreter) and receives an empty answer.

In order to check whether a query is erroneous, or even to help finding the error, it is sometimes useful to have test-cases, i.e., XML files which can produce some answer for the query. Then the test-cases and the original XML documents can be compared, and this can help finding the error. In our setting, such test-cases are obtained for free, thanks to the generate and test capabilities of logic programming. The general process can be described as follows:

1. Let Q' be the translation `parse_xquery` Q of query Q into $\mathcal{TOY}$.
2. Let $F_1, \ldots, F_k$ be the names of the XML documents occurring in Q'. That is, for each F_i, $1 \leq i \leq k$, there is an occurrence of an expression of the form `load_xml_file`(F_i) in Q' (which corresponds to expressions doc(F_i) in Q). Let Q'' be the result of replacing each doc(F_i) expression by a new variable D_i, for $i = 1 \ldots k$.
3. Let *"expected.xml"* be a document containing an expected answer for the query Q.
4. Let E the expression Q''==load_doc *"expected.xml"*.
5. Try the goal
   G $\equiv$ E, `write_xml_file` D_1 F'_1, ..., `write_xml_file` D_k F'_k

The idea is that the goal G looks for values of the logic variables D_i fulfilling the strict equality. The result is that after solving this goal, the D_i variables contain XML documents that can produce the expected answer for this query. Then each document is saved into a new file with name F'_i. For instance F'_i can consist of the original name F_i preceded by some suitable prefix *tc*. The process can be automatized, and the result is the code of Figure 5.

```
prepareTC (xp E)          = (xp E',L)
                            where (E',L) = prepareTCPath E
prepareTC (xmlExp X)      = (xmlExp X, [])
prepareTC (comp Q1 Q2)    = (comp Q1' Q2', L1++L2)
                              where (Q1',L1) = prepareTC Q1
                                    (Q2',L2) = prepareTC Q2
prepareTC (xfor X Q1 Q2)  = (xfor X Q1' Q2', L1++L2)
                              where (Q1',L1) = prepareTC Q1
                                    (Q2',L2) = prepareTC Q2
prepareTC (xif (Q1:=Q2) Q3) = (xif (Q1':=Q2') Q3',L1++(L2++L3))
                              where (Q1',L1) = prepareTC Q1
                                    (Q2',L2) = prepareTC Q2
                                    (Q3',L3) = prepareTC Q3
prepareTC (xif (cond Q1) Q2) = (xif (cond Q1) Q2, L1++L2)
                              where (Q1',L1) = prepareTC Q1
                                    (Q2',L2) = prepareTC Q2


prepareTCPath (var X)   = (var X, [])
prepareTCPath (X :/ S)  = (X :/ S, [])
prepareTCPath (doc F S) = (A :/ S, [write_xml_file A ("tc"++F)])

generateTC Q F = true <==  sxq Qtc == load_doc F, L==_
                where (Qtc,L) = prepareTC Q
```

Fig. 5. $\mathcal{TOY}$ transformation rules for SXQ

The code uses the list concatenation operator ++ which is defined in $\mathcal{TOY}$ as usual in functional languages such as Haskell. It is worth observing that if there are no test-case documents that can produce the expected result for the query, the call to generateTC will loop. The next example shows the generation of test-cases for the wrong query of Example 5.

Example 6. Consider the query of Example 5 , and suppose the user writes the following document "expected.xml":

```
<rev>
 <title>Some title</title>
 <review>The review</review>
 <publisher>Publisher</publisher>
</rev>
```

This is a possible expected answer for the query. Now we can try the goal:

```
Toy> Q == parse_xquery "for....", R == generateTC Q "expected.xml"
```

The first strict equality parses the query, and the second one generates the XML documents which constitute the test cases. In this example the test-cases obtained are:

```
                                    % revtc.xml
% bibtc.xml                         <reviews>
<bib>                                <entry>
 <book>                               <title>Some title</title>
  <title>Some title</title>           <review>The review </review>
 </book>                              <publisher>Publisher</publisher>
</bib>                                </entry>
                                    </reviews>
```

By comparing the test-case "revtc.xml" with the file "reviews.xml" we observe that the publisher is not in the reviews. Then it is easy to check that in the query the publisher is obtained from the reviews instead of from the *bib* document, and that this constitutes the error.

6   Conclusions

The paper shows the embedding of a fragment of the XQuery language for query-ing XML documents into the functional-logic language $\mathcal{TOY}$. Although only a small subset of XQuery consisting of *for*, *where/if* and *return* statements has been considered, the users of $\mathcal{TOY}$ can now perform simple queries such as *join* operations. The formal definition of the embedding allows us to prove the sound-ness of the approach with respect to the operational semantics of XQuery. The proposal respects the declarative nature of $\mathcal{TOY}$, exploiting its non-deterministic nature for obtaining the different results produced by XQuery expressions. An advantage of this approach with respect to the use of lists usually employed in functional languages is that our embedding allows the user to generate test-cases automatically when possible, which is useful for testing the query, or even for helping to find the error in the query. An extended version of this paper, includ-ing the proofs of the theoretical results and more detailed explanations about how to install $\mathcal{TOY}$ and run the prototype can be found in [2].

The most obvious future work would be introducing the *let* statement, which presents two novelties. The first is that they are *lazy*, that is, they are not evaluated if they are not required by the result. This part is easy to fulfill since we are in a lazy language. In particular, they could be introduced as local definitions (*where* statements in $\mathcal{TOY}$). The second novelty is more difficult to capture, and it is that the variables introduced by *let* represent an XML sequence. The natural representation in $\mathcal{TOY}$ would be a list, but the non-deterministic nature of our proposal does not allow us to collect all the results provided by an expression in a declarative way. A possible idea would be to use the functional-logic language Curry [8] and its encapsulated-search [10], or even the non-declarative *collect* primitive included in $\mathcal{TOY}$. In any case, this will imply a different theoretical framework and new proofs for the results. A different line for future work is the use of test cases for finding the error in the query using some variation of declarative debugging [14] that could be applied to this setting.

References

1. Almendros-Jiménez, J.M.: An Encoding of XQuery in Prolog. In: Bellahsène, Z., Hunt, E., Rys, M., Unland, R. (eds.) XSym 2009. LNCS, vol. 5679, pp. 145–155. Springer, Heidelberg (2009)
2. Almendros-Jiménez, J., Caballero, R., García-Ruiz, Y., Sáenz-Pérez, F.: A Declarative Embedding of XQuery in a Functional-Logic Language. Technical Report SIC-04/11, Facultad de Informática, Universidad Complutense de Madrid (2011), http://gpd.sip.ucm.es/rafa/xquery/
3. Baader, F., Nipkow, T.: Term Rewriting and All That. Cambridge University Press (1999)
4. Benedikt, M., Koch, C.: From XQuery to relational logics. ACM Trans. Database Syst. 34, 25:1–25:48 (2009)
5. Caballero, R., García-Ruiz, Y., Sáenz-Pérez, F.: Integrating XPath with the Functional-Logic Language Toy. In: Rocha, R., Launchbury, J. (eds.) PADL 2011. LNCS, vol. 6539, pp. 145–159. Springer, Heidelberg (2011)
6. Fegaras, L.: Propagating updates through XML views using lineage tracing. In: IEEE 26th International Conference on Data Engineering (ICDE), pp. 309–320 (March 2010)
7. González-Moreno, J., Hortalá-González, M., López-Fraguas, F., Rodríguez-Artalejo, M.: A Rewriting Logic for Declarative Programming. In: Riis Nielson, H. (ed.) ESOP 1996. LNCS, vol. 1058, pp. 156–172. Springer, Heidelberg (1996)
8. Hanus, M.: Curry: An Integrated Functional Logic Language (version 0.8.2 March 28, 2006) (2003), http://www.informatik.uni-kiel.de/~mh/curry/
9. Hanus, M.: Declarative processing of semistructured web data. Technical report 1103, Christian-Albrechts-Universität Kiel (2011)
10. Hanus, M., Steiner, F.: Controlling Search in Declarative Programs. In: Palamidessi, C., Meinke, K., Glaser, H. (eds.) ALP 1998 and PLILP 1998. LNCS, vol. 1490, pp. 374–390. Springer, Heidelberg (1998)
11. Fraguas, F.J.L., Hernández, J.S.: $\mathcal{TOY}$: A Multiparadigm Declarative System. In: Narendran, P., Rusinowitch, M. (eds.) RTA 1999. LNCS, vol. 1631, pp. 244–247. Springer, Heidelberg (1999)
12. Olteanu, D., Meuss, H., Furche, T., Bry, F.: XPath: Looking Forward. In: Chaudhri, A.B., Unland, R., Djeraba, C., Lindner, W. (eds.) EDBT 2002. LNCS, vol. 2490, pp. 109–127. Springer, Heidelberg (2002)
13. Seipel, D., Baumeister, J., Hopfner, M.: Declaratively Querying and Visualizing Knowledge Bases in XML. In: Seipel, D., Hanus, M., Geske, U., Bartenstein, O. (eds.) INAP/WLP 2004. LNCS (LNAI), vol. 3392, pp. 16–31. Springer, Heidelberg (2005)
14. Shapiro, E.: Algorithmic Program Debugging. ACM Distiguished Dissertation. MIT Press (1982)
15. W3C. Extensible Markup Language (XML) (2007)
16. W3C. XML Path Language (XPath) 2.0 (2007)
17. W3C. XML Query Use Cases (2007), http://www.w3.org/TR/xquery-use-cases/
18. W3C. XQuery 1.0: An XML Query Language (2007)
19. W3C. XQuery 1.0 and XPath 2.0 Formal Semantics, 2nd edn. (2010), http://www.w3.org/TR/xquery-semantics/
20. Walmsley, P.: XQuery. O'Reilly Media, Inc. (2007)

Automata-Based Computation of Temporal Equilibrium Models[*]

Pedro Cabalar[1] and Stéphane Demri[2]

[1] Department of Computer Science,
University of Corunna, Spain
cabalar@udc.es
[2] LSV, ENS de Cachan, CNRS, INRIA, France
demri@lsv.ens-cachan.fr

Abstract. Temporal Equilibrium Logic (TEL) is a formalism for temporal logic programming that generalizes the paradigm of Answer Set Programming (ASP) introducing modal temporal operators from standard Linear-time Temporal Logic (LTL). In this paper we solve some problems that remained open for TEL like decidability, bounds for computational complexity as well as computation of temporal equilibrium models for arbitrary theories. We propose a method for the latter that consists in building a Büchi automaton that accepts exactly the temporal equilibrium models of a given theory, providing an automata-based decision procedure and illustrating the ω-regularity of such sets. We show that TEL satisfiability can be solved in exponential space and it is hard for polynomial space. Finally, given two theories, we provide a decision procedure to check if they have the same temporal equilibrium models.

1 Introduction

Stable models. *Stable models* (or *answer sets*) have their roots in Logic Programming and in the search for a semantical interpretation of default negation [9]. They have given rise to a successful declarative paradigm, known as *Answer Set Programming* (ASP) [18,15], for practical knowledge representation. ASP has been applied to a wide spectrum of domains for solving several types of reasoning tasks: making diagnosis for the Space Shuttle [19], information integration of different data sources [13], distributing seaport employees in work teams [10] or automated music composition [3] to cite some examples. Some of these application scenarios frequently involve representing transition-based systems under linear time, so that discrete instants are identified with natural numbers. ASP offers interesting features for a formal treatment of temporal scenarios. For instance, it provides a high degree of *elaboration tolerance* [16], allowing a simple and natural solution to typical representational issues such as the *frame* problem and the *ramification* problem, see respectively [17] and [12].

---

[*] This research was partially supported by Spanish MEC project TIN2009-14562-C05-04 and Xunta de Galicia project INCITE08-PXIB105159PR.

G. Vidal (Ed.): LOPSTR 2011, LNCS 7225, pp. 57–72, 2012.
© Springer-Verlag Berlin Heidelberg 2012

Another interesting feature is that it allows a uniform treatment of different kinds of reasoning problems such as prediction, postdiction, planning, diagnosis or verification. However, since ASP is not a temporal formalism, it also involves some difficulties for dealing with temporal problems. In particular, since most ASP tools must deal with finite domains, this additionally requires fixing a finite path length with an obvious impossibility for solving problems such as proving the non-existence of a plan for a given planning scenario, or checking whether two temporal representations are *strongly equivalent* (i.e., they are interchangeable inside any context and for any path length).

Temporal Equilibrium Logic. To overcome these difficulties, in [1] a temporal extension of ASP, called *Temporal Equilibrium Logic* (TEL), was considered. This extension is an orthogonal combination of *linear-time temporal logic* (LTL) (see e.g. [22]) with the nonmonotonic formalism of *Equilibrium Logic* [20], probably the most general and best studied logical characterisation of ASP. TEL extends the stable model semantics to arbitrary LTL theories, that is, sets of formulae that combine atoms, the standard Boolean connectives, and the temporal operators X (read "next"), G (read "always"), F (read "eventually"), U (read "until") and R (read "release").

Towards arbitrary TEL theories. The definition of TEL has allowed studying problems like the aforementioned strong equivalence [1] of two temporal theories, but it had mostly remained as a theoretical tool, since there was no method for computing the temporal stable models of a temporal theory, at least until quite recently. In a first step in this direction, the paper [2] started from the normal form for TEL called *temporal logic programs* (TLPs) from [5] and showed that, when a syntactic subclass is considered (the so-called *splitable* TLPs), its temporal stable models can be obtained by a translation into LTL. This method has been implemented in a tool called STeLP [6] that uses an LTL model checker as a backend and provides the temporal stable models of a splitable TLP in terms of a Büchi automaton.

Although the splitable TLPs are expressive enough to capture most temporal scenarios treated in the ASP literature, a general method to compute the temporal equilibrium models for *arbitrary* TEL theories was not available until now. The interest for obtaining such a method is not only to cover the full expressiveness of this logic, but also to show its decidability and assess the complexity associated to its main reasoning tasks. In this sense, it is not convenient to use TLPs as a starting point since, despite of being a normal form for TEL, they are obtained by introducing new auxiliary atoms not present in the original propositional signature.

Our Contributions. In this paper we cover this gap and introduce an automata-based method to compute the temporal equilibrium models of an arbitrary temporal theory. We will pay a special attention to recall standard relationships between LTL and Büchi automata in order to facilitate the connection between ASP concepts and those from model-checking with temporal logics. More precisely, we propose automata-based decision procedures as follows:

1. We show that the satisfiability problem for the monotonic basis of TEL – the so-called logic of *Temporal Here-and-There* (THT) – can be solved in PSPACE by translation into the satisfiability problem for LTL. Whence, any decision procedure for LTL (automata-based, tableaux-based, resolution-based, etc.) can be used for THT. We are also able to demonstrate the PSPACE-hardness of the problem.
2. For any temporal formula, we effectively build a Büchi automaton that accepts exactly its temporal equilibrium models which allows to provide an automata-based decision procedure. We are able to show that TEL satisfiability can be solved in EXPSPACE and it is PSPACE-hard. Filling the gap is part of future work. Hence, we provide a symbolic representation for sets of temporal equilibrium models raising from temporal formulae.
3. Consequently, given two theories, we provide a decision procedure to check whether they have the same temporal equilibrium models (that is, *regular* equivalence, as opposed to *strong* equivalence).
4. Our proof technique can indeed be adapted to any extension of LTL provided that formulae can be translated into Büchi automata (as happens with LTL with past or LTL with fixed-points operators).

2   Temporal Equilibrium Logic

Let $AT = \{p, q, \ldots\}$ be a countably infinite set of *atoms*. A *temporal formula* is defined with the formal grammar below:

$$\varphi ::= p \mid \bot \mid \varphi_1 \wedge \varphi_2 \mid \varphi_1 \vee \varphi_2 \mid \varphi_1 \rightarrow \varphi_2 \mid X\varphi \mid \varphi_1 U \varphi_2 \mid \varphi_1 R \varphi_2$$

where $p \in AT$. We will use the standard abbreviations:

$$\neg\varphi \stackrel{\text{def}}{=} \varphi \rightarrow \bot$$
$$\top \stackrel{\text{def}}{=} \neg\bot$$
$$\varphi \leftrightarrow \varphi' \stackrel{\text{def}}{=} (\varphi \rightarrow \varphi') \wedge (\varphi' \rightarrow \varphi)$$

$$G\varphi \stackrel{\text{def}}{=} \bot R \varphi$$
$$F\varphi \stackrel{\text{def}}{=} \top U \varphi$$

The temporal connectives X, G, F, U and R have their standard meaning from LTL. A *theory* Γ is defined as a finite set of temporal formulae.

In the non-temporal case, Equilibrium Logic is defined by introducing a criterion for selecting models based on a non-classical monotonic formalism called the logic of *Here-and-There* (HT) [11], an intermediate logic between intuitionistic and classical propositional calculus. Similarly, TEL will be defined by first introducing a monotonic, intermediate version of LTL, we call the logic of *Temporal Here-and-There* (THT), and then defining a criterion for selecting models in order to obtain nonmonotonicity.

In this way, we will deal with two classes of models. An *LTL model* $\mathbf{H}$ is a map $\mathbf{H} : \mathbb{N} \rightarrow \mathcal{P}(AT)$, viewed as an ω-sequence of propositional valuations. By contrast, the semantics of THT is defined in terms of sequences of pairs of propositional valuations, which can be also viewed as a pair of LTL models. A *THT model* is a pair $\mathbf{M} = (\mathbf{H}, \mathbf{T})$ where $\mathbf{H}$ and $\mathbf{T}$ are LTL models and for

$i \geq 0$, we impose that $\mathbf{H}(i) \subseteq \mathbf{T}(i)$. $\mathbf{H}(i)$ and $\mathbf{T}(i)$ are sets of atoms standing for *here* and *there* respectively. A THT model $\mathbf{M} = (\mathbf{H}, \mathbf{T})$ is said to be *total* when $\mathbf{H} = \mathbf{T}$. The satisfaction relation $\models$ is interpreted as follows on THT models ($\mathbf{M}$ is a THT model and $k \in \mathbb{N}$):

1. $\mathbf{M}, k \models p \overset{\text{def}}{\Leftrightarrow} p \in \mathbf{H}(k)$.
2. $\mathbf{M}, k \models \varphi \wedge \varphi' \overset{\text{def}}{\Leftrightarrow} \mathbf{M}, k \models \varphi$ and $\mathbf{M}, k \models \varphi'$.
3. $\mathbf{M}, k \models \varphi \vee \varphi' \overset{\text{def}}{\Leftrightarrow} \mathbf{M}, k \models \varphi$ or $\mathbf{M}, k \models \varphi'$.
4. $\mathbf{M}, k \models \varphi \rightarrow \varphi' \overset{\text{def}}{\Leftrightarrow}$ for all $\mathbf{H}' \in \{\mathbf{H}, \mathbf{T}\}$, $(\mathbf{H}', \mathbf{T}), k \not\models \varphi$ or $(\mathbf{H}', \mathbf{T}), k \models \varphi'$.
5. $\mathbf{M}, k \models \mathsf{X}\varphi \overset{\text{def}}{\Leftrightarrow} \mathbf{M}, k+1 \models \varphi$.
6. $\mathbf{M}, k \models \varphi \mathsf{U} \varphi' \overset{\text{def}}{\Leftrightarrow}$ there is $j \geq k$ such that $\mathbf{M}, j \models \varphi'$ and for all $j' \in [k, j-1]$, $\mathbf{M}, j' \models \varphi$.
7. $\mathbf{M}, k \models \varphi \mathsf{R} \varphi' \overset{\text{def}}{\Leftrightarrow}$ for all $j \geq k$ such that $\mathbf{M}, j \not\models \varphi'$, there exists $j' \in [k, j-1]$, $\mathbf{M}, j' \models \varphi$.
8. never $\mathbf{M}, k \models \perp$.

A *model* for a theory Γ is a THT model $\mathbf{M}$ such that for every formula $\varphi \in \Gamma$, we have $\mathbf{M}, 0 \models \varphi$. A formula φ is *THT valid* $\overset{\text{def}}{\Leftrightarrow} \mathbf{M}, 0 \models \varphi$ for every THT model $\mathbf{M}$. Similarly, a formula φ is *THT satisfiable* $\overset{\text{def}}{\Leftrightarrow}$ there is a THT model $\mathbf{M}$ such that $\mathbf{M}, 0 \models \varphi$.

As we can see, the main difference with respect to LTL is the interpretation of implication (item 4), that must be checked in both components, $\mathbf{H}$ and $\mathbf{T}$, of $\mathbf{M}$. In fact, it is easy to see that when we take total models $\mathbf{M} = (\mathbf{T}, \mathbf{T})$, THT satisfaction $(\mathbf{T}, \mathbf{T}), k \models \varphi$ collapses to standard LTL satisfaction $\mathbf{T}, k \models \varphi$. We write $\mathbf{T}, k \models \varphi$ instead of $(\mathbf{T}, \mathbf{T}), k \models \varphi$ whenever convenient. For instance, item 4 in the above definition can be rewritten as:

4'. $\mathbf{M}, k \models \varphi \rightarrow \varphi' \overset{\text{def}}{\Leftrightarrow} (\mathbf{M}, k \models \varphi$ implies $\mathbf{M}, k \models \varphi')$ and $\mathbf{T}, k \models \varphi \rightarrow \varphi'$ (LTL satisfaction)

Note that $\mathbf{M}, k \models \neg p$ iff $\mathbf{M}, k \models p \rightarrow \perp$ iff $(\mathbf{M}, k \models p$ implies $\mathbf{M}, k \models \perp$ and $\mathbf{T}, k \models p$ implies $\mathbf{T}, k \models \perp$) iff $(p \notin \mathbf{H}(k)$ and $p \notin \mathbf{T}(k))$.

Similarly, a formula φ is *LTL valid* $\overset{\text{def}}{\Leftrightarrow} \mathbf{M}, 0 \models \varphi$ for every total THT model $\mathbf{M}$ whereas a formula φ is *LTL satisfiable* $\overset{\text{def}}{\Leftrightarrow}$ there is a total THT model $\mathbf{M}$ such that $\mathbf{M}, 0 \models \varphi$. We write $\mathrm{Mod}(\varphi)$ to denote the set of LTL models for φ (restricted to the set of atoms occurring φ denoted by $\mathrm{AT}(\varphi)$).

Obviously, any THT valid formula is also LTL valid, but not the other way around. For instance, the following are THT valid equivalences:

$$\neg(\varphi \wedge \psi) \leftrightarrow \neg\varphi \vee \neg\psi \qquad \mathsf{X}(\varphi \oplus \psi) \leftrightarrow \mathsf{X}\varphi \oplus \mathsf{X}\psi$$
$$\neg(\varphi \vee \psi) \leftrightarrow \neg\varphi \wedge \neg\psi \qquad \mathsf{X} \otimes \varphi \leftrightarrow \otimes \mathsf{X}\varphi$$

for any binary connective $\oplus$ and any unary connective $\otimes$. This means that De Morgan laws are valid, and that we can always shift the X operator to all the operands of any connective. On the contrary, the LTL valid formula $\varphi \vee \neg\varphi$ (known as *excluded middle* axiom) is not THT valid. This is inherited from the intermediate/intuitionistic nature of THT: in fact, the addition of this axiom makes THT collapse into LTL. By adding a copy of this axiom for any atom at any position of the models, we can force that THT models of any formula are total, as stated next.

Proposition 1. *Given a temporal formula φ built over the propositional atoms in $\mathrm{AT}(\varphi)$, for every THT model $(\mathbf{H}, \mathbf{T})$, the propositions below are equivalent:*

(I) $(\mathbf{H}, \mathbf{T}), 0 \models \varphi \wedge \bigwedge_{p \in \mathrm{AT}(\varphi)} \mathsf{G}(p \vee \neg p)$,
(II) $\mathbf{T}, 0 \models \varphi$ *in LTL, and for $i \geq 0$ and $p \in \mathrm{AT}(\varphi)$, we have $p \in \mathbf{H}(i)$ iff $p \in \mathbf{T}(i)$.*

As a consequence, we can easily encode LTL in THT, since LTL models of φ coincide with its total THT models. Let us state another property whose proof can be obtained by structural induction.

Proposition 2 (Persistence). *For any formula φ, any THT model $\mathbf{M} = (\mathbf{H}, \mathbf{T})$ and any $i \geq 0$, if $\mathbf{M}, i \models \varphi$, then $\mathbf{T}, i \models \varphi$.*

Corollary 1. $(\mathbf{H}, \mathbf{T}), i \models \neg\varphi$ *iff* $\mathbf{T}, i \not\models \varphi$ *in LTL.*

We proceed now to define an ordering relation among THT models, so that only the minimal ones will be *selected* for a temporal theory. Given two LTL models $\mathbf{H}$ and $\mathbf{H}'$, we say that $\mathbf{H}$ is *less than or equal to* $\mathbf{H}'$ (in symbols $\mathbf{H} \leq \mathbf{H}'$) $\overset{\text{def}}{\Leftrightarrow}$ for $k \geq 0$, we have $\mathbf{H}(k) \subseteq \mathbf{H}'(k)$. We write $\mathbf{H} < \mathbf{H}'$ if $\mathbf{H} \leq \mathbf{H}'$ and $\mathbf{H} \neq \mathbf{H}'$. The relations $\leq$ and $<$ can be lifted at the level of THT models. Given two THT models $\mathbf{M} = (\mathbf{H}, \mathbf{T})$ and $\mathbf{M}' = (\mathbf{H}', \mathbf{T}')$, $\mathbf{M} \leq \mathbf{M}' \overset{\text{def}}{\Leftrightarrow} \mathbf{H} \leq \mathbf{H}'$ and $\mathbf{T} = \mathbf{T}'$. Similarly, we write $\mathbf{M} < \mathbf{M}'$ if $\mathbf{M} \leq \mathbf{M}'$ and $\mathbf{M} \neq \mathbf{M}'$.

Definition 1 (Temporal Equilibrium Model). *A THT model $\mathbf{M}$ is a temporal equilibrium model (or TEL model, for short) of a theory Γ if $\mathbf{M}$ is a total model of Γ and there is no $\mathbf{M}' < \mathbf{M}$ such that $\mathbf{M}', 0 \models \Gamma$.*

Temporal Equilibrium Logic (TEL) is the logic induced by temporal equilibrium models and it is worth noting that any temporal equilibrium model of Γ is a *total* THT model of the form $(\mathbf{T}, \mathbf{T})$ (by definition). The corresponding LTL model $\mathbf{T}$ of Γ is said to be a *temporal stable model* of Γ.

When we restrict the syntax to non-modal theories and semantics to HT interpretations $\langle \mathbf{H}(0), \mathbf{T}(0) \rangle$ we talk about (non-temporal) equilibrium models, which coincide with stable models in their most general definition [8].

The TEL satisfiability problem consists in determining whether a temporal formula has a TEL model. As an example, consider the formula

$$\mathsf{G}(\neg p \to \mathsf{X}p) \tag{1}$$

Its intuitive meaning corresponds to the logic program consisting of rules of the form: $p(s(X)) \leftarrow not\ p(X)$ where time has been reified as an extra parameter $X = 0, s(0), s(s(0)), \ldots$. Notice that the interpretation of $\neg$ is that of default negation *not* in logic programming. In this way, (1) is saying that, at any situation $i \geq 0$, if there is no evidence on p, then p will become true in the next state $i + 1$. In the initial state, we have no evidence on p, so this will imply $\mathsf{X}p$. As a result $\mathsf{XX}p$ will have no applicable rule and thus will be false by default, and so on. It is easy to see that the unique temporal stable model of (1) is defined by the formula $\neg p \wedge \mathsf{G}(\neg p \leftrightarrow \mathsf{X}p)$.

It is worth noting that an LTL satisfiable formula may have no temporal stable model. As a simple example (well-known from non-temporal ASP) the logic program rule $\neg p \to p$, whose only (classical) model is $\{p\}$, has no stable models. This is because if we take a model $\mathbf{M} = (\mathbf{H}, \mathbf{T})$ where p holds in $\mathbf{T}$, then (Corollary 1) $\mathbf{M} \not\models \neg p$ and so $\mathbf{M} \models \neg p \to p$ true regardless $\mathbf{H}$, so we can take a strictly smaller $\mathbf{H} < \mathbf{T}$ whose only difference with respect to $\mathbf{T}$ is that p does not hold. On the other hand, if we take any $\mathbf{T}$ in which p does not hold, then $\mathbf{M} \models \neg p$ and so $\neg p \to p$ would make p true both in $\mathbf{H}$ and $\mathbf{T}$ reaching a contradiction. When dealing with logic programs, it is well-known that non-existence of stable models is always due to a kind of cyclic dependence on default negation like this.

In the temporal case, however, non-existence of temporal stable models may also be due to a lack of a finite justification for satisfying the criterion of minimal knowledge. As an example, take the formula $\mathsf{GF}p$, typically used in LTL to assert that property p occurs infinitely often. This formula has no temporal stable models[1]: all models must contain infinite occurrences of p and there is no way to establish a minimal $\mathbf{H}$ among them. Thus, formula $\mathsf{GF}p$ is LTL satisfiable but it has no temporal stable model. By contrast, forthcoming Proposition 4 states that for a large class of temporal formulae, LTL satisfiability is equivalent to THT satisfiability and TEL satisfiability.

3   Automata-Based Approach for LTL in a Nutshell

Before presenting our decision procedures, let us briefly recall what are the main ingredients of the automata-based approach. It consists in reducing logical problems into automata-based decision problems in order to take advantage of known results from automata theory. The most standard target problems on automata used in this approach are the nonemptiness problem (checking whether an automaton admits at least one accepting computation) and the inclusion problem (checking whether the language accepted by the automaton $\mathcal{A}$ is included in the language accepted by the automaton $\mathcal{B}$). In a pioneering work [4] Büchi introduced a class of automata showing that they are equivalent to formulae in monadic second-order logic (MSO) over $(\mathbb{N}, <)$.

In full generality, here are a few desirable properties of the approach. The reduction should be conceptually simple, see the translation from LTL formulae into alternating automata [27]. Formula structure is reflected directly in the transition formulae of alternating automata. The computational complexity

---

[1] Note that this does not mean that the property "infinitely often" cannot be represented in TEL. In fact, as we will see later, LTL can be embedded in TEL.

of the automata-based target problem should be well-characterized – see, for instance, the translation from PDL formulae into nondeterministic Büchi tree automata [28]. It is also highly desirable that not only the reduction is conceptually simple but also that it is semantically faithful so that the automata involve in the target instance are closely related to the instance of the original logical problem. Last but not least, preferrably, the reduction might allow obtaining the optimal complexity for the source logical problem.

3.1 Basics on Büchi Automata

We recall that a *Büchi automaton* $\mathcal{A}$ is a tuple $\mathcal{A} = (\Sigma, Q, Q_0, \delta, F)$ such that Σ is a finite *alphabet*, Q is a finite set of *states*, $Q_0 \subseteq Q$ is the set of *initial* states, the *transition relation* δ is a subset of $Q \times \Sigma \times Q$ and $F \subseteq Q$ is a set of *final* states. Given $q \in Q$ and $a \in \Sigma$, we also write $\delta(q,a)$ to denote the set of states q' such that $(q,a,q') \in \delta$.

A *run* ρ of $\mathcal{A}$ is a sequence $q_0 \overset{a_0}{\to} q_1 \overset{a_1}{\to} q_2 \dots$ such that for every $i \geq 0$, $(q_i, a_i, q_{i+1}) \in \delta$ (also written $q_i \overset{a_i}{\to} q_{i+1}$). The run ρ is *accepting* if $q_0 \in Q_0$ is initial and some state of F is repeated infinitely often in ρ: $\inf(\rho) \cap F \neq \emptyset$ where we let $\inf(\rho) = \{q \in Q : \forall i, \exists j > i, q = q_j\}$. The *label* of ρ is the word $\sigma = a_0 a_1 \dots \in \Sigma^\omega$. The automaton $\mathcal{A}$ *accepts* the language $\mathrm{L}(\mathcal{A})$ of ω-words $\sigma \in \Sigma^\omega$ such that there exists an accepting run of $\mathcal{A}$ on the word σ, i.e., with label σ.

Now, we introduce a standard generalization of the Büchi acceptance condition by considering conjunctions of classical Büchi conditions. A *generalized Büchi automaton* (GBA) is a structure $\mathcal{A} = (\Sigma, Q, Q_0, \delta, \{F_1, \dots, F_k\})$ such that $F_1, \dots, F_k \subseteq Q$ and Σ, Q, Q_0 and δ are defined as for Büchi automata. A run is defined as for Büchi automata and a run ρ of $\mathcal{A}$ is *accepting* iff the first state is initial and for $i \in [1, n]$, we have $\inf(\rho) \cap F_i \neq \emptyset$. It is known that every GBA $\mathcal{A}$ can be easily translated in logarithmic space into a Büchi automaton, preserving the language of accepted ω-words (see e.g. [21]). Moreover, the nonemptiness problem for GBA or BA is known to be NLogSpace-complete.

3.2 From LTL Formulae to Büchi Automata

We recall below how to define a Büchi automaton that accepts the linear models of an LTL formula. Given an LTL formula φ, we define its *closure* $cl(\varphi)$ to denote a finite set of formulae that are relevant to check the satisfiability of φ. For each LTL formula φ, we define its *main components* (if any) according to the table below:

formula φ	main components
p or $\neg p$	none
$\neg\neg\psi$ or $\mathsf{X}\psi$	ψ
$\neg\mathsf{X}\psi$	$\neg\psi$
$\psi_1 \mathsf{U} \psi_2$ or $\psi_1 \wedge \psi_2$ or $\psi_1 \mathsf{R} \psi_2$	ψ_1, ψ_2
$\neg(\psi_1 \mathsf{U} \psi_2)$ or $\neg(\psi_1 \wedge \psi_2)$ or $\neg(\psi_1 \mathsf{R} \psi_2)$	$\neg\psi_1, \neg\psi_2$

We write $cl(\varphi)$ to denote the least set of formulae such that $\varphi \in cl(\varphi)$ and $cl(\varphi)$ is closed under main components. It is routine to check that $\text{card}(cl(\varphi)) \le |\varphi|$ (the size of φ). Moreover, one can observe that if $\psi \in cl(\varphi)$, then for each immediate subformula (if any), either it belongs to $cl(\varphi)$ or its negation belongs to $cl(\varphi)$. A subset $\Gamma \subseteq cl(\varphi)$ is *consistent and fully expanded* whenever

- $\psi_1 \wedge \psi_2 \in \Gamma$ implies $\psi_1, \psi_2 \in \Gamma$,
- $\neg(\psi_1 \wedge \psi_2) \in \Gamma$ implies $\neg\psi_1 \in \Gamma$ or $\neg\psi_2 \in \Gamma$,
- $\neg\neg\psi \in \Gamma$ implies $\psi \in \Gamma$,
- Γ does not contain a contradictory pair ψ and $\neg\psi$.

The pair of consistent and fully expanded sets (Γ_1, Γ_2) is *one-step consistent* $\overset{\text{def}}{\Leftrightarrow}$

1. $\mathsf{X}\psi \in \Gamma_1$ implies $\psi \in \Gamma_2$ and $\neg\mathsf{X}\psi \in \Gamma_1$ implies $\neg\psi \in \Gamma_2$.
2. $\psi_1 \mathsf{U} \psi_2 \in \Gamma_1$ implies $\psi_2 \in \Gamma_1$ or $(\psi_1 \in \Gamma_1$ and $\psi_1 \mathsf{U} \psi_2 \in \Gamma_2)$,
3. $\neg(\psi_1 \mathsf{U} \psi_2) \in \Gamma_1$ implies $\neg\psi_2 \in \Gamma_1$ and $(\neg\psi_1 \in \Gamma_1$ or $\neg(\psi_1 \mathsf{U} \psi_2) \in \Gamma_2)$.
4. $\psi_1 \mathsf{R} \psi_2 \in \Gamma_1$ implies $\psi_2 \in \Gamma_1$ and $(\psi_1 \in \Gamma_1$ or $\psi_1 \mathsf{R} \psi_2 \in \Gamma_2)$,
5. $\neg(\psi_1 \mathsf{R} \psi_2) \in \Gamma_1$ implies $\neg\psi_2 \in \Gamma_1$ or $(\neg\psi_1 \in \Gamma_1$ and $\neg(\psi_1 \mathsf{R} \psi_2) \in \Gamma_2)$.

Given an LTL formula φ, let us build the generalized Büchi automaton $\mathcal{A}_\varphi = (\Sigma, Q, Q_0, \delta, \{F_1, \ldots, F_k\})$ where

- $\Sigma = \mathcal{P}(\text{AT}(\varphi))$ and Q is the set of consistent and fully expanded sets.
- $Q_0 = \{\Gamma \in Q : \varphi \in \Gamma\}$.
- $\Gamma \overset{a}{\to} \Gamma' \in \delta \overset{\text{def}}{\Leftrightarrow} (\Gamma, \Gamma')$ is one-step consistent, $(\Gamma \cap \text{AT}(\varphi)) \subseteq a$ and $\{p : \neg p \in \Gamma\} \cap a = \emptyset$.
- If the temporal operator U does not occur in φ, then $F_1 = Q$ and $k = 1$. Otherwise, suppose that $\{\psi_1 \mathsf{U} \psi_1', \ldots, \psi_k \mathsf{U} \psi_k'\}$ is the set of U-formulae from $cl(\varphi)$. Then, for every $i \in [1, k]$, $F_i = \{\Gamma \in Q : \psi_i \mathsf{U} \psi_i' \notin \Gamma$ or $\psi_i' \in \Gamma\}$.

It is worth observing that $\text{card}(Q) \le 2^{|\varphi|}$ and $\mathcal{A}_\varphi$ can be built in exponential time in $|\varphi|$.

Proposition 3. *[29]* $\text{Mod}(\varphi) = \text{L}(\mathcal{A}_\varphi)$, *i.e. for every* $a_0 a_1 \cdots \in \Sigma^\omega$, $a_0 a_1 \cdots \in \text{L}(\mathcal{A}_\varphi)$ *iff* $\mathbf{T}, 0 \models \varphi$ *where for all* $i \in \mathbb{N}$, $\mathbf{T}(i) = a_i$.

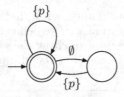

Fig. 1. Büchi automaton for models of $\mathsf{G}(\neg p \to \mathsf{X}p)$ (over the alphabet $\{\emptyset, \{p\}\}$)

Figure 1 presents a Büchi automaton recognizing the models for $\mathsf{G}(\neg p \to \mathsf{X}p)$ over the alphabet $\{\emptyset, \{p\}\}$. The automaton obtained from the above systematic construction would be a bit larger since $cl(\mathsf{G}(\neg p \to \mathsf{X}p))$ has about 2^4 subsets.

However, the systematic construction has the advantage to be generic. Other translations exist with other advantages, for instance to build small automata, see e.g. [7]. However, herein, we need to use the following properties (apart from the correctness of the reduction): (1) the size of each state of $\mathcal{A}_\varphi$ is linear in the size of φ, (2) it can be checked if a state is initial [resp. final] in linear space in the size of φ and (3) given two subsets X, X' of $cl(\varphi)$ and $a \in \Sigma$, one can check in linear space in the size of φ whether $X \xrightarrow{a} X'$ is a transition of $\mathcal{A}_\varphi$ (each transition of $\mathcal{A}_\varphi$ can be checked in linear space in the size of φ).

These are key points towards the PSPACE upper bound for LTL satisfiability since the properties above are sufficient to check the nonemptiness of $\mathcal{A}_\varphi$ in nondeterministic polynomial space in the size of φ (guess on-the-fly a prefix and a loop of length at most exponential) and then invoke Savitch Theorem [24] to eliminate nondeterminism. We will use similar arguments to establish that TEL satisfiability can be solved in EXPSPACE.

4   Building TEL Models with Büchi Automata

In this section, we provide an automata-based approach to determine whether a formula φ built over the atoms $\{p_1, \ldots, p_n\}$ has a TEL model. This is the place where starts our main contribution. To do so, we build a Büchi automaton $\mathcal{B}$ over the alphabet $\Sigma = \mathcal{P}(\{p_1, \ldots, p_n\})$ such that $L(\mathcal{B})$ is equal to the set of TEL models for φ. Moreover, nonemptiness can be checked in EXPSPACE, which allows to answer the open problem about the complexity of determining whether a temporal formula has a TEL model.

Each model $\mathbf{M} = \langle \mathbf{H}, \mathbf{T} \rangle$ restricted to the atoms in $\{p_1, \ldots, p_n\}$ can be encoded into an LTL model $\mathbf{H}'$ over the alphabet $\Sigma' = \mathcal{P}(\{p_1, \ldots, p_n, p'_1, \ldots, p'_n\})$ such that for $i \geq 0$, $\mathbf{H}'(i) = (\mathbf{T}(i) \cap \{p_1, \ldots, p_n\}) \cup \{p'_j : p_j \in \mathbf{H}(i), j \in [1, n]\}$. In that case, we write $\mathbf{H}' \approx \mathbf{M}$.

Lemma 1.

(I) *For every THT model $\mathbf{M} = \langle \mathbf{H}, \mathbf{T} \rangle$ restricted to atoms in $\{p_1, \ldots, p_n\}$, there is a unique LTL model $\mathbf{H}'$ such that $\mathbf{H}' \approx \mathbf{M}$.*

(II) *For every LTL model $\mathbf{H}' : \mathbb{N} \to \Sigma'$ such that $\mathbf{H}', 0 \models \bigwedge_{i \in [1,n]} \mathsf{G}(p'_i \to p_i)$, there is a unique THT model $\mathbf{M} = \langle \mathbf{H}, \mathbf{T} \rangle$ restricted to atoms s.t. $\mathbf{H}' \approx \mathbf{M}$.*

In (II), a THT model $\mathbf{M}$ can be built thanks to the satisfaction of the formula $\bigwedge_{i \in [1,n]} \mathsf{G}(p'_i \to p_i)$ by $\mathbf{H}'$, which guarantees that for all $i \in \mathbb{N}$, we have $\mathbf{H}(i) \subseteq \mathbf{T}(i)$. The proof is by an easy verification. This guarantees a clear isomorphism between two sets of models. In order to complete this model-theoretical correspondence, let us define the translation f between temporal formulae:

- f is homomorphic for conjunction, disjunction and temporal operators,
- $f(\bot) \stackrel{\text{def}}{=} \bot$, $f(p_i) \stackrel{\text{def}}{=} p'_i$ and $f(\psi \to \psi') \stackrel{\text{def}}{=} (\psi \to \psi') \wedge (f(\psi) \to f(\psi'))$.

Lemma 2. *Let φ be a temporal formula built over the atoms in $\{p_1, \ldots, p_n\}$ and $\mathbf{M}$ restricted to $\{p_1, \ldots, p_n\}$ and $\mathbf{H}'$ be models such that $\mathbf{H}' \approx \mathbf{M}$. For $l \geq 0$, we have $\mathbf{H}', l \models f(\psi)$ iff $\mathbf{M}, l \models \psi$ for every subformula ψ of φ.*

The proof is by an easy structural induction. So, there is a polynomial-time reduction from THT satisfiability into LTL satisfiability by considering the mapping $f(\cdot) \wedge \bigwedge_{i\in[1,n]} G(p'_i \to p_i)$.

Let $\mathcal{A}_1$ be the Büchi automaton such that $L(\mathcal{A}_1) = \text{Mod}(\varphi)$, following any construction similar to [29] (see Section 3.2). The set $L(\mathcal{A}_1)$ can be viewed as the set of total THT models of φ. Let φ' be the formula $f(\varphi) \wedge \bigwedge_{i\in[1,n]} G(p'_i \to p_i)$.

Lemma 3. *The set of LTL models for the formula φ' corresponds to the set of THT models for the temporal formula φ.*

For instance, taking the formula $\varphi = G(\neg p \to Xp)$, we can compute its THT models **M** by obtaining the corresponding LTL models (with atoms p and p') for the formula below:

$$\varphi' = f(\ G(\neg p \to Xp)\ ) \ \wedge\ G(p' \to p)$$
$$= G(\ (\neg p \to Xp) \wedge (\neg p \wedge \neg p' \to Xp')\ ) \ \wedge\ G(p' \to p)$$

Figure 2 presents a Büchi automaton for the models of the formula $f(G(\neg p \to Xp)) \wedge G(p' \to p)$ over the alphabet $\{\emptyset, \{p\}, \{p'\}, \{p, p'\}\}$. Hence, we provide a symbolic representation for the THT models of $G(\neg p \to Xp)$. For instance, reading the letter $\{p\}$ at position i corresponds to a pair $(\mathbf{H}(i), \mathbf{T}(i))$ with $p \notin \mathbf{H}(i)$ and $p \in \mathbf{T}(i)$. Similarly, reading the letter $\{p, p'\}$ at position i corresponds to a pair $(\mathbf{H}(i), \mathbf{T}(i))$ with $p \in \mathbf{H}(i)$ and $p \in \mathbf{T}(i)$. However, $\{p'\}$ cannot be read since $\mathbf{H}(i) \subseteq \mathbf{T}(i)$.

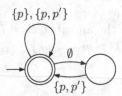

Fig. 2. Büchi automaton for models of $f(G(\neg p \to Xp)) \wedge G(p' \to p)$

Hence, φ is THT satisfiable iff φ' is LTL satisfiable. The map f shall be also useful to show Proposition 4 below, becoming a key step to obtain PSPACE-hardness results (see e.g. Theorem 2). Proposition 4 below states that for a large class of formulae, LTL satisfiability is equivalent to TEL satisfiability.

Proposition 4. *Let φ be temporal formula built over the connectives $\vee$, $\wedge$, $\to$, X and U and such that $\to$ occurs only in subformulae of the form $p \to \bot$ with $p \in AT$. The propositions below are equivalent: (I) φ is LTL satisfiable; (II) φ is THT satisfiable; (III) φ has a temporal stable model, i.e. φ is TEL satisfiable.*

Proof. Let $L^+(\vee, \wedge, X, U)$ be the class of temporal formulae involved in the statement. Every temporal formula φ in $L^+(\vee, \wedge, X, U)$ states a *guarantee property* in the sense of [30] (see also [14]). This means that

(P1) if $\mathbf{T}, 0 \models \varphi$ in the LTL sense, then there is $N \geq 0$ such that for any $\mathbf{T}'$ that only agrees with $\mathbf{T}$ on the positions in $[0, N]$, we also have $\mathbf{T}', 0 \models \varphi$ (φ states a guarantee property).

In particular, we can impose that for $l > N$, we have $\mathbf{T}'(l) = \emptyset$. Moreover, since the set of sequences of length $N + 1$ indexed by subsets of atoms occurring in φ is finite, if φ is LTL satisfiable, then there is an LTL model $\mathbf{T}$ such that $\mathbf{T}, 0 \models \varphi$ and no $\mathbf{T}' < \mathbf{T}$ verifies $\mathbf{T}', 0 \models \varphi$ (this property does not hold for every LTL formulae; consider $\mathsf{GF}p$). (II) implies (I) is by Proposition 2 and (III) implies (II) by definition of temporal stable model whereas (III) implies (I) is by definition of LTL satisfiability. It remains to show that (I) implies (III); we do it by *reductio ad absurdum*. Suppose that φ is LTL satisfiable. From the previous properties, we have seen that there is a minimal LTL model $\mathbf{T}$ such that $\mathbf{T}, 0 \models \varphi$. Suppose that there is $\mathbf{H} < \mathbf{T}$ such that $(\mathbf{H}, \mathbf{T}), 0 \models \varphi$. By minimality of $\mathbf{T}$, we have $\mathbf{H}, 0 \not\models \varphi$. Since $\mathbf{T}$ is minimal and has an infinite suffix of the form $\emptyset^\omega$, there is a finite amount of positions $l_1, \ldots, l_N$ such that $\mathbf{H}(l_i) \subset \mathbf{T}(l_i)$ (strict inclusion).

Let $\mathbf{H}'$ be the LTL model such that $\mathbf{H}' \approx (\mathbf{H}, \mathbf{T})$. We have seen that $\mathbf{H}', 0 \models f(\varphi)$ (Lemma 2). Let us define the map f', that we shall apply to φ, as a slight variant of f:

- f' is homomorphic for conjunction, disjunction and temporal operators,
- $f'(p_i) \stackrel{\text{def}}{=} (p_i' \wedge p_i)$ and $f'(\neg p_i) \stackrel{\text{def}}{=} (\neg p_i \wedge \neg p_i')$.

Since, $\mathbf{H}', 0 \models \bigwedge_{i \in [1,n]} \mathsf{G}(p_i' \to p_i)$, we also have $\mathbf{H}', 0 \models f'(\varphi)$. Let $\mathbf{H}''$ be the variant of $\mathbf{H}'$ such that for $l \notin \{l_1, \ldots, l_N\}$, $\mathbf{H}''(l) \stackrel{\text{def}}{=} \mathbf{H}'(l)$ and for $i \in [1, N]$ and $p \in \mathrm{AT}$, if $p \notin \mathbf{H}(l_i)$ and $p \in \mathbf{T}(l_i)$, then $p, p' \notin \mathbf{H}''(l_i)$, otherwise $(p, p' \in \mathbf{H}''(l_i) \stackrel{\text{def}}{\Leftrightarrow} p \in \mathbf{T}(l_i))$. Observe that in $\mathbf{H}''$ the valuations on $\{p_1, \ldots, p_n\}$ and $\{p_1', \ldots, p_n'\}$ agree and the 'atomic' formulae in $f'(\varphi)$ are the form either $(p_i' \wedge p_i)$ or $(\neg p_i' \wedge \neg p_i)$. Consequently, $\mathbf{H}'', 0 \models f'(\varphi)$ and by Lemma 2, $\mathbf{T}', 0 \models \varphi$ where $\mathbf{T}'(l) = \mathbf{H}''(l) \cap \{p_1, \ldots, p_n\}$ for $l \geq 0$. By construction of $\mathbf{T}'$, we have $\mathbf{T}' < \mathbf{T}$, leading to a contradiction. Thus, $(\mathbf{T}, \mathbf{T})$ is a temporal stable model for φ. □

Corollary 2. *THT satisfiability problem is* PSPACE-*complete.*

Proof. The translation f requires only polynomial-time and since LTL satisfiability is PSPACE-complete [25], we get that THT satisfiability is in PSPACE. It remains to show the PSPACE lower bound.

To do so, we can just observe that, as proved by Proposition 1, LTL satisfiability (which is PSPACE-complete) can be encoded into THT satisfiability using the translation from Proposition 1, which can be performed in linear time. Indeed, it just adds a formula $\mathsf{G}(p \vee \neg p)$ per each atom $p \in \mathrm{AT}$. □

We can strengthen the mapping φ' to obtain not only THT models of φ but also to constrain them to be strictly non-total (that is $\mathbf{H} < \mathbf{T}$) as follows

$$\varphi'' \stackrel{\text{def}}{=} \varphi' \wedge \bigvee_{i \in [1,n]} \mathsf{F}((p_i' \to \bot) \wedge p_i)$$

φ'' characterizes the non-total THT models of the formula φ. The generalized disjunction ensures that at some position j, $\mathbf{H}(j) \subset \mathbf{T}(j)$ (strict inclusion).

Lemma 4. *The set of LTL models for the formula φ'' corresponds to the set of non-total THT models for the temporal formula φ.*

The proof is again by structural induction. Let $\mathcal{A}_2$ be the Büchi automaton such that $\mathrm{L}(\mathcal{A}_2) = \mathrm{Mod}(\varphi'')$, following again any construction similar to [29] (see Section 3.2). $\mathrm{L}(\mathcal{A}_2)$ contains exactly the non-total THT models of φ.

Let $h : \Sigma' \to \Sigma$ be a map (renaming) between the two finite alphabets such that $h(a) = a \cap \{p_1, \ldots, p_n\}$. h consists in erasing the atoms from $\{p'_1, \ldots, p'_n\}$. h can be naturally extended as an homomorphism between finite words, infinite words and as a map between languages. Similary, given a Büchi automaton $\mathcal{A}_2 = (\Sigma', Q, Q_0, \delta, F)$, we write $h(\mathcal{A}_2)$ to denote the Büchi automaton $(\Sigma, Q, Q_0, \delta', F)$ such that $q \xrightarrow{a} q' \in \delta' \overset{\text{def}}{\Leftrightarrow}$ there is $b \in \Sigma'$ such that $q \xrightarrow{b} q' \in \delta$ and $h(b) = a$. Obviously, $\mathrm{L}(h(\mathcal{A}_2)) = h(\mathrm{L}(\mathcal{A}_2))$. Indeed, the following propositions imply each other:

1. $a_0 a_1 \cdots \in \mathrm{L}(\mathcal{A}_2)$,
2. $h(a_0)h(a_1) \cdots \in h(\mathrm{L}(\mathcal{A}_2))$ (by definition of h on languages),
3. $h(a_0)h(a_1) \cdots \in \mathrm{L}(h(\mathcal{A}_2))$ (by definition of h on $\mathcal{A}_2$).

The inclusion $\mathrm{L}(h(\mathcal{A}_2)) \subseteq h(\mathrm{L}(\mathcal{A}_2))$ can be shown in a similar way. So, $\mathrm{L}(h(\mathcal{A}_2))$ can be viewed as the set of total THT models for φ having a strictly smaller THT model.

Proposition 5. *φ has a TEL model iff $\mathrm{L}(\mathcal{A}_1) \cap (\Sigma^\omega \setminus \mathrm{L}(h(\mathcal{A}_2))) \neq \emptyset$.*

Proof. A TEL model $\mathbf{M} = (\mathbf{H}, \mathbf{T})$ for φ satisfies the following properties:

1. $\mathbf{M}, 0 \models \varphi$ and $\mathbf{H} = \mathbf{T}$.
2. For no $\mathbf{H}' < \mathbf{H}$, we have $(\mathbf{H}', \mathbf{T}), 0 \models \varphi$.

We have seen that $\mathrm{L}(\mathcal{A}_1)$ contains exactly the LTL models of φ, i.e. the total THT models satisfying φ. For taking care of condition (2.), by construction, $\mathcal{A}_2$ accepts the non-total THT models for φ whereas $\mathrm{L}(h(\mathcal{A}_2))$ contains the total THT models for φ having a strictly smaller THT model satisfying φ, the negation of (2.). Hence, $(\mathbf{T}, \mathbf{T})$ is a TEL model for φ iff $\mathbf{T} \in \mathrm{L}(\mathcal{A}_1)$ and $\mathbf{T} \notin \mathrm{L}(h(\mathcal{A}_2))$.   □

Hence, the set of TEL models for a given φ forms an ω-regular language.

Proposition 6. *For each temporal formula φ, one can effectively build a Büchi automaton that accepts exactly the TEL models for φ.*

Proof. The class of languages recognized by Büchi automata (the class of ω-regular languages) is effectively closed under union, intersection and complementation. Moreover, it is obviously closed under the renaming operation. Since $\mathcal{A}_1$, $\mathcal{A}_2$, and $h(\mathcal{A}_2)$ are Büchi automata, one can build a Büchi automaton $\mathcal{A}'$ such that $\mathrm{L}(\mathcal{A}') = \Sigma^\omega \setminus \mathrm{L}(h(\mathcal{A}_2))$. Similarly, one can effectively build a Büchi automaton $\mathcal{B}_\varphi$ such that $\mathrm{L}(\mathcal{B}_\varphi) = \mathrm{L}(\mathcal{A}_1) \cap \mathrm{L}(\mathcal{A}')$. Complementation can be performed using the constructions in [26] or in [23] (if optimality is required). Roughly speaking, complementation induces an exponential blow-up.   □

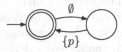

Fig. 3. Büchi automaton for stable models of $G(\neg p \to Xp)$

Figure 3 presents a Büchi automaton accepting the (unique) temporal equilibrium model for φ. The next step consists in showing that the nonemptiness check can be done in exponential space.

Proposition 7. *Checking whether a TEL formula has a TEL model can be done in* ExpSpace.

Proof. Let φ be a temporal formula and $\mathcal{A}_1$ and $\mathcal{A}_2$ be Büchi automata such that $L(\mathcal{A}_1) \cap (\Sigma^\omega \setminus L(h(\mathcal{A}_2)))$ accepts exactly the TEL models for φ. We shall show that nonemptiness of the language can be tested in exponential space. With the construction of $\mathcal{A}_1$ using [29], we have seen that

1. the size of each state of $\mathcal{A}_1$ is linear in the size of φ (written $|\varphi|$),
2. it can be checked if a state is initial [resp. final] in linear space in $|\varphi|$,
3. each transition of $\mathcal{A}_1$ can be checked in linear space in $|\varphi|$,
4. $\mathcal{A}_1$ has a number of states exponential in $|\varphi|$.

Similarly, let us observe the following simple properties:

(a) φ'' is of linear size in $|\varphi|$; (b) Automaton $\mathcal{A}_2$ can be built from the formula φ'' using the construction in [29]; (c) $\mathcal{A}_2$ and $h(\mathcal{A}_2)$ have the same sets of states, initial states and final states and checking whether a transition belongs to $h(\mathcal{A}_2)$ is not more complex than checking whether a transition belongs to $\mathcal{A}_2$. So,

1. the size of each state of $h(\mathcal{A}_2)$ is linear in $|\varphi|$,
2. it can be checked if a state is initial [resp. final] in linear space in $|\varphi|$,
3. each transition of $h(\mathcal{A}_2)$ can be checked in linear space in $|\varphi|$.
4. $h(\mathcal{A}_2)$ has a number of states exponential in $|\varphi|$.

Using the complementation construction from [26] (the construction in [23] would be also fine) to complement $h(\mathcal{A}_2)$, one can obtain a Büchi automaton $\mathcal{A}'$ such that $L(\mathcal{A}') = \Sigma^\omega \setminus L(h(\mathcal{A}_2))$ and

1. the size of each state of $\mathcal{A}'$ is exponential in $|\varphi|$,
2. it can be checked if a state is initial [resp. final] in exponential space in $|\varphi|$,
3. each transition of $\mathcal{A}'$ can be checked in exponential space in $|\varphi|$.
4. $\mathcal{A}'$ has a number of states doubly exponential in $|\varphi|$.

Indeed, $h(\mathcal{A}_2)$ is already of exponential size in $|\varphi|$. So, using the above-mentioned property, one can check on-the-fly whether $L(\mathcal{A}_1) \cap L(\mathcal{A}')$ is nonempty by guessing a synchronized run of length at most double exponential (between the automata $\mathcal{A}_1$ and $\mathcal{A}'$) and check that it satisfies the acceptance conditions of both automata. At any stage of the algorithm, at most 2 product states need to be stored and this requires exponential space. Similarly, counting until a double exponential value requires only an exponential amount of bits. Details are omitted but the very algorithm is based on standard arguments for checking on-the-fly graph accessibility and checking nonemptiness of the intersection of two languages accepted by Büchi automata (similar arguments are used in [26, Lemma 2.10]). By Savitch Theorem [24], nondeterminism can be eliminated, providing the promised EXPSPACE upper bound.   □

Theorem 1. *Checking whether a formula has a TEL model is* PSPACE-*hard.*

Proof. We can use again the linear encoding in Proposition 1 and observe that any THT model $(\mathbf{T}, \mathbf{T})$ of $\psi = \varphi \wedge \bigwedge_{p \in \mathrm{AT}(\varphi)} \mathsf{G}(p \vee \neg p)$ will also be a TEL model of φ, since there are no non-total models for ψ and thus $(\mathbf{T}, \mathbf{T})$ will always be minimal. But then $\mathbf{T} \models \varphi$ in LTL iff $(\mathbf{T}, \mathbf{T}) \models \psi$ in THT iff $(\mathbf{T}, \mathbf{T})$ is a TEL model of ψ. Thus LTL satisfiability can be reduced to TEL satisfiability and so the latter problem is PSPACE-hard.   □

Theorem 2. *Checking whether two temporal formulae have the same TEL models is decidable in* EXPSPACE *and it is* PSPACE-*hard.*

Proof. Let $\varphi \in \mathrm{L}^+(\vee, \wedge, \mathsf{X}, \mathsf{U})$ and $\psi = \mathsf{GF}p_1$. We recall that ψ has no temporal equilibrium model. The propositions below are equivalent: (a) φ is LTL satisfiable, (b) φ has a temporal equilibrium model and (c) φ and ψ have distinct sets of temporal equilibrium models. Since LTL satisfiability for the fragment $\mathrm{L}^+(\vee, \wedge, \mathsf{X}, \mathsf{U})$ is PSPACE-hard, coPSPACE= PSPACE and ((a) iff (c)), then the equivalence problem with temporal equilibrium models is PSPACE-hard.

Let φ and ψ be two temporal formulae built over the same set of atoms and $\mathcal{A}_1^\varphi$, $\mathcal{A}_2^\varphi$, $\mathcal{A}_1^\psi$ and $\mathcal{A}_2^\psi$ be Büchi automata such that $\mathrm{L}(\mathcal{A}_1^\varphi) \cap (\varSigma^\omega \setminus \mathrm{L}(h(\mathcal{A}_2^\varphi)))$ recognizes the temporal equilibrium models for φ and $\mathrm{L}(\mathcal{A}_1^\psi) \cap (\varSigma^\omega \setminus \mathrm{L}(h(\mathcal{A}_2^\psi)))$ recognizes the temporal equilibrium models for ψ. So, φ and ψ have distinct sets of temporal equilibrium models iff one of the sets below is non-empty.

(I) $\mathrm{L}(\mathcal{A}_1^\varphi) \cap (\varSigma^\omega \setminus \mathrm{L}(h(\mathcal{A}_2^\varphi))) \cap (\varSigma^\omega \setminus \mathrm{L}(\mathcal{A}_1^\psi))$,
(II) $\mathrm{L}(\mathcal{A}_1^\varphi) \cap (\varSigma^\omega \setminus \mathrm{L}(h(\mathcal{A}_2^\varphi))) \cap \mathrm{L}(h(\mathcal{A}_2^\psi))$,
(III) $\mathrm{L}(\mathcal{A}_1^\psi) \cap (\varSigma^\omega \setminus \mathrm{L}(h(\mathcal{A}_2^\psi))) \cap (\varSigma^\omega \setminus \mathrm{L}(\mathcal{A}_1^\varphi))$,
(IV) $\mathrm{L}(\mathcal{A}_1^\psi) \cap (\varSigma^\omega \setminus \mathrm{L}(h(\mathcal{A}_2^\psi))) \cap \mathrm{L}(h(\mathcal{A}_2^\varphi))$.

A nondeterministic algorithm running in exponential space is designed as follows:

1. Guess which sets among (I)–(IV) is tested for nonemptiness.
2. Run a nondeterministic algorithm in exponential space by synchronizing the transitions of the three automata (one or two of them are designed by complementation) as done in the proof of Proposition 7.

Again, elimination of nondeterminism can be performed thanks to Savitch Theorem [24]. So, non equivalence problem is in EXPSPACE. Since coEXPSPACE= EXPSPACE (simply because EXPSPACE refers to a deterministic class of Turing machines), the equivalence problem is in EXPSPACE.   □

5 Concluding Remarks

We have introduced an automata-based method for computing the temporal equilibrium models of an arbitrary temporal theory, under the syntax of Linear-time Temporal Logic (LTL). This construction has allowed us solving several open problems about Temporal Equilibrium Logic (TEL) and its monotonic basis Temporal Here-and-There (THT). In particular, we were able to prove that THT satisfiability can be solved in PSPACE and is PSPACE-hard whereas TEL satisfiability is decidable (something not proven before) being solvable in EXPSPACE and at least PSPACE-hard (filling the gap is part of future work). Our method consists in constructing a Büchi automaton that captures all the temporal equilibrium models of an arbitrary theory. This also implies that the set of TEL models of any theory is ω-regular.

A recent approach [2,6] has developed a tool, called STeLP, that also captures TEL models of a theory in terms of a Büchi automaton. Our current proposal, however, has some important advantages. First, STeLP restricts the input syntax to so-called *splitable temporal logic programs*, a strict subclass of a normal form for TEL that further requires the introduction of auxiliary atoms for removing U and R operators, using a structure preserving transformation. On the contrary, our current method has no syntactic restrictions and directly works on the alphabet of the original theory, for which no transformation is required prior to the automaton construction. Second, once the STeLP input is written in the accepted syntax, it translates the input program into LTL by the addition of a set of formulae (the so-called *loop formulae*) whose number is, in the worst case, exponential on the size of the input. Future work includes the implementation of our current method as well a comparison in terms of efficiency with respect to the tool STeLP.

References

1. Aguado, F., Cabalar, P., Pérez, G., Vidal, C.: Strongly Equivalent Temporal Logic Programs. In: Hölldobler, S., Lutz, C., Wansing, H. (eds.) JELIA 2008. LNCS (LNAI), vol. 5293, pp. 8–20. Springer, Heidelberg (2008)
2. Aguado, F., Cabalar, P., Pérez, G., Vidal, C.: Loop Formulas for Splitable Temporal Logic Programs. In: Delgrande, J.P., Faber, W. (eds.) LPNMR 2011. LNCS, vol. 6645, pp. 80–92. Springer, Heidelberg (2011)
3. Boenn, G., Brain, M., De Vos, M., Ffitch, J.: ANTON: Composing Logic and Logic Composing. In: Erdem, E., Lin, F., Schaub, T. (eds.) LPNMR 2009. LNCS, vol. 5753, pp. 542–547. Springer, Heidelberg (2009)
4. Büchi, R.: On a decision method in restricted second-order arithmetic. In: Intl. Congress on Logic, Method and Philosophical Science 1960, pp. 1–11 (1962)
5. Cabalar, P.: A Normal Form for Linear Temporal Equilibrium Logic. In: Janhunen, T., Niemelä, I. (eds.) JELIA 2010. LNCS, vol. 6341, pp. 64–76. Springer, Heidelberg (2010)
6. Cabalar, P., Diéguez, M.: STeLP – A Tool for Temporal Answer Set Programming. In: Delgrande, J.P., Faber, W. (eds.) LPNMR 2011. LNCS, vol. 6645, pp. 370–375. Springer, Heidelberg (2011)
7. Demri, S., Gastin, P.: Specification and verification using temporal logics. In: Modern Applications of Automata Theory. IIsc Research Monographs, vol. 2. World Scientific (2011) (to appear)

8. Ferraris, P.: Answer Sets for Propositional Theories. In: Baral, C., Greco, G., Leone, N., Terracina, G. (eds.) LPNMR 2005. LNCS (LNAI), vol. 3662, pp. 119–131. Springer, Heidelberg (2005)
9. Gelfond, M., Lifschitz, V.: The stable model semantics for logic programming. In: ICLP 1988, pp. 1070–1080. MIT Press, Cambridge (1988)
10. Grasso, G., Iiritano, S., Leone, N., Lio, V., Ricca, F., Scalise, F.: An ASP-Based System for Team-Building in the Gioia-Tauro Seaport. In: Carro, M., Peña, R. (eds.) PADL 2010. LNCS, vol. 5937, pp. 40–42. Springer, Heidelberg (2010)
11. Heyting, A.: Die formalen Regeln der intuitionistischen Logik. Sitzungsberichte der Preussischen Akademie der Wissenschaften, Physikalisch-mathematische Klasse, 42–56 (1930)
12. Kautz, H.: The logic of persistence. In: AAAI 1986, pp. 401–405 (1986)
13. Leone, N., Eiter, T., Faber, W., Fink, M., Gottlob, G., Greco, G.: Boosting information integration: The INFOMIX system. In: Proc. of the 13th Italian Symposium on Advanced Database Systems, SEBD 2005, pp. 55–66 (2005)
14. Manna, Z., Pnueli, A.: A hierarchy of temporal properties. In: PODC 1990, pp. 377–408. ACM Press (1990)
15. Marek, V., Truszczyński, M.: Stable models and an alternative logic programming paradigm, pp. 169–181. Springer (1999)
16. McCarthy, J.: Elaboration tolerance. In: Proc. of the 4th Symposium on Logical Formalizations of Commonsense Reasoning (Common Sense 1998), London, UK, pp. 198–217 (1998)
17. McCarthy, J., Hayes, P.: Some philosophical problems from the standpoint of artificial intelligence. Machine Intelligence Journal 4, 463–512 (1969)
18. Niemelä, I.: Logic programs with stable model semantics as a constraint programming paradigm. Annals of Mathematics and Artificial Intelligence 25, 241–273 (1999)
19. Nogueira, M., Balduccini, M., Gelfond, M., Watson, R., Barry, M.: An A-Prolog Decision Support System for the Space Shuttle. In: Ramakrishnan, I.V. (ed.) PADL 2001. LNCS, vol. 1990, pp. 169–183. Springer, Heidelberg (2001)
20. Pearce, D.: A New Logical Characterisation of Stable Models and Answer Sets. In: Dix, J., Przymusinski, T.C., Moniz Pereira, L. (eds.) NMELP 1996. LNCS(LNAI), vol. 1216, pp. 57–70. Springer, Heidelberg (1997)
21. Perrin, D., Pin, J.-E.: Infinite Words: Automata, Semigroups, Logic and Games. Elsevier (2004)
22. Pnueli, A.: The temporal logic of programs. In: FOCS 1977, pp. 46–57. IEEE (1977)
23. Safra, S.: Complexity of Automata on Infinite Objects. PhD thesis, The Weizmann Institute of Science, Rehovot (1989)
24. Savitch, W.J.: Relationships between nondeterministic and deterministic tape complexities. JCSS 4(2), 177–192 (1970)
25. Sistla, A., Clarke, E.: The complexity of propositional linear temporal logic. JACM 32(3), 733–749 (1985)
26. Sistla, A., Vardi, M., Wolper, P.: The complementation problem for Büchi automata with applications to temporal logic. TCS 49, 217–237 (1987)
27. Vardi, M.: Alternating Automata: Unifying Truth and Validity Checking for Temporal Logics. In: McCune, W. (ed.) CADE 1997. LNCS, vol. 1249, pp. 191–206. Springer, Heidelberg (1997)
28. Vardi, M., Wolper, P.: Automata-theoretic techniques for modal logics of programs. JCSS 32, 183–221 (1986)
29. Vardi, M., Wolper, P.: Reasoning about infinite computations. I & C 115, 1–37 (1994)
30. Černá, I., Pelánek, R.: Relating Hierarchy of Temporal Properties to Model Checking. In: Rovan, B., Vojtáš, P. (eds.) MFCS 2003. LNCS, vol. 2747, pp. 318–327. Springer, Heidelberg (2003)

Simplifying Questions in Maude Declarative Debugger by Transforming Proof Trees*

Rafael Caballero, Adrián Riesco, Alberto Verdejo, and Narciso Martí-Oliet

Facultad de Informática, Universidad Complutense de Madrid, Spain

Abstract. Declarative debugging is a debugging technique that abstracts the execution details that in general may be difficult to follow in declarative languages to focus on results. It relies on a data structure representing the wrong computation, the *debugging tree*, which is traversed by asking questions to the user about the correctness of the computation steps related to each node. Thus, the complexity of the questions is an important factor regarding the applicability of the technique. In this paper we present a transformation for debugging trees for Maude specifications that ensures that any subterm occurring in a question has been previously replaced by the most reduced form that it has taken during the computation, thus ensuring that questions become as simple as possible.

Keywords: declarative debugging, Maude, proof tree transformation.

1 Introduction

Declarative debugging [15], also called *algorithmic debugging*, is a debugging technique that abstracts the execution details, that may be difficult to follow in general in declarative languages, to focus on results. This approach, that has been used in logic [17], functional [11], and multi-paradigm [8] languages, is a two-phase process [10]: first, a data structure representing the computation, the so-called *debugging tree*, is built; in the second phase this tree is traversed following a *navigation strategy* and asking to an external oracle about the correctness of the computation associated to the current node until a *buggy node*, an incorrect node with all its children correct, is found. The structure of the debugging tree must ensure that buggy nodes are associated to incorrect fragments of code, that is, finding a buggy node is equivalent to finding a bug in the program. Note that, since the oracle used to navigate the tree is usually the user, the number and complexity of the questions are the main issues when discussing the applicability of the technique.

Maude [4] is a high-level language and high-performance system supporting both equational and rewriting logic computation. Maude modules correspond to specifications in rewriting logic [9], a simple and expressive logic which allows the representation of many models of concurrent and distributed systems. This logic is an extension of equational logic, that in the Maude case corresponds to membership equational logic (MEL) [1], which, in addition to equations, allows the statement of membership axioms

---

* Research supported by MEC Spanish projects *DESAFIOS10* (TIN2009-14599-C03-01) and *STAMP* (TIN2008-06622-C03-01), Comunidad de Madrid programs *PROMETIDOS* (S2009/TIC1465) and *PROMESAS* (S-0505/TIC/0407), and UCM-BSCH-GR58/08-910502.

G. Vidal (Ed.): LOPSTR 2011, LNCS 7225, pp. 73–89, 2012.
© Springer-Verlag Berlin Heidelberg 2012

characterizing the elements of a sort. Rewriting logic extends MEL by adding rewrite rules, that represent transitions in a concurrent system.

In previous papers we have faced the problem of declarative debugging of Maude specifications both for *wrong answers* (incorrect results obtained from a valid input), and for missing answers (incomplete results obtained from a valid input); a complete description of the system can be found in [14]. Conceptually debugging trees in Maude are obtained in two steps. First a *proof tree* for the erroneous result (either a wrong or missing answer) in a suitable semantic calculus is considered. Then this tree is pruned by removing those nodes that correspond to logic inference steps that does not depend on the program and are consequently valid. The result is an *abbreviated proof tree* (APT) which has the property of requiring less questions to find the error in the program. Moreover, the terms in the APT nodes appear in their most reduced forms (for instance function calls have been replaced by their results). Although unnecessary from the theoretical point of view, this property of containing terms in their most reduced form has been required since the earlier works in declarative debugging (see Section 2) since otherwise the debugging process becomes unfeasible in practice due to the complexity of the questions performed to the user.

However, the situation changes when debugging Maude specifications with the strat attribute [4], that directs the evaluation order and can prevent some arguments from being reduced, that is, this attribute introduces a particular notion of laziness, making some subterms to be evaluated later than they would be in a "standard" Maude computation. For this reason we will use in this paper a slightly different notion of normal form that takes into account strat: a term is in normal form if neither the term nor its subterms has been further reduced in the current computation. When dealing with specifications with this attribute the APT no longer contains the terms in their most reduced forms, and thus the questions performed by the tool become too involved.

The purpose of this work is to define a program transformation that converts an arbitrary proof tree T built for a specification with the strat attribute into a proof tree T' whose APT contains all the subterms in their most reduced form. Since T' is also a proof tree for the same computation the soundness and completeness of the technique obtained in previous papers remain valid. Note that this improvement, described for the equational subset of Maude (where strat is applied) improves the questions asked in the debugging of both wrong and missing answers, including system modules, because reductions are used by all the involved calculi. Note that, although we present here a transformation for arbitrary proof trees, our tool builds the debugging trees in such a way that some of this transformations are not needed (more specifically, we do not need the "canonical" transformation we will see later). We prefer to build the proof tree and then transform it to make the approach conservative: the user can decide whether he wants to use the transformation or not.

The rest of the paper is organized as follows: the following section introduces some related work and shows the contributions of our approach with respect to related proposals. Section 3 introduces Maude functional modules, the debugging trees used to debug this kind of modules, and the trees we want to obtain to improve the debugging process. Section 4 presents the transformations applied to obtain these trees and the theoretical results that ensure that the transformation is safe. Finally, we present the conclusions and discuss some related ongoing work.

The source code of the debugger, examples, and much more information is available at http://maude.sip.ucm.es/debugging/. Detailed proofs of the results shown in this paper and extended information about the transformations can be found in [2].

2   Related Work

Since the introduction of declarative debugging [15] the main concerns with respect to this technique were the complexity of the questions performed to the user, and also that the process can become very tedious, and thus error-prone. The second point is related to the number of questions and has been addressed in different ways [14,16]: nodes whose correction only depends on the correction of their children are removed; statements and modules can be trusted, and thus the corresponding nodes can be removed from the debugging tree; a database can be used to prevent debuggers from asking the same question twice; trees can be compressed [5], which consists in removing from the debugging tree the children of nodes that are related to the same error as the father, in such a way that the father will provide all the debugging information; a different approach consists in *adding* nodes to the debugging tree to balance it and thus traverse it more efficiently [7]; finally, other techniques reduce the number of questions by allowing complex answers, that direct the debugging process in a more specific direction, e.g. [8] provides an answer to point out a specific subterm as erroneous.

This paper faces the first concern, the complexity of the questions, considering the case of Maude specifications including the strat attribute. This attribute can be used to alter the execution order, and thus the same subterm can be found in different forms in the tree. The unpredictability of the execution order was already considered in the first declarative debuggers proposed for lazy functional programming. In [12] the authors proposed two ways of constructing the debugging trees. The first one was based on source code transformations and the introduction of an impure primitive employed for ensuring that all the subterms take the most reduced form (or a special symbol denoting unevaluated calls). This idea was implemented in Buddha [13], a declarative debugger for Haskell, and in the declarative debugger of the functional-logic language Toy [3]. The second proposal was to change the underlying language implementation, which offers better performance. This technique was exploited in [11], where an implementation based on graph reduction was proposed for the language Haskell.

In this paper we address a similar problem from a different point of view. We are interested in proving formally the adequacy of the proposal and thus we propose a transformation at the level of the proof trees, independent of the implementation. The transformation takes an arbitrary proof tree and generates a new proof tree. We prove that the transformed tree is a valid proof tree with respect to rewriting logic calculus underlying Maude and that the subterms in questions are in their most reduced form.

3   Declarative Debugging in Maude

We present here Maude and the debugging trees used to debug Maude specifications.

3.1 Maude

For our purposes in this paper we are interested in the equational subset of Maude, which corresponds to specifications in MEL [1]. Maude functional modules [4], introduced with syntax fmod ... endfm, are executable MEL specifications and their semantics is given by the corresponding initial membership algebra in the class of algebras satisfying the specification. In a functional module we can declare sorts (by means of keyword sort(s)); subsort relations between sorts (subsort); operators (op) for building values of these sorts, giving the sorts of their arguments and result, and which may have attributes such as being associative (assoc) or commutative (comm), for example; memberships (mb) asserting that a term has a sort; and equations (eq) identifying terms. Both memberships and equations can be conditional (cmb and ceq). The executability requirements for equations and memberships are confluence, termination, and sort-decreasingness [4].

We illustrate the features described before with an example. The LAZY-LISTS module below specifies lists with a lazy behavior. At the beginning of the module we define the sort NatList for lists of natural numbers, which has Nat as a subsort, indicating that a natural number constitutes the singleton list:

```
(fmod LAZY-LISTS is
  pr NAT .
  sort NatList .
  subsort Nat < NatList .
```

Lists are built with the operator nil for empty lists and with the operator _ _ for bigger lists, which is associative and has nil as identity. It also has the attribute strat(0) indicating that only reductions at the top (the position 0) are allowed:

```
op nil : -> NatList [ctor] .
op _ _ : NatList NatList -> NatList [ctor assoc id: nil strat(0)] .
```

Next, we define a function from that generates a potentially infinite list starting from the number given as argument. Note that the attribute strat(0) in _ _, used in the right-hand side of the equation, does not permit reductions in the subterms of N from(s(N)), thus preventing an infinite computation because no equations can be applied to from(s(N)):

```
op from : Nat -> NatList .
eq [f] : from(N) = N from(s(N)) .
```

where f is a label identifying the equation. The module also contains a function take that extracts the number of elements indicated by the first argument from the list given as the second argument. Since the strat(0) attribute in _ _ prevents the list from evolving, we take the first element of the list and apply the function to the rest of the list in a matching condition, thus separating the terms built with _ _ into two different terms and allowing the lazy lists to develop all the needed elements:

```
op take : Nat NatList -> NatList .
ceq [t1] : take(s(N), N' NL) = N' NL' if NL' := take(N, NL) .
eq [t2] : take(N, NL) = 0 [owise] .
```

where owise stands for otherwise, indicating that the equation is used when no other equation can be applied. Finally, the function head extracts the first element of a list, where ~> indicates that the function is partial:

(Reflexivity)   **(Congruence)**

$$\frac{}{t \to t} \text{ Rf} \qquad \frac{t_1 \to t_1' \quad \cdots \quad t_n \to t_n'}{f(t_1, \ldots, t_n) \to f(t_1', \ldots, t_n')} \text{ Cong}$$

(Transitivity)   **(Replacement)**

$$\frac{t_1 \to t' \quad t' \to t_2}{t_1 \to t_2} \text{ Tr} \qquad \frac{\{\theta(u_i) \downarrow \theta(u_i')\}_{i=1}^n \quad \{\theta(v_j) : s_j\}_{j=1}^m}{\theta(t) \to \theta(t')} \text{ Rep}$$
$$\text{if } t \to t' \Leftarrow \bigwedge_{i=1}^n u_i = u_i' \wedge \bigwedge_{j=1}^m v_j : s_j$$

Fig. 1. Semantic calculus for reductions

```
  op head : NatList ~> Nat .
  eq [h] : head(N NL) = N .
endfm)
```

We can now introduce the module in Maude and reduce the following term:

```
Maude> (red take(2 * head(from(1)), from(1)) .)
result NatList : 1 2 0
```

However, instead of returning the first two elements of the list, it appends 0 to the result. The unexpected result of this computation indicates that there is some error in the program. The following sections show how to build a debugging tree for this reduction, how to improve it, and how to use this improved tree to debug the specification.

3.2 Debugging Trees

Debugging trees for Maude specifications [14] are conceptually built in two steps:[1] first, a proof tree is built with the proof calculus in Figure 1, which is a modification of the calculus in [1], where we use the notation $t \downarrow t'$ to indicate that t and t' are reduced to the same term (which is used for both equality conditions of the form $t = t'$ and matching conditions $t := t'$, where t may contain new variables) and we assume that the equations are terminating and confluent and hence they can be oriented from left to right, and that replacement inferences keep the label of the applied statement in order to point it out as wrong when a buggy node is found. In the second step a pruning function, called *APT*, is applied to the proof tree in order to remove those nodes whose correctness only depends on the correctness of their children (and thus they are useless for the debugging process) and to improve the questions asked to the user. This transformation can be found in [14].

Figure 1 describes the part of MEL we will use throughout this paper, the extension to full MEL is straightforward and can be found in [2]. The figure shows that the proof trees can infer *judgments* of the form $t \to t'$, indicating that t is reduced to t' by using equations. The inference rules in this calculus are reflexivity, that proves that a term can be reduced to itself; congruence, that allows to reduce the subterms; transitivity, used to compose reductions; and replacement, that applies a equation to a term if a substitution

---

[1] The implementation applies these two steps at once.

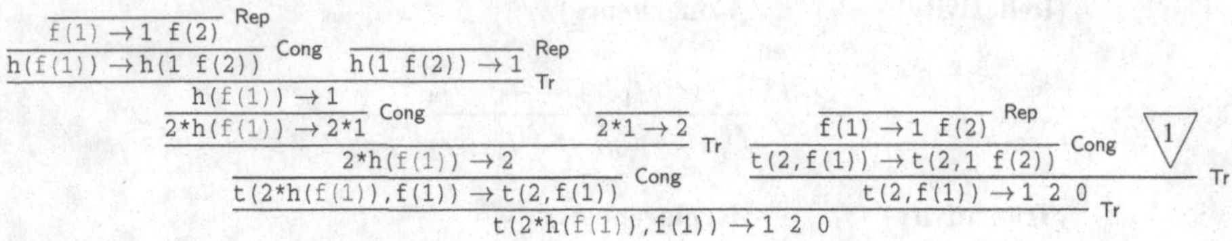

$$
\dfrac{
  \dfrac{
    \dfrac{
      \dfrac{\overline{f(1)} \to 1\ f(2)}{h(f(1)) \to h(1\ f(2))}\ \text{Rep}\ \ \text{Cong}
      \quad \overline{h(1\ f(2)) \to 1}\ \text{Rep}
    }{h(f(1)) \to 1}\ \text{Tr}
  }{2*h(f(1)) \to 2*1}\ \text{Cong}
  \quad \overline{2*1 \to 2}\ \text{Tr}
}{2*h(f(1)) \to 2}
$$

Fig. 2. Proof tree for the reduction on Section 3.1

θ making the term match the lefthand side of the equation and fulfilling the conditions is found. It is easy to see that the only inference rule whose correctness depends on the specification is replacement; intuitively, nodes inferred with this rule will be the only ones kept by *APT*. Thus, *APT* removes some nodes from the tree and can attach the debugging information to some others in order to ease the questions asked to the user, but cannot modify the judgments in the nodes, since it would require to modify the whole structure of the tree, as we will see later.

We show in Figures 2 and 3 the proof tree associated to the reduction presented in the previous section, obtained following Maude execution strategies,[2] where t stands for take, h for head, and f for from. The left child of the root of the tree in Figure 2 obtains the number of elements that must be extracted from the list, while the right child unfolds the list one step further and takes the element thus obtained, repeating this operation until all the required elements have been taken. Note that Maude cannot reduce the second argument of t, f(1), to its normal form (with respect to the tree) 1 2 3 f(4) with three consecutive replacement steps because the attribute strat(0) prevents it.

From the point of view of declarative debugging, this tree is not very satisfactory, because it contains nodes like t(2,1 f(2)) → 1 2 0, the root of the tree in Figure 3, where the subterm f(2) is not fully reduced, which forces the user to obtain its expected result and then (mentally) substitute it in the node in order to answer the question about the correction of the node. We show the APT corresponding to this tree in Figure 4; note that a transformation like *APT* cannot improve this kind of questions because there is no node with the information we want to use, and thus the node (†) described above is kept and will be used in the debugging process. Intuitively, we would like to gather all the replacements related to the same term so we can always ask about terms with the subterms in normal form, like t(2,1 2 ⊥), where ⊥ is a special symbol indicating that a term could be further reduced but its value is not necessary. The next section explains how to transform proof trees in order to obtain questions with this form.

When examining a proof tree we are interested in distinguishing whether two syntactically identical terms are copies of the same term or not. The reason is that it is more natural for the user to have each copy in its more reduced form, without considering the reductions of other copies of the same term (as happens with the term f(1) in the

---

[2] Actually, the value 3 in Figure 3 has been computed to mimic Maude's behavior. Once it has obtained take(0, f(3)) it tries to reduce its subterms, obtaining 3 although it will be never used. All the transformations in this paper also work if this term is not computed.

$$\cfrac{\cfrac{}{(\spadesuit)\ \text{f(2)} \to 2\ \text{f(3)}}\ \text{Rep}\qquad \cfrac{\cfrac{\cfrac{}{\text{f(3)} \to 3\ \text{f(4)}}\ \text{Rep}}{\text{t(0,f(3))} \to \text{t(0,3\ f(4))}}\ \text{Cong}\qquad \cfrac{}{\text{t(0,3\ f(4))} \to 0}\ \text{Rep}}{\cfrac{\text{t(0,f(3))} \to 0}{\text{t(1,2\ f(3))} \to 2\ 0}\ \text{Rep}}\ \text{Tr}}{\cfrac{\text{t(1,f(2))} \to \text{t(1,2\ f(3))}}{\cfrac{(\lozenge)\ \text{t(1,f(2))} \to 2\ 0}{\text{t(2,1\ f(2))} \to 1\ 2\ 0}\ \text{Rep}}\ \text{Tr}}\ \text{Cong}$$

Fig. 3. Proof tree for the subtree $\bigtriangledown$ on Figure 2

$$\cfrac{\cfrac{}{\text{f(1)} \to 1\ \text{f(2)}}\ \text{Rep}\quad \cfrac{}{\text{h(1\ f(2))} \to 1}\ \text{Rep}\quad \cfrac{}{\text{f(1)} \to 1\ \text{f(2)}}\ \text{Rep}\qquad \cfrac{\cfrac{}{\text{f(2)} \to 2\ \text{f(3)}}\ \text{Rep}\qquad \cfrac{\cfrac{}{\text{f(3)} \to 3\ \text{f(4)}}\ \text{Rep}\quad \cfrac{}{\text{t(0,3\ f(4))} \to 0}\ \text{Rep}}{\text{t(1,2\ f(3))} \to 2\ 0}\ \text{Rep}}{(\dagger)\ \text{t(2,1\ f(2))} \to 1\ 2\ 0}\ \text{Tr}}{\text{t(2*h(f(1)),f(1))} \to 1\ 2\ 0}$$

Fig. 4. Abbreviated proof tree for the proof tree in Figure 2

example above; one of these terms is reduced to 1 f(2) while the second one is reduced to 1 2 3 f(4)). We achieve this goal by "painting" related terms in a proof tree with the same color. Hence the same term can be repeated in several places in a proof tree, but only those copies coming from the same original term will have the same color. We refer to colored terms as *c-terms* and to trees with colored terms in their nodes as *c-trees*. When talking about colored trees, $t_1 = t_2$ means that t_1 and t_2 are equally colored. Therefore talking about two occurrences of a c-term t implicitly means that there are two copies of the same term equally colored. Intuitively, all the terms in the lefthand side of the root have different colors; the replacement inference rule introduces new colors, while the reflexivity, transitivity, and congruence rules propagate them. More details can be found in [2].

It is worth observing that computation trees represent a particular computation that has already taken place in Maude. This means that we can be sure that the debugged program satisfies the constraints required by Maude functional modules: equations must be terminating, confluent, and sort-decreasing. Other requirements such as left-linearity or constructor-based rules are not required in these modules. The details of how to carry out a computation correspond to Maude and not to the debugger, which only represents computations. During the tree construction process, the debugger already knows the appropriate substitutions used in the associated computation as well as the places where they must be applied for each computation step; our algorithms just modify the tree taking into account this information. A subtle detail that allows us to move the computations forward is that the lefthand side of Maude equations must be a pattern, and thus "frozen" terms (i.e., terms that are not built with constructors and cannot be reduced because of the strat attribute) such as N . from(N') cannot be used. Notice that in Maude, term sharing is introduced incrementally by equational simplification, because it analyzes righthand sides of equations to identify its shared subterms [6]. More details can be found in [2].

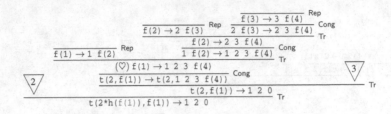

Fig. 5. Proof tree for the reduction on Section 3.1

$$\dfrac{\dfrac{2\ 3\ f(4) \rightarrow 2\ 3\ f(4)}{t(1,\ 2\ 3\ f(4)) \rightarrow t(1,2\ 3\ f(4))}\ \text{Cong} \qquad \dfrac{\dfrac{\dfrac{3\ f(4) \rightarrow 3\ f(4)}{t(0,\ 3\ f(4)) \rightarrow t(0,3\ f(4))}\ \text{Cong}\quad t(0,3\ f(4)) \rightarrow 0\ \text{Rep}}{\dfrac{t(0,3\ f(4)) \rightarrow 0}{t(1,2\ 3\ f(4)) \rightarrow 2\ 0}\ \text{Rep}}\ \text{Tr}}{\dfrac{t(1,\ 2\ 3\ f(4)) \rightarrow 2\ 0}{t(2,1\ 2\ 3\ f(4)) \rightarrow 1\ 2\ 0}\ \text{Rep}}}$$

Fig. 6. Proof tree for the subtree ▽ on Figure 5

3.3   The Lists Example Revisited

As explained in the previous section, the debugging tree in Figure 4 presents the drawback of containing nodes of the form t(2,1 f(2)) → 1 2 0, whose correction is difficult to state because the subterms must be mentally reduced by the user in order to compute the final result. We give in this section the intuitions motivating the transformations in the next section, transforming the trees and in Figures 2 and 3, that give rise to the proof trees in Figures 5 and 6. The tree ▽ in Figure 5 has the same left premise as the one in Figure 2, which shows the importance of coloring the terms in the proof trees, because the algorithm distinguishes between the two f(1) thanks to their different colors. The part of the tree depicted in Figure 5 shows how the reduction of the subterms is "anticipated" by the algorithm in the previous section and thus the node (♡) performs the reduction of the second argument of t, f(1), to its normal form (with respect to the tree), and all the replacement steps that were needed to reach it are contained in this subtree. The tree ▽ in Figure 6 shows the other part of the transformations: we get rid of the relocated replacement inferences by using reflexivity steps.

The APT of our transformed proof tree is depicted in Figure 7. It has removed all the useless information like reflexivity and congruence inferences, and has associated the replacement inferences, that contain debugging information, to the transitivity inferences below them, returning a debugging tree where the lefthand side of all the reductions have their subterms in normal form, as expected because the transformation works driven by the *APT* transformation.

Since this transformation has been implemented in our declarative debugger, we can start a debugging session to find the error in the specification described in Section 3.1. The debugging process starts with the command:

```
Maude> (debug take(2 * head(from(1)), from(1)) -> 1 2 0 .)
```

$$\cfrac{\cfrac{}{\text{f(1)} \to 1 \ \text{f(2)}}\text{Rep} \quad \cfrac{}{\text{h(1 f(2))} \to 1}\text{Rep} \quad \cfrac{\cfrac{\cfrac{}{\text{f(3)} \to 3 \ \text{f(4)}}\text{Rep}}{\text{f(2)} \to 2 \ 3 \ \text{f(4)}}\text{Rep}}{\text{f(1)} \to 1 \ 2 \ 3 \ \text{f(4)}}\text{Rep} \quad \cfrac{\cfrac{\cfrac{(\bullet)\,\text{t(0,3 f(4))} \to 0}{\text{t(1,2 3 f(4))} \to 2 \ 0}\text{Rep}}{(\ddagger)\,\text{t(2,1 2 3 f(4))} \to 1 \ 2 \ 0}\text{Rep}}{}\text{Rep}}{\text{t(2*h(f(1)),f(1))} \to 1 \ 2 \ 0}\text{Tr}$$

Fig. 7. Abbreviated proof tree for the transformed tree

This command builds the tree shown in Figure 7,[3] which is traversed following the navigation strategy divide and query [16], that selects in each case a node rooting a subtree with approximately half the size of the whole tree, and the first question, associated with the node ($\ddagger$) in Figure 7, is:

```
Is this reduction (associated with the equation t1) correct?
take(2,1 2 3 ?:NatList) -> 1 2 0
Maude> (no .)
```

where we must interpret ?:NatList as a term that has not reached a normal form (in the sense it is not built with operators with the ctor attribute) but whose value is irrelevant to compute the final result. The answer is (no .) because we expected to take only 1 2. Note that this node is the transformed version of the node (†) in Figure 4, that would perform the question:

```
Is this reduction (associated with the equation t1) correct?
take(2,1 from(2)) -> 1 2 0
```

which is more difficult to answer because we have to think first about the reduction of from(2) and then use the result to reason about the complete reduction. With the answer given above the subtree rooted by ($\ddagger$) in Figure 7 is considered as the current one and the next questions are:

```
Is this reduction (associated with the equation t1) correct?
take(1,2 3 ?:NatList) -> 2 0
Maude> (no .)
Is this reduction (associated with the equation t2) correct?
take(0,3 ?:NatList) -> 0
Maude> (no .)
```

We answer (no .) in both cases for the same reason as in the previous case. With these answers we have discovered that the node ($\bullet$) is wrong. Hence, since it has no children, it is the buggy node and is associated with a wrong statement:

```
The buggy node is: take(0,3 ?:NatList) -> 0
with the associated equation: t2
```

In fact, the equation t2 returns 0 but it should return nil.

4   Transforming Debugging Trees

In this section we present the transformation that ensures that the abbreviated proof tree contains every term reduced as much as possible. This transformation is a two-step process. First, a sequence of three tree transformations prepares the proof tree for the

---

[3] Actually, it builds the tree in Figure 4 and then transforms it into the tree in Figure 7.

second phase. We call the trees obtained by this transformation *canonical trees*. The second phase takes a canonical tree as input and applies the algorithm that replaces terms by its most reduced forms. The result is a proof tree for the same computation whose APT verifies that every term is reduced as much as possible.

4.1 Reductions

We need to formally define the concepts of reduction and number of steps, which will be necessary to ensure that a tree is in its most reduced form.

Definition 1. *Let T be an APT, and t, t' two c-terms. We say that $t \to t'$ is a reduction w.r.t. T if there is a node $N \in T$ of the form $t_1 \to t_2$ verifying:*

- *$pos(t, t_1) \neq \emptyset$, where $pos(t, t_1)$ is the set of positions of t containing t_1.*
- *$t' = t[t_1 \mapsto t_2]$, where $t[t_1 \mapsto t_2]$ represents the replacement of every occurrence of the c-term t_1 by t_2 in t.*

In this case we also say that t is reducible (w.r.t. T). A reduction chain *for t will be a sequence of reductions $t_0 = t \to t_1 \to t_2 \to \cdots \to t_n$ s.t. each $t_i \to t_{i+1}$ is a reduction and that t_n cannot be further reduced w.r.t. T.*

Definition 2. *Let T be an APT. Then:*

- *The number of reductions of a term t w.r.t. T, denoted as $reduc(t, T)$ is the sum of the length of all the possible different reduction chains of t w.r.t. T.*
- *The number of reductions of a node of the form $N = f(t_1, \ldots, t_n) \to t$ w.r.t. T, denoted as $reduc(N, T)$ is defined as $(\sum_{i=1}^{n} reduc(t_i, T)) + reduc(t, T)$.*

In this definition the length of a reduction chain $t_0 \to \cdots \to t_n$ is defined as n. Remember that the aim of this paper is to present a technique to put together these reductions chains, transforming appropriately the proof tree, and using colors to distinguish terms; when dealing with commutative or associativity, we will assume flatten terms with the subterms ordered in an alphabetical fashion. Moreover, our technique assumes that there is only one normal form for each c-term in the tree.

Definition 3. *We say that an occurrence of a c-term t occurring in an APT T is in* normal form *w.r.t. T if there is no reduction for any c-subterm of t in T.*

Definition 4. *Let T be an APT. We say that T is* confluent *if every c-term t occurring in T has a unique normal form with respect to T.*

Note that this notion of confluence is different from the usual notion of confluence required in Maude functional modules: it requires all the copies of a (colored) term, that can be influenced by the `strat` attribute, to be reduced to the same term. In the rest of the paper we assume that, unless stated otherwise, all the APTs are colored and confluent. With these definitions we are ready to define the concept of *norm*:

(InsCong$_1$)

$$InsCong\left(\frac{T_1 \ldots T_m}{f(t_1,\ldots,t_n) \to t}\text{Rep}\right) =$$

$$\cfrac{\cfrac{}{t_1 \to t_1}\text{Rf} \ldots \cfrac{}{t_n \to t_n}\text{Rf}}{\cfrac{f(t_1,\ldots,t_n) \to f(t_1,\ldots,t_n)}{f(t_1,\ldots,t_n) \to t}\text{Cong} \quad \cfrac{InsCong(T_1)\ldots InsCong(T_m)}{f(t_1,\ldots,t_n) \to t}\text{Rep}}\text{Tr}$$

(InsCong$_2$)

$$InsCong\left(\frac{T_1 \ldots T_m}{aj}\text{R}\right) = \frac{InsCong(T_1)\ldots InsCong(T_m)}{aj}\text{R}$$

aj any judgment, R any inference rule, $n > 0$

Fig. 8. Insert Congruences (*InsCong*)

Definition 5. *Let T be a proof tree, and $T' = APT(T)$. The norm of T, represented by $\| T \|$, is the sum of the lengths of all the reduction chains that can be applied to terms in T'. More formally, given the reduc function in Definition 2:*

$$\| T \| = \sum_{\substack{N \in T' \\ N \neq root(T')}} reduc(N, T')$$

Thus, the norm is the number of reductions that can be performed in the corresponding APT. Our goal is to obtain proof trees with associated norm 0, ensuring that the questions performed to the user contain terms as reduced as possible. This is the purpose of the proof tree transformations in the following section, which start with some initial proof tree and produces an equivalent proof tree with norm 0.

4.2 Canonical Trees

Canonical trees are obtained from proof trees as explained in the following definition.

Definition 6. *We define the canonical form of a proof tree T, which will be denoted from now on as $Can(T)$, as*

$$Can(T) = RemInf(NTr(InsCong(T)))$$

where InsCong (insert congruences), NTr (normalize transitivities), and RemInf (remove superfluous inferences) are defined in Figures 8, 9, and 10, respectively.

It is assumed that the rules of each transformation are applied top-down. The first transformation, *InsCong*, prepares the proof tree for allowing reductions on the arguments t_i of judgments of the form $f(t_1,\ldots,t_n) \to t$ by introducing congruence inferences before these judgments take place. Initially no reduction is applied, and each argument is simply reduced to itself using a reflexivity inference. Replacing these reflexivities by non-trivial reductions for the arguments is the role of the algorithm introduced in the

(NTr₁)

$$NTr\left(\dfrac{\dfrac{T_{t_1 \to t_2} \quad T_{t_2 \to t_3}}{t_1 \to t_3}\text{Tr} \quad T_{t_3 \to t_4}}{t_1 \to t_4}\text{Tr}\right) = NTr\left(\dfrac{NTr(T_{t_1 \to t_2}) \quad NTr\left(\dfrac{T_{t_2 \to t_3} \quad T_{t_3 \to t_4}}{t_2 \to t_4}\text{Tr}\right)}{t_1 \to t_4}\text{Tr}\right)$$

(NTr₂)

$$NTr\left(\dfrac{T_1 \dots T_n}{aj}\text{R}\right) = \dfrac{NTr(T_1) \dots NTr(T_n)}{aj}\text{R} \qquad aj \text{ any judgment, R any inference rule}$$

Fig. 9. Normalize Transitivities (*NTr*)

next section. The next transformation, *NTr*, takes care of righthand sides. The idea is that transitivity inferences occurring as left premises of other transitivity are associated to intermediate, not fully-reduced computations. Thus, *NTr* ensures that righthand sides can be completely reduced by the algorithm in the next section. Finally, *RemInf* eliminates some superfluous steps involving reflexivities, and combines consecutive congruences in a "bigger step" single congruence, which avoids the production of unnecessary intermediate results in the proof tree. This last process is done with the help of an auxiliary transformation *merge* (Figure 11), that combines two trees by using a transitivity.

A proof tree in canonical form is also a proof tree proving the same judgment.

Proposition 1. *Let T be a proof tree. Then Can(T) is a proof tree with the same root.*

Moreover, applying these transformations cannot produce an increase of the norm:

Proposition 2. *Let T be a proof tree and $T' = Can(T)$. Then $\| T \| \geq \| T' \|$.*

4.3   Reducing the Norm of Canonical Trees

We describe in this section the main transformation applied to the proof trees. This transformation relies on the following proposition, that declares that in any proof tree in canonical form there exist (1) a node with a reduction $t_1 \to t_1'$ such that t_1' is in normal form, that will be used to further reduce the terms, (2) a node that contains a reduction $t_2 \to t_2'$, with $t_1 \in t_2'$ (t_1 is a subterm of t_2'), which means that t_2' can be further reduced by the previous reduction, and (3) a node such that it is not affected by the transformations in the previous nodes. We will use node (1) to improve the reductions in node (2); this transformation will only affect the nodes in the subtree that has (3) as root:

Proposition 3. *Let T be a confluent c-proof tree in canonical form such that $\| T \| > 0$. Then T contains:*

1. *A node related to a judgment $t_1 \to t_1'$ such that:*
   - *It is either the consequence of a transitivity inference with a replacement as left premise, or the consequence of a replacement inference which is not the left premise of a transitivity.*
   - *t_1' is in normal form w.r.t. T.*

$$(\textbf{RemInf}_1) \quad \mathit{RemInf}\left(\dfrac{\dfrac{T_{t_1 \to t_1'} \ldots T_{t_n \to t_n'}}{f(t_1,\ldots,t_n) \to f(t_1',\ldots,t_n')}\,\mathsf{Cong} \qquad \dfrac{T_{t_1' \to t_1''} \ldots T_{t_n' \to t_n''}}{f(t_1',\ldots,t_n') \to f(t_1'',\ldots,t_n'')}\,\mathsf{Cong}}{f(t_1,\ldots,t_n) \to f(t_1'',\ldots,t_n'')}\,\mathsf{Tr} \right) =$$

$$\dfrac{\mathit{RemInf}\left(\mathit{merge}\left(T_{t_1 \to t_1'},\ T_{t_1' \to t_1''}\right)\right) \ldots \mathit{RemInf}\left(\mathit{merge}\left(T_{t_n \to t_n'},\ T_{t_n' \to t_n''}\right)\right)}{f(t_1,\ldots,t_n) \to f(t_1'',\ldots,t_n'')}\,\mathsf{Cong}$$

$$(\textbf{RemInf}_2) \quad \mathit{RemInf}\left(\dfrac{\dfrac{T_{t_1 \to t_1'} \ldots T_{t_n \to t_n'}}{f(t_1,\ldots,t_n) \to f(t_1',\ldots,t_n')}\,\mathsf{Cong} \quad \dfrac{\dfrac{T_{t_1' \to t_1''} \ldots T_{t_n' \to t_n''}}{f(t_1',\ldots,t_n') \to f(t_1'',\ldots,t_n'')}\,\mathsf{Cong} \quad T_{f(t_1'',\ldots,t_n'') \to e}}{f(t_1',\ldots,t_n') \to e}\,\mathsf{Tr}}{f(t_1,\ldots,t_n) \to e}\,\mathsf{Tr} \right) =$$

$$\mathit{RemInf}\left(\dfrac{\dfrac{\mathit{merge}\left(T_{t_1 \to t_1'},\ T_{t_1' \to t_1''}\right) \ldots \mathit{merge}\left(T_{t_n \to t_n'},\ T_{t_n' \to t_n''}\right)}{f(t_1,\ldots,t_n) \to f(t_1'',\ldots,t_n'')}\,\mathsf{Cong} \qquad \mathit{RemInf}(T_{f(t_1'',\ldots,t_n'') \to e})}{f(t_1,\ldots,t_n) \to e}\,\mathsf{Tr} \right)$$

$$(\textbf{RemInf}_3) \quad \mathit{RemInf}\left(\dfrac{\dfrac{F}{t \to t_1}\,\mathsf{Cong} \quad \dfrac{\overline{t_1 \to t_1}\,\mathsf{Rf} \quad T_{t_1 \to t'}}{t_1 \to t'}\,\mathsf{Tr}}{t \to t'}\,\mathsf{Tr} \right) = \mathit{RemInf}\left(\dfrac{\dfrac{F}{t \to t_1}\,\mathsf{Cong} \quad T_{t_1 \to t'}}{t \to t'}\,\mathsf{Tr} \right)$$

$$(\textbf{RemInf}_4) \quad \mathit{RemInf}\left(\dfrac{T^{\mathsf{Rf}} \quad T'}{aj}\,\mathsf{Tr} \right) = \mathit{RemInf}\left(\dfrac{T' \quad T^{\mathsf{Rf}}}{aj}\,\mathsf{Tr} \right) = \mathit{RemInf}(T'),\ aj \text{ any judgment}$$

$$(\textbf{RemInf}_5) \quad \mathit{RemInf}\left(\dfrac{T_1 \ldots T_n}{aj}\,\mathsf{R} \right) = \dfrac{\mathit{RemInf}(T_1) \ldots \mathit{RemInf}(T_n)}{aj}\,\mathsf{R},\ aj \text{ any judgment, } \mathsf{R} \text{ any inference rule}$$

Fig. 10. Remove superfluous inferences (*RemInf*)

(Merge₁)

$$merge\left(\frac{T_{t\to t_1}\ T_{t_1\to t'}}{t\to t'}\mathsf{Tr}, T_{t'\to t''}\right) = \frac{T_{t\to t_1}\ merge\left(T_{t_1\to t'}, T_{t'\to t''}\right)}{t\to t''}\mathsf{Tr}$$

(Merge₂)

$$merge\left(T_{t\to t'}, T_{t'\to t''}\right) = \frac{T_{t\to t'}\ T_{t'\to t''}}{t\to t''}\mathsf{Tr}$$

Fig. 11. Merge Trees

2. A node related to a judgment $t_2 \to t_2'$ with $t_1 \in t_2'$.
3. A node related to a judgment $t_3 \to t_3'$ consequence of a transitivity step, with $t_1 \notin t_3'$.

Algorithm 1 presents the transformation in charge of reducing the norm of the proof trees until it reaches 0. It first selects a node N_{ible} (from *reducible node*), that contains a term that has been further reduced during the computation,[4] a node N_{er} (from *reducer node*) that contains the reduction needed by the terms in N_{ible}, and a node p_0 limiting the range of the transformation. Note that we can distinguish two parts in the subtree rooted by the node in p_0, the left premise, where N_{ible} is located, and the right premise, where N_{er} is located. Then, we create some copies of these nodes in order to use them after the transformations. For example, the first step of the loop for the proof tree in Figures 2 and 3 would set N_{ible} to $f(2) \to 2\ f(3)$ and N_{er} to $f(3) \to 3\ f(4)$; they are located in the subtree rooted by p_0, the node ($\Diamond$).

Step 6 replaces the proof of the reduction $t_1 \to t_1'$ by reflexivity steps $t_1' \to t_1'$. Since the algorithm is trying to use this reduction before its current position, a natural consequence will be to transform all the appearances of t_1 in the path between the old and the new position by t_1', what means that in this particular place we would obtain the reduction $t_1' \to t_1'$ inferred, by Proposition 3, by either a transitivity or a replacement rule, and with the appropriate proof trees as children. Since this would be clearly incorrect, the whole tree is replaced by a reflexivity. In our example, the replacement $f(3) \to 3\ f(4)$ (N_{er}) would be transformed into the reflexivity step $3\ f(4) \to 3\ f(4)$.

Step 7 replaces all the occurrences of t_1 by t_1' in the right premise of p_0, as explained in the previous step. In this way, the right premise of p_0 is a new subtree where t_1 has been replaced by t_1' and all the proofs related to $t_1 \to t_1'$ have been replaced by reflexivity steps $t_1' \to t_1'$. Note that intuitively these steps are correct because t_1' is required to be in normal form, the tree is confluent, and the norm of this tree is 0, that is, all the possible reductions of terms with the same color have been previously modified by the algorithm to create a $t_1 \to t_1'$ proof. In our example, the appearances of $f(3)$ in the right premise of the node ($\Diamond$) are replaced by $3\ f(4)$; this subtree is already a proof tree.

Step 8 replaces the occurrences of t_1 by t_1' in the left premise of p_0. We apply this transformation only in the righthand sides because they are in charge of transmitting the information, and in this way we prevent the algorithm from changing correct values (inherited perhaps from the root). This substitution can be used thanks to the position p_0, which ensures that only the righthand sides are affected. In our example, we substitute the term $f(3)$ by $3\ f(4)$ in the left child of ($\Diamond$).

---

[4] We select the first one in post-order to ensure that this node is the one that generated the term.

Step 9 combines the reduction in N_{ible} with the reduction in N_{er} (actually, it merges their copies, since the previous transformations have modified them). If the term t_1 we are further reducing corresponds to the term t_2' in the lefthand side of the judgment in N_{ible}, then it is enough to use a transitivity to "join" the two subtrees. In other case, the term we are reducing is a subterm of t_2' and thus we must use a congruence inference rule to reduce it, using again a transitivity step to infer the new judgment. This last step would generate, in our example, a node combining the replacement ($\spadesuit$) and the one in N_{er} in a transitivity step, giving rise to the node $f(2) \rightarrow 2\ 3\ f(4)$; in this way the left child of ($\Diamond$), and consequently the tree, becomes a proof tree again.

Finally, these transformations make the trees to lose their canonical form, and hence the canonical form of the tree is computed again in step 10.

Algorithm 1. *Let T be a proof tree in canonical form.*

1. *Let $T_r = T$*
2. *Loop while $\|\, T_r \,\| > 0$*
3. *Let $N_{er} = t_1 \rightarrow t_1'$ be a node satisfying the conditions of item 1 in Proposition 3, $N_{ible} = t_2 \rightarrow t_2'$ the first node in T's post-order verifying the conditions of item 2 in Proposition 3, and p_0 the position of the subtree of T rooted by the first (furthest from the root) ancestor of N_{ible} satisfying item 3 in Proposition 3, such that the right premise of the node in p_0, T_{rp}, has $\|\, T_{rp} \,\| = 0$.*
4. *Let C_{er} be a copy of the tree rooted by N_{er}.*
5. *Let C_{ible} be a copy of the tree rooted by N_{ible} and p_{ible} the position of N_{ible}.*
6. *Let T_1 be the result of replacing in T all the subtrees rooted by N_{er} by a reflexivity inference step with conclusion $t_1' \rightarrow t_1'$.*
7. *Let T_2 be the result of substituting all the occurrences of the c-term t_1 by t_1' in the right premise of the subtree at position p_0 in T_1.*
8. *Let T_3 be the result of substituting all the occurrences of the c-term t_1 with t_1' in the righthand sides of the left premise of the subtree at position p_0 in T_2.*
9. *Let T_4 be the result of replacing the subtree at position p_{ible} in T_3 by the following subtree:*

    (a) if $t_2' = t_1$.

    $$\frac{C_{ible} \quad C_{er}}{t_2 \rightarrow t_1'}\ \text{Tr}$$

    (b) if $t_2' \neq t_1$.

    $$\frac{C_{ible} \quad \dfrac{C_{er}}{t_2' \rightarrow t_2'[t_1 \mapsto t_1']}\ \text{Cong}}{t_2 \rightarrow t_2'[t_1 \mapsto t_1']}\ \text{Tr}$$

10. *Let T_r be the result of normalizing T_4.*
11. *End Loop*

The next theorem is the main result of this paper. It states that after applying the algorithm we obtain a proof tree for the same computation whose nodes are as reduced as possible. Thus, the declarative debugging tool that uses this tree as debugging tree will ask questions in its most simplified form.

Theorem 1. *Let T be a proof tree in canonical form. Then the result of applying Algorithm 1 to T is a proof tree T_r such that $root(T_r) = root(T)$ and $\| T_r \| = 0$.*

Observe that we have improved the "quality" of the information in the nodes without increasing the number of questions, since the transformations do not introduce new replacement inferences in the APT.

5   Concluding Remarks and Ongoing Work

One of the main criticisms to declarative debugging is the high complexity of the questions performed to the user. Thus, if the same computation can be represented by different debugging trees, we must choose the tree containing the simplest questions. In Maude, an improvement in this direction is to ensure that the judgments involving reductions are presented to the user with the terms reduced as much as possible. We have presented a transformation that allows us to produce debugging trees fulfilling this property starting with any valid proof tree for a wrong computation. The result is a debugging tree with questions as simple as possible without increasing the number of questions, which is specially useful when dealing with the strat attribute. Moreover, the theoretical results supporting the debugging technique presented in previous papers remain valid since we have proved that our transformation transforms proof trees into proof trees for the same computation.

Although for the sake of simplicity we have focused in this paper on the equational part of Maude, this transformation has been applied to all the judgments $t \to t'$ appearing in the debugging of both wrong (including system modules) and missing answers. However, our calculus for missing answers also considers judgments $t \to_{norm} t'$, indicating that t' is the normal form of t; when facing the strat attribute, the inferences for these judgments have the same problem shown here; we are currently working to define a transformation for this kind of judgment.

References

1. Bouhoula, A., Jouannaud, J.-P., Meseguer, J.: Specification and proof in membership equational logic. Theoretical Computer Science 236, 35–132 (2000)
2. Caballero, R., Martí-Oliet, N., Riesco, A., Verdejo, A.: Improving the debugging of membership equational logic specifications. Technical Report SIC-02-11, Dpto. Sistemas Informáticos y Computación, Universidad Complutense de Madrid (March 2011),
   http://maude.sip.ucm.es/debugging/
3. Caballero, R., Rodríguez-Artalejo, M.: DDT: A Declarative Debugging Tool for Functional-Logic Languages. In: Kameyama, Y., Stuckey, P.J. (eds.) FLOPS 2004. LNCS, vol. 2998, pp. 70–84. Springer, Heidelberg (2004)
4. Clavel, M., Durán, F., Eker, S., Lincoln, P., Martí-Oliet, N., Meseguer, J., Talcott, C.: All About Maude - A High-Performance Logical Framework. LNCS, vol. 4350. Springer, Heidelberg (2007)
5. Davie, T., Chitil, O.: Hat-Delta: One right does make a wrong. In: 7th Symposium on Trends in Functional Programming, TFP 2006 (2006)

6. Eker, S.: Term rewriting with operator evaluation strategies. In: Proceedings of the 2nd International Workshop on Rewriting Logic and its Applications, WRLA 1998. Electronic Notes in Theoretical Computer Science, vol. 15, pp. 311–330 (1998)

7. Insa, D., Silva, J., Riesco, A.: Balancing execution trees. In: Gulías, V.M., Silva, J., Villanueva, A. (eds.) Proceedings of the 10th Spanish Workshop on Programming Languages, PROLE 2010, pp. 129–142. Ibergarceta Publicaciones (2010)

8. MacLarty, I.: Practical declarative debugging of Mercury programs. Master's thesis, University of Melbourne (2005)

9. Meseguer, J.: Conditional rewriting logic as a unified model of concurrency. Theoretical Computer Science 96(1), 73–155 (1992)

10. Naish, L.: A declarative debugging scheme. Journal of Functional and Logic Programming 1997(3) (1997)

11. Nilsson, H.: How to look busy while being as lazy as ever: the implementation of a lazy functional debugger. Journal of Functional Programming 11(6), 629–671 (2001)

12. Nilsson, H., Sparud, J.: The evaluation dependence tree as a basis for lazy functional debugging. Automated Software Engineering 4, 121–150 (1997)

13. Pope, B.: Declarative Debugging with Buddha. In: Vene, V., Uustalu, T. (eds.) AFP 2004. LNCS, vol. 3622, pp. 273–308. Springer, Heidelberg (2005)

14. Riesco, A., Verdejo, A., Martí-Oliet, N., Caballero, R.: Declarative debugging of rewriting logic specifications. Journal of Logic and Algebraic Programming (2011) (to appear)

15. Shapiro, E.Y.: Algorithmic Program Debugging. ACM Distinguished Dissertation. MIT Press (1983)

16. Silva, J.: A Comparative Study of Algorithmic Debugging Strategies. In: Puebla, G. (ed.) LOPSTR 2006. LNCS, vol. 4407, pp. 143–159. Springer, Heidelberg (2007)

17. Tessier, A., Ferrand, G.: Declarative Diagnosis in the CLP Scheme. In: Deransart, P., Hermenegildo, M.V., Maluszynski, J. (eds.) DiSCiPl 1999. LNCS, vol. 1870, pp. 151–174. Springer, Heidelberg (2000)

Clones in Logic Programs
and How to Detect Them

Céline Dandois* and Wim Vanhoof

University of Namur - Faculty of Computer Science
21 rue Grandgagnage, 5000 Namur Belgium
{cda,wva}@info.fundp.ac.be

Abstract. In this paper, we propose a theoretical framework that allows us to capture, by program analysis, the notion of code clone in the context of logic programming. Informally, two code fragments are considered as cloned if they implement the same functionality. Clone detection can be advantageous from a software engineering viewpoint, as the presence of code clones inside a program reveals redundancy, broadly considered a "bad smell". In the paper, we present a detailed definition of a clone in a logic program and provide an efficient detection algorithm able to identify an important subclass of code clones that could be used for various applications such as program refactoring and plagiarism recognition. Our clone detection algorithm is not tied to a particular logic programming language, and can easily be instantiated for different such languages.

Keywords: logic programming languages, code clone, code duplication, clone detection algorithm.

1 Introduction

Certain characteristics of a program's source code can be recognized as constituting a "bad smell", meaning an indication of poor design quality [6]. Among these characteristics, an important place is taken by *code duplication*, also called *code cloning* [6]. Intuitively, two code fragments can be considered cloned if they implement an identical or sufficiently similar functionality, independent of them being textually equal or not. Even if the scientific community is divided about the need of suppressing code clones (they are not always considered harmful), it is generally accepted that they should, at least, be detected [11] and techniques for clone detection can be applied in various domains like program refactoring [6], plagiarism detection [8] and virus recognition [18]. Since duplication is intrinsically linked to semantic equivalence, its detection is an undecidable problem that can however be approximated by automatic program analysis.

Clone detection has received a substantial amount of attention during recent years, mainly in the context of the imperative and object-oriented programming paradigms. Several overviews synthesizing the state-of-the-art of this research

---

* F.R.S.-FNRS Research Fellow.

G. Vidal (Ed.): LOPSTR 2011, LNCS 7225, pp. 90–105, 2012.

field can be found in [11,12]. As it is pointed out in [11], there is no generally accepted definition for what constitutes a clone and the latter's definition is often bound to a particular detection technique. In spite of the abundance of work on the subject, only few works consider code cloning for declarative programming languages. Two recent papers have focused on the functional programming paradigm by presenting an abstract syntax tree (AST)-based approach dedicated to Erlang [9] and Haskell programs [3]. In the context of logic programming, [15] essentially motivates the need for clone detection in logic programs, while [16] outlines a basic code duplication analysis of logic programs. Dealing with refactorings for Prolog programs, [13] proposes transformations handling two particular cases of duplication and is, as such, related to our own work. Some language-independent detection techniques were developed too but, as it was reported in [12], they suffer from an important loss in precision since they cannot be tuned for the target language.

This paper proposes a detailed study of code cloning in logic programs by expanding and formalizing the general ideas that were presented in [16]. In the remainder of this paper, we first establish the notion of code clone in a logic program. We basically consider two predicate definitions cloned if one is a (partial) renaming of the other modulo a permutation of the arguments, clauses and body atoms. We secondly present an efficient algorithm for detecting an important subclass of these code clones.

2   Theory of Duplication

2.1   Preliminaries

In what follows, we assume the reader to be familiar with the basic logic programming concepts as they are found, for example, in [1]. To simplify the presentation of the different concepts, we restrict ourselves to *definite programs*, i.e. programs without negation. In particular, we consider a logic program to be defined as a set of *predicates*, representing relations between data objects manipulated by the program. Each predicate p/m, with name p and arity m, is defined by a set of *clauses* of the form $H \leftarrow B_1, \ldots, B_s$ with H, an *atom*, i.e. a predicate name followed by its corresponding number of arguments, *head* of the clause, and $B_1, \ldots, B_s$, a conjunction of atoms, *body* of the clause.

Without loss of generality, we make the hypothesis that clauses are converted to a flattened form such that each atom is of the form $q(X_1, \ldots, X_n)$, $X = Y$ or $X = f(X_1, \ldots, X_n)$ (with $X, Y, X_1, \ldots, X_n$ different variables, q a predicate name and f a functor name). This can be easily done by replacing the terms in the original clauses by new variables and creating (and simplifying) explicit unifications for variables in the body of the clause. We suppose that in this flattened form all clauses defining a predicate share the same identical head and that all other variables are uniquely renamed. For example, the definition of append/2 in flattened form would be as follows:

```
append(X,Y,Z) ← X=[], Z=Y
append(X,Y,Z) ← X=[Xh|Xt], append(Xt,Y,Zt), Z=[Xh|Zt]
```

In our definitions, we will formally refer to a predicate definition by a couple $(H, \langle C_1, \ldots, C_k \rangle)$ where H is the head shared by all clauses of the predicate and $\langle C_1, \ldots, C_k \rangle$ the sequence of clause bodies as they occur in the definition. Likewise, the body of a clause or, more generally, any conjunction, will be represented as a sequence of atoms $\langle B_1, \ldots, B_s \rangle$. For any conjunction of atoms C (possibly a single atom), we use $vars(C)$ to denote the set of variables occurring in C. For any *finite mapping* $m = \{(e_1, e'_1), \ldots, (e_n, e'_n)\}$[1], we use $dom(m)$ and $img(m)$ to represent, respectively, the *domain* and *image* of the mapping, and for an element $e \in dom(m)$ we use $m(e)$ to represent the element associated to e in the mapping. A *variable renaming* ρ is a finite bijective mapping from a set of variables to another set of variables, distinct from the original ones. Given a conjunction of atoms C, $C\rho$ represents the result of simultaneously replacing in C every occurrence of a variable $X \in dom(\rho)$ by $\rho(X)$.

2.2   Definition of a Clone

In order to ease the definition of a clone in a logic program, we first define a *predicate mapping* as a way of expressing a link between the syntactical constructs of two predicate definitions.

Definition 1. *Given two predicates p/m and q/n defined by $p = (p(X_1, \ldots, X_m), \langle C_1, \ldots, C_k \rangle)$ and $q = (q(Y_1, \ldots, Y_n), \langle C'_1, \ldots, C'_l \rangle)$, a predicate mapping from p to q is defined as a triple $\langle \phi, \chi, \{\psi^{i,j} \mid (i,j) \in \chi\} \rangle$ where ϕ, χ are bijective mappings between subsets of, respectively, the argument and clause positions of p and q, and each $\psi^{i,j}$ a bijective mapping between subsets of the atom positions of C_i and C'_j. Given a predicate mapping $\langle \phi, \chi, \{\psi^{i,j} \mid (i,j) \in \chi\} \rangle$, we will call its components ϕ, χ, and each of the $\psi^{i,j}$ respectively an* argument mapping, clause mapping *and* body mapping.

A predicate mapping merely defines a possibly incomplete correspondence between the arguments and clauses of two predicate definitions and – for each pair of associated clauses – the body atoms. Note that a predicate mapping is a purely structural concept, with no meaning attached.

Informally, a *clone* in a logic program refers to a mapping between two predicate definitions such that the parts of the definitions that participate in the mapping are syntactically identical modulo a variable renaming.

Definition 2. *Given two predicates p/m and q/n defined by $p = (p(X_1, \ldots, X_m), \langle C_1, \ldots, C_k \rangle)$ and $q = (q(Y_1, \ldots, Y_n), \langle C'_1, \ldots, C'_l \rangle)$, a clone from p to q is a predicate mapping $\langle \phi, \chi, \{\psi^{i,j} \mid (i,j) \in \chi\} \rangle$ from p to q such that $\chi \neq \emptyset$ and, for each pair of clauses $C_i = \langle B_1, \ldots, B_s \rangle$ and $C'_j = \langle B'_1, \ldots, B'_t \rangle$ with $(i,j) \in \chi$, $\psi^{i,j} \neq \emptyset$ and there exists a renaming ρ with $dom(\rho) \subseteq vars(C_i) \cup \{X_u \mid 1 \leq u \leq m\}$ and $img(\rho) \subseteq vars(C'_j) \cup \{Y_v \mid 1 \leq v \leq n\}$ such that:*

---

[1] We will sometimes use the equal notation $m = \{e_1/e'_1, \ldots, e_n/e'_n\}$ for a mapping.

1. $(u, v) \in \phi \Leftrightarrow (X_u, Y_v) \in \rho$
2. *for each atom* $B_w \in C_i$ *with* $w \in dom(\psi^{i,j})$:
   - *if* B_w *is an atom other than a recursive call,*
     then $B_w \rho = B'_{\psi^{i,j}(w)}$
   - *if* B_w *is a recursive call, say* $p(Z_1, \ldots, Z_m)$,
     then $B'_{\psi^{i,j}(w)}$ *is a recursive call, say* $q(Z'_1, \ldots, Z'_n)$,
     and $(u, v) \in \phi \Leftrightarrow (Z_u, Z'_v) \in \rho$.

The renaming ρ *will be called* induced by ϕ *and* $\psi^{i,j}$.

The duplication relation embodied by our concept of clone from the predicate p to the predicate q is such that at least one atom from a clause of p has to present some similarity with an atom from a clause of q. However, there is no constraint on the minimal size of the argument mapping. Two body atoms are considered similar if they differ only in a variable renaming. Concerning recursive calls, their mapping is conditioned by a permutation of their arguments guided by the argument mapping. If the latter is empty, recursive calls are mapped unconditionally, regardless of their arguments.

Example 1. Let us consider the following predicate definitions, in which the body atoms of each clause are indexed by their position, to facilitate later reference:
```
abs_list(A,B) ← A=[]₁, B=[]₂
abs_list(A,B) ← A=[Ah|At]₁, abs(Ah,Bh)₂, abs_list(At,Bt)₃, B=[Bh|Bt]₄

add1&sqr_list(Y,X) ← X=[]₁, Y=[]₂
add1&sqr_list(Y,X) ← X=[Xh|Xt]₁, N is Xh+1₂, Yh is N*N₃,
        add1&sqr_list(Yt,Xt)₄, Y=[Yh|Yt]₅
```

The predicates `abs_list` and `add1&sqr_list` generate, from a given list, the list where each element is, respectively, the absolute value of the original element and the original element added to 1 and then squared. The predicate mapping $c = \langle \phi, \chi, \{\psi^{1,1}, \psi^{2,2}\}\rangle$ with $\phi = \{(1,2),(2,1)\}$, $\chi = \{(1,1),(2,2)\}$, $\psi^{1,1} = \{(1,1),(2,2)\}$ and $\psi^{2,2} = \{(1,1),(3,4),(4,5)\}$ is a clone according to our definition. The induced renamings associated to ϕ and $\psi^{1,1}$ and $\psi^{2,2}$ are, respectively, $\rho_1 = \{A/X, B/Y\}$ and $\rho_2 = \{A/X, B/Y, Ah/Xh, At/Xt, Bh/Yh, Bt/Yt\}$. This clone reflects a partial correspondence between both predicates and, in a perspective of refactoring, this common functionality could be generalized in a higher-order predicate *maplist*/3, performing a certain operation (given as third argument) on each element of an input list.

As shown by Example 1, the renamings induced by the argument mapping and each body mapping share a common part, i.e. the renaming induced by the argument mapping. Furthermore, if a body mapping induces a match between two arguments, then the latter have to be mapped by the argument mapping. But two arguments may be linked by the argument mapping without being involved in any body mapping. As a consequence, two clones may vary due to the argument mapping alone. Note that our definition of a clone is general enough to represent duplication between two different predicate definitions, duplication within a single predicate definition $(p = q)$, as well as duplication within a single clause of a single definition $(p = q, \chi = \{(i, i)\}$ for some $i)$.

2.3   Hierarchy of Clones

Given two predicates p and q, we use $Clones(p, q)$ to denote the finite set of all distinct clones from p to q. We can define the following natural partial order relation over $Clones(p, q)$.

Definition 3. *Given two predicates p and q and two clones $c_1 = \langle \phi_1, \chi_1, \{\psi_1^{i,j}|$ $(i, j) \in \chi_1\} \rangle$ and $c_2 = \langle \phi_2, \chi_2, \{\psi_2^{i,j}|(i, j) \in \chi_2\} \rangle$ in $Clones(p, q)$, c_1 is a subclone of c_2, denoted $c_1 \subseteq c_2$, iff $(\phi_1 \subseteq \phi_2) \wedge (\chi_1 \subseteq \chi_2) \wedge (\forall(i, j) \in \chi_1 : \psi_1^{i,j} \subseteq \psi_2^{i,j})$.*

Although stating a preference relation over the set of clones is difficult since this notion is intrinsically linked to the programmer's assessment [17], the order relation defined above allows us to focus on the *maximal* clones in the sense that none of these can be extended into another valid clone. For two predicates p and q, we denote by $\mathcal{C}(p, q)$ the finite unique set of maximal elements of $Clones(p, q)$.

Example 2. The mapping c from Example 1 is a maximal element of the set $Clones(\texttt{abs\_list}, \texttt{add1\&sqr\_list})$. One can verify that the only other maximal element of this set is $\langle \phi, \chi, \{\psi^{1,1}, \psi^{2,2}\} \rangle$ with $\phi = \{(1, 1), (2, 2)\}$, $\chi = \{(1, 1), (2, 2)\}$, $\psi^{1,1} = \{(1, 2), (2, 1)\}$ and $\psi^{2,2} = \{(1, 5), (3, 4), (4, 1)\}$.

How clones can be further classified, most probably in function of the desired application, is a topic for further research. We will now present how we can compute the set of maximal clones between two predicate definitions.

3   Analysis of Duplication

3.1   Preliminaries

The analysis we propose focuses on a subset of the maximal clones that may exist between two predicates. Indeed, our clone detection algorithm is parameterized with respect to a given clause mapping and only computes clones involving monotonically increasing body mappings. This choice is justified as follows.

First, since clauses can be seen as isolated rules or computation units, finding clones involving different clause mappings boils down to iterating the search algorithm over these different clause mappings, which can be easily implemented. Moreover, practical logic programming languages or systems may or may not consider clause order as part of the definition of a predicate. Hence, the clause mappings that should be considered for finding valid clones will depend on the language or system under consideration.

Second, practical logic languages - an example being Prolog [4] - often consider the order of the individual atoms in a rule as part of the definition. For such languages, it makes sense to restrict the search for clones to monotonically increasing body mappings as they will consider the conjunction a,b to be semantically different from b,a. Even for more declarative languages that abstract from the order of the atoms – an example being Mercury [14] – the atoms in a conjunction can often be rearranged in some standardized and semantics preserving way such that if two conjunctions a,b and b,a have the same semantics, they will be reordered in the same way [5].

3.2   Clone Detection Algorithm

A body mapping ψ whose $dom(\psi)$ and $img(\psi)$ are intervals, i.e. sets composed only of successive values, will be called a *block mapping*. As a special case, a block mapping ψ whose $|dom(\psi)| = |img(\psi)| = 1$ will be called an *atomic block mapping*. A block mapping, respectively a body mapping, ψ between two conjunctions is maximal if there does not exist a block mapping, respectively a body mapping, ψ' between the two conjunctions such that $\psi \subset \psi'$ [2]. Obviously, a body mapping can be seen as built from individual block mappings and a block mapping from individual atomic block mappings. The converse is not necessarily true: not all sets of (atomic) block mappings can be combined into a valid body mapping. This principle is the basis of our algorithm: from two conjunctions, the set of maximal block mappings is generated by combining the atomic block mappings, and the set of maximal body mappings is generated by combining the maximal block mappings. This combination operation will be refined later.

Given an argument mapping ϕ between two argument sets $\{X_1, \ldots, X_m\}$ and $\{Y_1, \ldots, Y_n\}$, and a body mapping ψ between two conjunctions that contain no recursive call, if ρ_1 is the renaming induced by ϕ and ρ_2 the one by ψ, ψ is said *conform to* ϕ if $\rho_1 \cup \rho_2$ is a renaming such that $(u,v) \in \phi \Leftrightarrow (X_u, Y_v) \in \rho_1 \cup \rho_2$. This is denoted by ψ_ϕ and corresponds to the first condition of the definition of a clone. By extension, a *set of body mappings* S, not involving recursive calls, is said *conform to* ϕ if each of its elements is conform to ϕ. This is denoted by S_ϕ.

Finally, we define two special union operators intended to be used on sets of sets: $\uplus$, whose goal is to create a set where each element is the union between two elements of the two original sets : $A \uplus B = \{a \cup b \mid a \in A, \ b \in B\}$, and $\bigsqcup$, whose goal is to keep only the maximal elements among all elements of the two original sets: $A \bigsqcup B = (A \cup B) \setminus \{x | \exists x' \in (A \cup B) \setminus \{x\} : x \subseteq x'\}$.

Figure 1 gives an overview of our clone detection algorithm, which takes as input two predicates p/m and q/n and a clause mapping χ. The algorithm performs the following operations for each argument mapping ϕ (of size from 0 to m, if we suppose, without loss of generality, $m \leq n$):

The first operation consists in comparing each pair (C_i, C'_j) of clauses associated by the clause mapping χ. This comparison will be performed by three successive procedures. First, the recursive calls of the clauses are transformed to make it possible for them to be treated just like the other atoms of the clause bodies. Concretely, the procedure *transform_recursive_calls* replaces each recursive call in C_i, say $p(Z_1, \ldots, Z_m)$, respectively in C'_j, say $q(Z'_1, \ldots, Z'_n)$, by an

---

[2] Note that we use the symbol $\subset$ to represent strict inclusion.

ALGORITHM($LoClo(p, q, \chi)$)

    <u>Input:</u>   predicates p and q, clause mapping χ

    <u>Output:</u> set Res of maximal clones from p to q with a clause mapping

                subset of χ and monotonically increasing body mappings

1 $Res = \emptyset$

2 $\forall \phi$:

3     $\chi' = W = \emptyset$

4     $\forall (i, j) \in \chi$, with C_i in p and C'_j in q:

5        $(Cr_i, Cr'_j) = transform\_recursive\_calls(\phi, C_i, C'_j)$

6        $\mathcal{BM}^{i,j}_\phi = compute\_block\_mappings(\phi, Cr_i, Cr'_j)$

7        $\mathcal{M}^{i,j}_\phi = compute\_body\_mappings(\mathcal{BM}^{i,j}_\phi)$

8   **If** $\mathcal{M}^{i,j}_\phi \neq \emptyset$

        Then

9          $\chi' = \chi' \bigcup \{(i, j)\}$

10       $W = W \biguplus \{\{\psi^{i,j}_\phi\} | \psi^{i,j}_\phi \in \mathcal{M}^{i,j}_\phi\}$

11   **If** $\chi' \neq \emptyset$

      Then

12      $Res = Res \bigsqcup \{< \phi, \chi', w > | w \in W\}$

13 **Return** Res

Fig. 1. Clone detection algorithm

atom whose predicate name is rec, where "rec" is a special name, different from all other predicate names occurring in the program, and where the arguments are a subset of the actual arguments, determined as $\{Z_u | u \in dom(\phi)\}$, respectively $\{Z'_v | v \in img(\phi)\}$. The positions of the new arguments are also modified: in the recursive calls of C_i, they are written in increasing order with respect to their index u, while in the recursive calls of C'_j, the new position of the argument Z'_v with $(u, v) \in \phi$ is equal to $|\{(x, y) | (x, y) \in \phi \land x \leq u\}|$. The result of transforming the clauses C_i and C'_j in this way is denoted by Cr_i and Cr'_j.

Example 3. The predicates `abs_list` and `add1&sqr_list` from Example 1 allow for seven different argument mappings. In case $\phi = \{(1, 2)\}$, their recursive calls are transformed into, respectively, $rec(At)$ and $rec(Xt)$. We see that the variable Xt, which was originally the second argument, is now the first one. In case $\phi = \emptyset$, both recursive calls will simply become rec.

Then, a second procedure computes, from the given pair of transformed clauses, its set of all maximal monotonically increasing block mappings conform to ϕ, denoted by $\mathcal{BM}^{i,j}_\phi$. This operation, which will not be detailed due to lack of space, may be performed by a dynamic programming algorithm with an upper bound for the running time of $S \times T \times L$ where S and T are the lengths of the clause bodies in terms of number of atoms and L is the length of the longest block mapping. Thirdly, this set $\mathcal{BM}^{i,j}_\phi$ serves as a basis to generate the set of all maximal monotonically increasing body mappings conform to ϕ, denoted by $\mathcal{M}_\phi^{i,j}$. This important operation will be refined later. If $\mathcal{M}_\phi^{i,j}$ is not empty, i.e. if the clauses i and j share some similarity in accordance with ϕ, then the sets χ' and W are updated. These sets, both initialized as empty sets before the loop examining the pairs of clauses, represent, respectively, all the pairs of clauses which share some similarity in accordance with ϕ (thus $\chi' \subseteq \chi$) and all the sets of body mappings, each of them leading to the creation of a clone.

Finally, once all pairs of clauses have been considered, if χ' is not empty, i.e. if at least one pair of clauses presents some similarity in accordance with ϕ, the clones containing ϕ are build from χ' and W. They are placed in the set *Res*, from which all subclones are suppressed.

Example 4. Let us reconsider the predicates abs_list and add1&sqr_list from Example 1 and, for the sake of illustration, we fix $\chi = \{(1,1),(2,2)\}$ and $\phi = \emptyset$. It follows that $\mathcal{BM}_\phi^{1,1} = \mathcal{M}_\phi^{1,1} = \emptyset$ and $\mathcal{BM}_\phi^{2,2} = \mathcal{M}_\phi^{2,2} = \{\psi_\phi^{2,2}\}$, where $\psi_\phi^{2,2} = \{(3,4)\}$. Only the recursive calls are mapped by the body mapping since the empty argument mapping prevents it from creating a link between one of the arguments A and B and one of the arguments Y and X. After having compared all pairs of clauses, it results that $\chi' = \{(2,2)\}$ and $W = \{\{\psi_\phi^{2,2}\}\}$. This allows to generate the clone $< \emptyset, \chi', \{\psi_\phi^{2,2}\} >$. More interestingly, for the same χ and $\phi = \{(1,2),(2,1)\}$, we obtain $\mathcal{BM}_\phi^{1,1} = \mathcal{M}_\phi^{1,1} = \{\psi_\phi^{1,1}\}$, where $\psi_\phi^{1,1} = \{(1,1),(2,2)\}$, and $\mathcal{BM}_\phi^{2,2} = \{\{(1,1)\},\{(3,4),(4,5)\}\}$. These two block mappings will be combined to give $\mathcal{M}_\phi^{2,2} = \{\psi_\phi^{2,2}\}$, where $\psi_\phi^{2,2} = \{(1,1),(3,4),(4,5)\}$. In this case, we have $\chi' = \chi$ and $W = \{\{\psi_\phi^{1,1}, \psi_\phi^{2,2}\}\}$ from which the clone $< \phi, \chi, \{\psi_\phi^{1,1}, \psi_\phi^{2,2}\} >$ is created. It is a maximal clone between both predicates, as stated in Example 1. As the first clone of this example is a subclone of the second one, only the latter will be kept in the solution set *Res*.

The hard part of the algorithm consists in combining the block mappings of $\mathcal{BM}_\phi^{i,j}$ in order to construct $\mathcal{M}_\phi^{i,j}$. This set can be computed by an iterative process as follows. The process starts with a single arbitrary block mapping from $\mathcal{BM}_\phi^{i,j}$. Then, the following instructions are repeated until $\mathcal{BM}_\phi^{i,j} = \emptyset$: one of the remaining block mappings is selected from $\mathcal{BM}_\phi^{i,j}$ and successively combined with each body mapping currently in $\mathcal{M}_\phi^{i,j}$. As we will detail below, such a combination operation may result in several new body mappings. The latter replace the original body mapping that was used for combination in $\mathcal{M}_\phi^{i,j}$, at the same time keeping only the maximal ones in the set. The order in which the block mappings are selected from $\mathcal{BM}_\phi^{i,j}$ does not influence the result nor the complexity of the algorithm.

Let us discuss now how a block and a body mapping – or in general, any two body mappings – can be combined to obtain a (set of) new body mapping(s). We first introduce the concept of conflict between atomic block mappings.

Definition 4. *Given two atomic block mappings $\psi = \{(i,j)\}$ and $\psi' = \{(k,l)\}$ from the conjunctions $C = B_1, ..., B_s$ and $C' = B_1', ..., B_t'$, there exists an atomic conflict between ψ and ψ' if at least one of the following conditions holds:*

○ $(i = k \wedge j \neq l) \vee (i \neq k \wedge j = l)$: ψ and ψ' overlap and combining them would not respect the bijection constraint

○ $(i < k \wedge j > l) \vee (i > k \wedge j < l)$: ψ and ψ' cross and combining them would not respect the monotonic increase constraint

○ there does not exist a renaming ρ such that $(B_i, B_k)\rho = (B'_j, B'_l)$: the conjunctions targeted by the combination of ψ and ψ' are not syntactically equal modulo a variable renaming

Based on this definition, we define the unique undirected *conflict graph of two body mappings* ψ and ψ' as the graph $CG(\psi, \psi') = (V, E)$ where the set of vertices $V = \{\{a\}|a \in \psi \vee a \in \psi'\}$ and the set of edges $E = \{\{v, v'\}|v, v' \in V \wedge \exists$ an atomic conflict between v and $v'\}$. Each vertex represents thus an atomic block mapping and each edge an atomic conflict between the atomic block mappings corresponding to the vertices. Slightly abusing notation, we introduce the following terminology. Given a conflict graph $CG = (V, E)$ and an atomic block mapping $v \in V$, a *direct conflict of* v is an edge $e = \{v, v'\} \in E$ while an *indirect conflict of* v is an edge $e = \{v', v''\} \in E$ such that $\exists\{v, v'\} \in E$ or $\exists\{v, v''\} \in E$.

Our algorithm combines two body mappings by first organizing all their atomic conflicts into a conflict graph. Then, this graph is partitioned into connected components, each of them representing a conflict (sub)graph that contains atomic conflicts that are mutually dependent in the sense that they involve the same atomic block mappings. Each connected component will then progressively be pruned by exploring how each conflict can be resolved until one obtains a set of conflict graphs with an empty edge set, i.e. a set of body mappings that are valid combinations of (sub)mappings of the two original body mappings represented by the initial connected component. The principle of pruning requires two special operators over a conflict graph. The operator . serves at fixing one atomic block mapping and eliminating from the graph the parts conflicting with it. Formally, given a conflict graph $CG = (V, E)$ and an atomic block mapping $v \in V$, $CG.v = (V', E')$, where $V' = V \setminus \{x|\{v, x\} \in E\}$ and $E' = E \setminus \{\{x, y\}|\{x, y\}$ is a direct or indirect conflict of $v\}$. The second operator - serves at eliminating two conflicting atomic block mappings from the graph. Formally, given a conflict graph $CG = (V, E)$ and an atomic conflict $e = \{v, v'\} \in E$, $CG\text{-}e = (V', E')$, where $V' = V \setminus \{v, v'\}$ and $E' = E \setminus \{\{x, y\}|\{x, y\}$ is a direct conflict of v or $v'\}$.

The algorithm transforming a connected component G of a conflict graph is presented in Figure 2. In the algorithm, we use $d(v)$ to refer to the *degree of a vertex* v, i.e. the number of incident edges in the graph. As base case, an input graph containing no atomic conflict corresponds to a body mapping. As general case, we select in the graph an atomic conflict between atomic block mappings, ideally both but if not, one of them, characterized by a degree of value 1. If such an edge does not exist, the atomic conflict between the two atomic block mappings with the maximal sum of degrees is selected. This manner of selecting atomic conflicts according to the degree of their associated vertices aims at minimizing the running time of the algorithm, since the number of recursive calls to the algorithm may vary from 1 to 3 as it will be explained below.

ALGORITHM($resolve(G)$)
 <u>Input:</u> conflict graph $G = (V, E)$
 <u>Output:</u> set Res of maximal monotonically increasing body mappings
   constructed from non-conflicting atomic block mappings of G
1 **If** $E = \emptyset$
2 **Then** $Res = \{ \bigcup_{v \in V} v \}$
 Else
3 **If** $\exists e = \{v, v'\} \in E : d(v) = 1 \wedge d(v') = 1$
4 **Then** $Res = e \uplus resolve(G\text{-}e)$
  Else
5   **If** $\exists e = \{v, v'\} \in E : d(v) = 1 \vee d(v') = 1$
6   **Then** $Res = resolve(G.v) \bigcup resolve(G.v')$
   Else
7   **Select** $e = \{v, v'\} \in E : d(v) + d(v')$ maximal
8    $Res = resolve(G.v) \bigcup resolve(G.v') \bigcup resolve(G\text{-}e)$
9 **Return** Res

Fig. 2. Conflict graph resolution algorithm

The solution set Res will be computed differently depending on the type of the selected atomic conflict. In the first case (steps 3 and 4), the atomic conflict $e = \{v, v'\}$ is isolated from the others since both its atomic block mappings are not involved in any other conflict. To resolve e, we will thus separate the conflicting atomic block mappings and construct on the one hand, the set of solution body mappings containing v and the other hand, the set of solution body mappings containing v'. Concretely, v and v' are added separately to each element that will be found by continuing the search with the resolution of the initial conflict graph G without e. In the second case (steps 5 and 6), one of both atomic block mappings forming the atomic conflict e is involved in other atomic conflicts. Again, resolving e means keeping either v or v' but the rest of the graph can no more be resolved independently of them. For that reason, the procedure $resolve$ is called recursively, on the one hand, on the initial conflict graph G where v is fixed so that it will be part of all the subsequently found solution body mappings, and on the other hand, on G where v' is fixed, which ensures that no solution from one recursion branch will be subset of a solution from the other one. The set Res is the union of both solution sets. Finally, the third case (steps 7 and 8) is identical to the second one, except that a third recursion branch is needed, where none of the atomic block mappings v and v' is kept in the solution body mappings. This avoids loosing some solutions. Indeed, it is possible that a certain atomic block mapping conflicting with v and another one conflicting with v' are not conflicting when put together. This third recursion branch allows them to be joined in at least one solution body mapping.

To combine two body mappings ψ and ψ', we thus compute $\uplus_{g \in S} resolve(g)$, where S is the set of connected components of $CG(\psi, \psi')$. Each resulting body mapping is thus a union between $|S|$ smaller mutually non-conflicting body mappings, each from one solution set of a different connected component. The union operator ensures that no doubles are present in the resulting set.

Example 5. From the following two generic conjunctions of atoms $p(W)_1, q(X)_2,$
$r(Y)_3$, $r(Z)_4, s(X,Y,Z)_5$ and $p(A)_1, q(B)_2, r(C)_3, s(A,B,C)_4$, two block mappings
can be extracted: $\psi = \{(1,1),(2,2),(3,3)\}$ and $\psi' = \{(4,3),(5,4)\}$. Trying
to combine them leads to the conflict graph $CG(\psi,\psi') = (V,E)$, with $V =$
$\{\{(1,1)\},\{(2,2)\},\{(3,3)\},\{(4,3)\},\{(5,4)\}\}$ (to ease the reading, we will reference each atomic block mapping constituting a vertex by $v_i, 1 \le i \le 5$, respecting the order in which they are written) and $E = \{\{v_1,v_5\},\{v_2,v_5\},\{v_3,v_4\},$
$\{v_3,v_5\}\}$. Let us abbreviate $CG(\psi,\psi')$ to CG. This graph contains only one
connected component, i.e. itself. To resolve it, the first atomic conflict $\{v_1,v_5\}$
can be selected since $d(v_1) = 1$ and $d(v_5) = 3$. It is the second recursive case
of our procedure *resolve*. Two atomic (sub)graphs have thus to be treated:
$CG.v_1 = (\{v_1,v_2,v_3,v_4\},\{\{v_3,v_4\}\})$ and $CG.v_5 = (\{v_4,v_5\},\emptyset)$. $CG.v_1$ contains
only one atomic conflict in which both atomic block mappings are of degree 1. It
is the first recursion case, where the graph $(CG.v_1)$-$\{v_3,v_4\} = (\{v_1,v_2\},\emptyset)$ has
to be resolved. As its set of edges is empty, we fall in the base case of the procedure: there is no more atomic conflict and the graph represents the body mapping
$\{(1,1),(2,2)\}$. The latter will be united with the vertices v_3 and v_4 separately
to give the set of body mappings $\{\{(1,1),(2,2),(3,3)\},\{(1,1),(2,2),(4,3)\}\}$,
solution set of $CG.v_1$. The resolution of the graph $CG.v_5$ is direct, since its
set of atomic conflicts is empty, and produces the singleton set of body mappings $\{\{(4,3),(5,4)\}\}$. Finally, the solution set of the conflict graph CG is
the union between $resolve(CG.v_1)$ and $resolve(CG.v_5)$ which gives three body
mappings corresponding to the following non-conflicting subconjunction pairs:
(1) $p(W)_1, q(X)_2, r(Y)_3$ and $p(A)_1, q(B)_2, r(C)_3$, (2) $p(W)_1, q(X)_2, r(Z)_4$ and $p(A)_1,$
$q(B)_2, r(C)_3$, and (3) $r(Z)_4, s(X,Y,Z)_5$ and $r(C)_3, s(A,B,C)_4$.

3.3   Complexity

It can be easily seen that the algorithm computes all maximal clones between two
given predicates having monotonically increasing body mappings and a clause
mapping that is a subset of the given input clause mapping. The complexity of
the algorithm is determined by the computation of all argument mappings. Then,
the examination of all pairs of clauses depends on the size of the clause mapping
given as input. The time consuming operations consist in the computation of
the block mappings between two clause bodies and, especially, in the resolution
of the conflict graph in the process of combining the block mappings into body
mappings. The latter operation is exponential in nature given the multiple ways
in which conflicts may be resolved. While this complexity did not turn out to
be problematic in our preliminary evaluation (see Section 3.4), it remains to be
studied whether it is still manageable in practice.

We will now define some supplementary notions closely linked to the complexity of our algorithm. Our definition of a clone in a logic program allows us
to highlight a particular kind of clone.

Definition 5. *Given two predicates p/m and q/n defined respectively by k and l clauses, an* exact clone *from p to q is a clone $\langle \phi, \chi, \{\psi^{i,j} | (i,j) \in \chi\} \rangle$ such that $|\phi| = m = n$, $|\chi| = k = l$, and $\forall (i,j) \in \chi$, with $C_i = B_1, \dots, B_s$ in p and $C'_j = B'_1, \dots, B'_t$ in q, $|\psi^{i,j}| = s = t$. A clone which is not exact is called* partial.*

An exact clone occurs typically when q is a "copy-pasted" instance of p, possibly renamed but without any other modification. Partial clones are of course much more frequent than exact clones.

Hence, we may characterize the duplication relation between two predicates p and q by discerning three main classes of duplication (according to our algorithm, we consider only clones with monotonically increasing body mappings):

1. *Empty duplication:* $C(p,q) = \emptyset$
2. *Exact duplication:* $C(p,q) \supseteq \{c | c$ is an exact clone from p to $q\} \neq \emptyset$
   In this class of duplication, it is possible that several different exact clones exist between both predicates (at least one is required). When $C(p,q)$ contains only one exact clone, the relation is called *perfect exact duplication.*
3. *Partial duplication:* $C(p,q) = \{c | c$ is a partial clone from p to $q\} \neq \emptyset$
   When $C(p,q)$ is a singleton, the relation is called *perfect partial duplication.*

The empty and perfect duplication relations are treated faster than the two others since they do not involve any atomic conflict to be solved (see Section 3.4). An interesting topic of research would be to investigate the relative frequency of the different kinds of duplication relations in a logic program.

3.4   Tests

To test the feasibility of our approach (performance-wise), we created - simply as a proof of concept - five basic test cases representing the following situations: no duplication (1), exact duplication (2), perfect exact duplication (3), partial duplication (4), and perfect partial duplication (5). These test cases are composed of 2 generic predicates of arity 3, defined by 3 clauses of 10 atoms. A sixth test case was set up with 20 predicates taken from a real Prolog program (the clone detection algorithm itself), having from 1 to 4 arguments and from 1 to 4 clauses per predicate, and from 1 to 21 atoms per clause body. While the predicates have not been chosen with the aim of detecting interesting clones, the latter test case is nevertheless useful as it may hint at the performance of the algorithm on arbitrary real-size predicates. For every test case, all pairs of predicates and all pairs of clauses, each from a different predicate definition, were examined in order to detect all existing clones between the predicates. The algorithm was run on an Intel Dual Core 2.5GHz 4GB DDR3 computer. To illustrate our approach and to reflect its complexity, we give 6 metrics: the total number of atomic conflicts that the algorithm had to resolve, the maximal number of atomic conflicts that occurred during the combination phase of two (block) mappings, the total number of found clones, the number of found clones which vary only by the argument mapping (i.e. such that there exists at least one other clone with the same clause mapping and the same set of body mappings) and the sizes of the shortest and the longest found clones (the size of a clone being computed in number of atoms, as the sum of the sizes of the body mappings). The results are reported in Figure 3, the last column showing the runtimes.

First of all, we can observe a clear division between the runtimes of the first five test cases, in spite of being similar in size. Indeed, the test cases 1, 3 and 5 were analyzed over ten times faster than the test cases 2 and 4. This is due to the construction of the algorithm, which performs better when there is no atomic conflict. Comparatively, the test case 6 has a runtime per pair of predicates from 0 to 32780 ms (in average, 525 ms), in function of the number of clones found. Secondly, we can see that the number of clones is linked to the total number of atomic conflicts. More numerous the conflicts, the more the clause bodies are split up into smaller body mappings, possibly leading to an abundance of (small) clones. However, it is interesting to note that the maximal number of atomic conflicts that the algorithm had to solve at once seems constant, and quite small. This is positive since the algorithm is exponential in this number. Finally, the number of found clones is generally particularly high. In test case 6, this number spreads from 1 to 3560 clones per pair of predicates (in average, 322 clones). This is explained because, with respect to our definition of a clone, every syntactic similarity is taken into account without any filtering process.

Test case	Tot. nb of atom. conf.	Max. nb of atom. conf.	Nb of clones	Nb of clones varying by ϕ	(Min.,Max.) clone size	Runtime in ms
1	0	0	0	0	(0,0)	9
2	1366	7	215	106	(2,30)	250
3	0	0	1	0	(30,30)	21
4	1556	6	278	137	(2,22)	359
5	0	0	1	0	(9,9)	19
6	36040	6	22859	16316	(1,11)	99740

Fig. 3. Results of tests on the algorithm

Concerning the sixth test case, we may add that 1 pair of predicates is characterized by a perfect partial duplication relation, while 70 pairs belong to the category of partial duplication relations. In total, those 71 pairs of predicates containing clones constitute 37% of all pairs of predicates. Among the 20 predicates forming the test case, only 5 are not involved in a clone and every other is involved in clones with several different predicates.

Two supplementary metrics were measured on the sixth test case: the relative size of a clone (computed as twice its size divided by the sum of the sizes of both predicates, themselves computed as the total number of atoms in their clause bodies) and the total number of gaps (a gap in a clone being a maximal set of successive non-cloned atoms in a clause body). The first metric represents the degree of coverage of the clone on the predicates while the second one represents its degree of splitting. The extrema of the relative size are 3.85% (24 clones, one half with 4 gaps and the other with 5 gaps) and 75.86% (1 clone, with 5 gaps), and the interval from 15% to 20% is the most filled (7637 clones, with from 3 to 16 gaps). The extrema of the number of gaps are 2 (2 clones, with a relative size of 50%) and 19 (10 clones, of which 2 with a relative size of 28.57% and 8 with a relative size of 32.56%), and 13 is the highest frequency of gaps (4642 clones, with relative sizes from 14.29% to 54.55%). In general, the found clones are thus small and/or contain many gaps. There is nevertheless no systematic link between both metrics. Moreover, a link between the runtime and the values of these metrics, aiming at evaluating a single clone, cannot be easily and clearly established since the algorithm works with sets of clones.

The prototype that was implemented as well as the six test cases can be found on http://www.info.fundp.ac.be/~cda/LoClo/loclo.html. Be aware that in its current form the prototype performs an unguided search for all existing clones and is as such not adapted to, nor intended for, use as a software engineering tool on real-world logic programs.

4   Discussion and Future Work

The definition of a clone and the clone detection algorithm we propose are, to our best knowledge, the most general and the first to be fully formalized in logic programming literature. As proof of concept, we made an implementation of the algorithm that we ran on a limited series of tests. This illustrates the feasibility of our technique and the obtained results are encouraging. All clones between two predicates are detected in a non-naive way. It can be noted that our algorithm is generic. Indeed, it could easily be adapted to a particular logic programming language. Besides, it could be tuned and optimized by introducing parameters guiding the search process. For example, one could want to impose a minimal relative size and/or a maximal number of gaps to the detected clones (according to some appropriate notion of size) in order to avoid small and potentially meaningless clones. Or, in a typed and moded context, the variable renamings induced by the argument and body mappings could be restricted to pairs of variables of the same type and/or mode.

As it is based on syntactic equality, our clone definition may be related to the notion of *structural clone* [11], which depends on the design structures of a certain language and is particularly suited for refactoring. This way of approximating semantic similarity is expedient in logic programming. Indeed, contrary to an imperative program which represents a step-by-step algorithm manipulating data objects, a logic program basically specifies a number of relations that hold between the data objects. In this sense, its arguably simple syntax is closely linked to its semantics. Following from our clone definition, our clone detection technique may be considered AST-based-like. Indeed, it corresponds to a search of two (possibly partially) isomorphic subgraphs in the AST of the given predicates, where leaves representing variable names are discarded but respecting however the constraint of a consistent variable renaming. This guarantees better precision compared to naive comparison for subtree equality. Some existing AST-based tools take also renamings into account, like CReN [7], CloneDR [2],

Wrangler [9], HaRe [3]. Except for the first one, these tools are able to find exact and partial clones. Regarding logic programming, this is a crucial concern and our algorithm is dedicated to detecting partial clones.

Moreover, [13] presents two refactorings to remove code duplication from logic programs (identifying and removing identical subsequences of goals in different clause bodies and eliminating copy-pasted predicates with identical definitions). Both refactorings are implemented in the tool ViPReSS. The detection phase employed in [13] (which is not formally defined but said to be comparable to the problem of determining longest common subsequences) is thus specifically targeted towards identifying opportunities for applying one of the presented refactorings. Our work generalizes this problem since our formal definition of code clone (no such definition is present in the cited work) covers not only the opportunities of both refactorings but also other situations and applications, which can all be detected by our formal algorithm.

Among the topics we see for further work, we cite a more thorough evaluation of our approach. Our algorithm could be compared to other existing AST-based techniques, including those mentioned above, and to ViPReSS. A detailed formal study of the different classes of duplication spotted by our algorithm could also be undertaken. Besides, our current definition of a clone in a logic program could be generalized to accept an abstraction of constants and data structures. Another way to extend the definition concerns its granularity. Currently, the range of the duplication relation is two predicates. Instead, it could be valuable to consider a set of duplicated predicates, forming then a *class of clones* [10], possibly with a certain threshold of similarity to be defined. In a perspective of refactoring, this would allow to refactor all predicates inside a family at the same time instead of separately. Furthermore, in [19] is exposed the concept of *chained clone* which involves several procedures linked by a caller-callee relation, and the concept of *chained clone set* which is an equivalent class of chained clone. These notions could be applied to our case, by combining "basic" clones, formed by a pair of predicates, into more complex clones, formed by a pair of sets of predicates. Finally, it could be interesting to characterize as precisely as possible the link between our structural definition of a clone and semantic similarity, which could lead to improve the definition for capturing duplicates partly, or even non, syntactically equal.

Ackowledgement. We thank the anonymous reviewers for their constructive remarks which helped to improve the paper.

References

1. Apt, K.R.: Logic programming. In: Handbook of Theoretical Computer Science, Volume B: Formal Models and Sematics, pp. 493–574. Elsevier (1990)
2. Baxter, I.D., Yahin, A., Moura, L., Sant'Anna, M., Bier, L.: Clone detection using abstract syntax trees. In: Proceedings of the International Conference on Software Maintenance (CSM 1998), pp. 368–377. IEEE (1998)

3. Brown, C., Thompson, S.: Clone detection and elimination for Haskell. In: Proceedings of the 2010 SIGPLAN Workshop on Partial Evaluation and Program Manipulation (PEPM 2010), pp. 111–120. ACM (2010)

4. Clocksin, W.F., Mellish, C.S.: Programming in Prolog: Using the ISO Standard. Springer (2003)

5. Degrave, F., Vanhoof, W.: Towards a Normal Form for Mercury Programs. In: King, A. (ed.) LOPSTR 2007. LNCS, vol. 4915, pp. 43–58. Springer, Heidelberg (2008)

6. Fowler, M., Beck, K., Brant, J., Opdyke, W., Roberts, D.: Refactoring: improving the design of existing code. Addison-Wesley (1999)

7. Jablonski, P., Hou, D.: CReN: a tool for tracking copy-and-paste code clones and renaming identifiers consistently in the IDE. In: Proceedings of the 2007 OOPSLA Workshop on Eclipse Technology eXchange (ETX 2007), pp. 16–20. ACM (2007)

8. Lancaster, T., Finta, C.: A comparison of source code plagiarism detection engines. Computer Science Education 14(2), 101–112 (2004)

9. Li, H., Thompson, S.: Clone detection and removal for Erlang/OTP within a refactoring environment. In: Proceedings of the 2009 SIGPLAN Workshop on Partial Evaluation and Program Manipulation (PEPM 2009), pp. 169–178. ACM (2009)

10. Rieger, M., Ducasse, S., Lanza, M.: Insights into system-wide code duplication. In: Proceedings of the 11th Working Conference on Reverse Engineering (WCRE 2004), pp. 100–109. IEEE (2004)

11. Roy, C.K., Cordy, J.R.: A survey on software clone detection research. Tech. Rep. 2007-541, School of Computing, Queen's University at Kingston, Canada (2007)

12. Roy, C.K., Cordy, J.R., Koschke, R.: Comparison and evaluation of code clone detection techniques and tools: A qualitative approach. Science of Computer Programming 74(7), 470–495 (2009)

13. Serebrenik, A., Schrijvers, T., Demoen, B.: Improving Prolog programs: Refactoring for Prolog. Theory and Practice of Logic Programming (TPLP) 8, 201–215 (2008), other version consulted: https://lirias.kuleuven.be/bitstream/123456789/164765/1/technical_note.pdf

14. Somogyi, Z., Henderson, F., Conway, T.: The execution algorithm of Mercury, an efficient purely declarative logic programming language. Journal of Logic Programming 29, 17–64 (1996)

15. Vanhoof, W.: Searching Semantically Equivalent Code Fragments in Logic Programs. In: Etalle, S. (ed.) LOPSTR 2004. LNCS, vol. 3573, pp. 1–18. Springer, Heidelberg (2005)

16. Vanhoof, W., Degrave, F.: An Algorithm for Sophisticated Code Matching in Logic Programs. In: Garcia de la Banda, M., Pontelli, E. (eds.) ICLP 2008. LNCS, vol. 5366, pp. 785–789. Springer, Heidelberg (2008)

17. Walenstein, A., Jyoti, N., Li, J., Yang, Y., Lakhotia, A.: Problems creating task-relevant clone detection reference data. In: Proceedings of the 10th Working Conference on Reverse Engineering (WCRE 2003), pp. 285–294. IEEE (2003)

18. Walenstein, A., Lakhotia, A.: The software similarity problem in malware analysis. In: Duplication, Redundancy, and Similarity in Software. IBFI (2007)

19. Yoshida, N., Higo, Y., Kamiya, T., Kusumoto, S., Inoue, K.: On refactoring support based on code clone dependency relation. In: Proceedings of the 11th International Software Metrics Symposium (METRICS 2005), pp. 16–25. IEEE (2005)

Using Real Relaxations during Program Specialization

Fabio Fioravanti[1], Alberto Pettorossi[2], Maurizio Proietti[3], and Valerio Senni[2,4]

[1] Dipartimento di Scienze, University 'G. D'Annunzio', Pescara, Italy
fioravanti@sci.unich.it
[2] DISP, University of Rome Tor Vergata, Rome, Italy
{pettorossi,senni}@disp.uniroma2.it
[3] CNR-IASI, Rome, Italy
maurizio.proietti@iasi.cnr.it
[4] LORIA-INRIA, Villers-les-Nancy, France
valerio.senni@loria.fr

Abstract. We propose a program specialization technique for locally stratified CLP($\mathbb{Z}$) programs, that is, logic programs with linear constraints over the set $\mathbb{Z}$ of the integer numbers. For reasons of efficiency our technique makes use of a relaxation from integers to reals. We reformulate the familiar unfold/fold transformation rules for CLP programs so that: (i) the applicability conditions of the rules are based on the satisfiability or entailment of constraints over the set $\mathbb{R}$ of the real numbers, and (ii) every application of the rules transforms a given program into a new program with the same perfect model constructed over $\mathbb{Z}$. Then, we introduce a strategy which applies the transformation rules for specializing CLP($\mathbb{Z}$) programs with respect to a given query. Finally, we show that our specialization strategy can be applied for verifying properties of infinite state reactive systems specified by constraints over $\mathbb{Z}$.

1 Introduction

Reactive systems are often composed of processes that make use of possibly unbounded data structures. In order to specify and reason about this type of systems, several formalisms have been proposed, such as unbounded counter automata [27] and vector addition systems [26]. These formalisms are based on linear constraints over variables ranging over the set $\mathbb{Z}$ of the integer numbers.

Several tools for the verification of properties of systems with unbounded integer variables have been developed in recent years. Among these we would like to mention ALV [36], FAST [6], LASH [25], and TReX [1]. These tools use sophisticated solvers for constraints over the integers which are based on automata-theoretic techniques [22] or techniques for proving formulas of Presburger Arithmetic [32].

Also constraint logic programming is a very powerful formalism for specifying and reasoning about reactive systems [20]. In fact, many properties of counter

G. Vidal (Ed.): LOPSTR 2011, LNCS 7225, pp. 106–122, 2012.

automata and vector addition systems, such as *safety* properties and, more generally, temporal properties, can be easily translated into constraint logic programs with linear constraints over the integers, called CLP($\mathbb{Z}$) programs [21].

Unfortunately, dealing with constraints over the integers is often a source of inefficiency and, in order to overcome this limitation, many verification techniques are based on the interpretation of the constraints over the set $\mathbb{R}$ of the real numbers, instead of the set $\mathbb{Z}$ of the integer numbers [7,13]. This extension of the domain of interpretation is sometimes called *relaxation*.

The relaxation from integers to reals, also called the *real relaxation*, has several advantages: (i) many constraint solving problems (in particular, the satisfiability problem) have lower complexity if considered in the reals, rather than in the integers [33], (ii) the class of linear constraints over the reals is closed under *projection*, which is an operation often used during program verification, while the class of linear constraints over the integers is not, and (iii) many highly optimized libraries are actually available for performing various operations on constraints over the reals, such as satisfiability testing, projection, widening, and convex hull, that are often used in the field of static program analysis [10,11] (see, for instance, the Parma Polyhedral Library [3]).

Relaxation techniques can be viewed as *approximation* techniques. Indeed, if a property holds for all real values of a given variable then it holds for all integer values, but not vice versa. This approximation technique can be applied to the verification of reactive systems. For instance, if a safety property $\varphi =_{def} \forall x \in \mathbb{R}$ (*reachable*$(x) \rightarrow$ *safe*(x)) holds, then it also holds when replacing the set $\mathbb{R}$ by the set $\mathbb{Z}$. However, if $\neg\varphi =_{def} \exists x \in \mathbb{R}$ (*reachable*$(x) \land \neg$*safe*(x)) holds, then we cannot conclude that $\exists x \in \mathbb{Z}$ (*reachable*$(x) \land \neg$*safe*(x)) holds.

Now, as indicated in the literature (see, for instance, [17,19,29,30,31]) the verification of infinite state reactive systems can be done via program specialization and, in particular, in [19] we proposed a technique consisting of the following two steps: (*Step* 1) the specialization of the constraint logic program that encodes the given reactive system, with respect to the query that encodes the property to be verified, and (*Step* 2) the construction of the *perfect model* of the specialized program.

In this paper we propose a variant of the verification technique introduced in [19]. This variant is based on the specialization of locally stratified CLP($\mathbb{Z}$) programs and uses a relaxation from the integers to the reals.

In order to do so, we need: (i) a suitable reformulation of the familiar unfold/fold transformation rules for CLP programs [14,18] so that: (i.1) the applicability conditions of the rules are based on the satisfiability or entailment of constraints over the reals $\mathbb{R}$, and (i.2) every application of the rules transforms a given program into a new program with the same perfect model constructed over the integers $\mathbb{Z}$, called *perfect $\mathbb{Z}$-model*, and then (ii) the introduction of a transformation strategy which applies the reformulated transformation rules for specializing a given CLP($\mathbb{Z}$) program with respect to a given query.

There are two advantages of the verification technique we consider here. The first advantage is that, since our specialization strategy manipulates constraints over the reals, it may exploit efficient techniques for checking satisfiability and entailment, for computing projection, and for more complex constructions, such as the widening and the convex hull operations over sets of constraints. The second advantage is that, since we use equivalence preserving transformation rules, that is, rules which preserve the perfect $\mathbb{Z}$-model, the property to be verified holds in the initial program if and only if it holds in the specialized program and, thus, we may apply to the specialized program any other verification technique we wish, including techniques based on constraints over the integers.

The rest of the paper is structured as follows. In Section 2 we introduce some basic notions concerning constraints and CLP programs. In Section 3 we present the rules for transforming CLP($\mathbb{Z}$) programs and prove that they preserve equivalence with respect to the perfect model semantics. In Section 4 we present our specialization strategy and in Section 5 we show its application to the verification of infinite state reactive systems. Finally, in Section 6 we discuss related work in the field of program specialization and verification of infinite state systems.

2   Constraint Logic Programs over Integers and Reals

We will consider CLP($\mathbb{Z}$) programs, that is, constraint logic programs with linear constraints over the set $\mathbb{Z}$ of the integer numbers. An *atomic constraint* is an inequality either of the form $r \geq 0$ or of the form $r > 0$, where r is a linear polynomial with integer coefficients. A *constraint* is a conjunction of atomic constraints. The equality $t_1 = t_2$ stands for the conjunction $t_1 \geq t_2 \wedge t_2 \geq t_1$. A clause of a CLP($\mathbb{Z}$) program is of the form $A \leftarrow c \wedge B$, where A is an atom, c is a constraint, and B is a conjunction of (positive or negative) literals. For reasons of simplicity and without loss of generality, we also assume that the arguments of all literals are variables, that is, the literals are of the form $p(X_1, \ldots, X_n)$ or $\neg p(X_1, \ldots, X_n)$, with $n \geq 0$, where p is a predicate symbol not in $\{>, \geq, =\}$ and $X_1, \ldots, X_n$ are distinct variables ranging over $\mathbb{Z}$.

Given a constraint c, by $vars(c)$ we denote the set of variables occurring in c. By $\forall(c)$ we denote the universal closure $\forall X_1 \ldots \forall X_n \, c$, where $vars(c) = \{X_1, \ldots, X_n\}$. Similarly, by $\exists(c)$ we denote the existential closure $\exists X_1 \ldots \exists X_n \, c$. Similar notation will also be used for literals, goals, and clauses.

For the constraints over the integers we assume the usual interpretation which, by abuse of language, we denote by $\mathbb{Z}$. A $\mathbb{Z}$-*model* of a CLP($\mathbb{Z}$) program P is defined to be a model of P which agrees with the interpretation $\mathbb{Z}$ for the constraints. We assume that programs are *locally stratified* [2] and, similarly to the case of logic programs without constraints, for a locally stratified CLP($\mathbb{Z}$) program P we can define its unique *perfect $\mathbb{Z}$-model* (or, simply, *perfect model*), denoted $M_{\mathbb{Z}}(P)$ (see [2] for the definition of the perfect model of a logic program).

We say that a constraint c is $\mathbb{Z}$-*satisfiable* if $\mathbb{Z} \models \exists(c)$. We also say that a constraint c $\mathbb{Z}$-*entails* a constraint d, denoted $c \sqsubseteq_{\mathbb{Z}} d$, if $\mathbb{Z} \models \forall(c \to d)$.

For the constraints over the reals we assume the usual interpretation which, by abuse of language, we denote by $\mathbb{R}$. A constraint c is $\mathbb{R}$-*satisfiable* if $\mathbb{R} \models \exists(c)$.

A constraint c $\mathbb{R}$-*entails* a constraint d, denoted $c \sqsubseteq_\mathbb{R} d$, if $\mathbb{R} \models \forall(c \rightarrow d)$. The $\mathbb{R}$-*projection* of a constraint c onto the set X of variables is a constraint c_p such that: (i) $vars(c_p) \subseteq X$ and (ii) $\mathbb{R} \models \forall(c_p \leftrightarrow \exists Y_1 \ldots \exists Y_k \ c)$, where $\{Y_1, \ldots, Y_k\} = vars(c)-X$. Recall that the set of constraints over $\mathbb{Z}$ is not closed under projection.

The following lemma states some simple relationships between $\mathbb{Z}$-satisfiability and $\mathbb{R}$-satisfiability, and between $\mathbb{Z}$-entailment and $\mathbb{R}$-entailment.

Lemma 1. *Let c and d be constraints and X be a set of variables.*
(i) If c is $\mathbb{Z}$-satisfiable, then c is $\mathbb{R}$-satisfiable. (ii) If $c \sqsubseteq_\mathbb{R} d$, then $c \sqsubseteq_\mathbb{Z} d$.
(iii) If c_p is the $\mathbb{R}$-projection of c on X, then $c \sqsubseteq_\mathbb{Z} c_p$.

3 Transformation Rules with Real Relaxations

In this section we present a set of transformation rules that can be used for specializing locally stratified CLP($\mathbb{Z}$) programs. The applicability conditions of the rules are given in terms of constraints interpreted over the set $\mathbb{R}$ and, as shown by Theorem 1, these rules preserve the perfect $\mathbb{Z}$-model semantics.

The rules we will consider are those needed for specializing constraint logic programs, as indicated in the Specialization Strategy of Section 4. Note, however, that the correctness result stated in Theorem 1 can be extended to a larger set of rules (including the *negative unfolding* rule [18,34]) or to more powerful rules (such as the *definition* rule with m (≥ 1) clauses, and the *multiple positive folding* [18]).

Before presenting these rules, we would like to show through an example that, if we consider different domains for the interpretation of the constraints and, in particular, if we apply the relaxation from the integers to the reals, we may derive different programs with different intended semantics.

Let us consider, for instance, the following constraint logic program P:

1. $p \leftarrow Y > 0 \wedge Y < 1$ 2. $q \leftarrow$

If we interpret the constraints over the reals, since $\mathbb{R} \models \exists Y (Y > 0 \wedge Y < 1)$, program P can be transformed into program $P_\mathbb{R}$:

1'. $p \leftarrow$ 2. $q \leftarrow$

If we interpret the constraints over the integers, since $\mathbb{Z} \models \neg \exists Y (Y > 0 \wedge Y < 1)$, program P can be transformed into program $P_\mathbb{Z}$:

2. $q \leftarrow$

Programs $P_\mathbb{R}$ and $P_\mathbb{Z}$ are not equivalent because they have different perfect $\mathbb{Z}$-models (which in this case coincide with their least Herbrand models). Thus, when we apply a relaxation we should proceed with some care. In particular, we will admit a transformation rule only when its applicability conditions interpreted over $\mathbb{R}$ imply the corresponding applicability conditions interpreted over $\mathbb{Z}$.

The transformation rules are used to construct a *transformation sequence*, that is, a sequence $P_0, \ldots, P_n$ of programs. We assume that P_0 is locally stratified. A transformation sequence $P_0, \ldots, P_n$ is constructed as follows. Suppose that we have constructed a transformation sequence $P_0, \ldots, P_k$, for $0 \leq k \leq n-1$. The next program P_{k+1} in the transformation sequence is derived from program P_k by the application of a transformation rule among R1–R5 defined below.

Our first rule is the *Constrained Atomic Definition* rule (or *Definition Rule*, for short), which is applied for introducing a new predicate definition.

R1. Constrained Atomic Definition. Let us consider a clause, called a *definition clause*, of the form:

$\delta:\ newp(X_1, \ldots, X_h) \leftarrow c \wedge p(X_1, \ldots, X_h)$

where: (i) *newp* does not occur in $\{P_0, \ldots, P_k\}$, (ii) $X_1, \ldots, X_h$ are distinct variables, (iii) c is a constraint with $vars(c) \subseteq \{X_1, \ldots, X_h\}$, and (iv) p occurs in P_0. By *constrained atomic definition* from program P_k we derive the program $P_{k+1} = P_k \cup \{\delta\}$. For $k \geq 0$, $Defs_k$ denotes the set of clauses introduced by the definition rule during the transformation sequence $P_0, \ldots, P_k$. In particular, $Defs_0 = \emptyset$.

R2. (Positive) Unfolding. Let $\gamma :\ H \leftarrow c \wedge G_L \wedge A \wedge G_R$ be a clause in program P_k and let

$\gamma_1: K_1 \leftarrow c_1 \wedge B_1 \quad \ldots \quad \gamma_m: K_m \leftarrow c_m \wedge B_m \qquad (m \geq 0)$

be all clauses of (a renamed apart variant of) program P_k such that, for $i = 1, \ldots, m$, the constraint $c \wedge c_i\rho_i$ is $\mathbb{R}$-satisfiable, where ρ_i is a renaming substitution such that $A = K_i\rho_i$ (recall that all atoms in a CLP($\mathbb{Z}$) program have distinct variables as arguments).

By *unfolding clause γ w.r.t. the atom A* we derive the clauses

$\eta_1 :\ H \leftarrow c \wedge c_1\rho_1 \wedge G_L \wedge B_1\rho_1 \wedge G_R$

$\quad \ldots$

$\eta_m :\ H \leftarrow c \wedge c_m\rho_m \wedge G_L \wedge B_m\rho_m \wedge G_R$

and from program P_k we derive the program $P_{k+1} = (P_k - \{\gamma\}) \cup \{\eta_1, \ldots, \eta_m\}$.

Note that if $m = 0$ then, by unfolding, clause γ is deleted from P_k.

Example 1. Let P_k be the following CLP($\mathbb{Z}$) program:

1. $p(X) \leftarrow X > 1 \wedge q(X)$
2. $q(Y) \leftarrow Y > 2 \wedge Z = Y - 1 \wedge q(Z)$
3. $q(Y) \leftarrow Y < 2 \wedge 5\,Z = Y \wedge q(Z)$
4. $q(Y) \leftarrow Y = 0$

Let us unfold clause 1 w.r.t. the atom $q(X)$. We have the renaming substitution $\rho = \{Y/X\}$, which unifies the atoms $q(X)$ and $q(Y)$, and the following three constraints:

(a) $X > 1 \wedge X > 2 \wedge Z = X - 1$,   derived from clauses 1 and 2,
(b) $X > 1 \wedge X < 2 \wedge 5\,Z = X$,   derived from clauses 1 and 3,
(c) $X > 1 \wedge X = 0$,   derived from clauses 1 and 4.

Only (a) and (b) are $\mathbb{R}$-satisfiable, and only (a) is $\mathbb{Z}$-satisfiable. By unfolding clause 1 w.r.t. $q(X)$ we derive the following clauses:

1.a $p(X) \leftarrow X > 1 \wedge X > 2 \wedge Z = X - 1 \wedge q(Z)$
1.b $p(X) \leftarrow X > 1 \wedge X < 2 \wedge 5\,Z = X \wedge q(Z)$

Now we introduce two versions of the folding rule: *positive folding* and *negative folding*, depending on whether folding is applied to positive or negative literals in the body of a clause.

R3. Positive Folding. Let $\gamma\colon H \leftarrow c \wedge G_L \wedge A \wedge G_R$ be a clause in P_k and let $\delta\colon K \leftarrow d \wedge B$ be a clause in (a renamed apart variant of) $Defs_k$. Suppose that there exists a renaming substitution ρ such that: (i) $A = B\,\rho$, and (ii) $c \sqsubseteq_{\mathbb{R}} d\,\rho$. By *folding γ using δ* we derive the clause $\eta\colon H \leftarrow c \wedge G_L \wedge K\rho \wedge G_R$ and from program P_k we derive the program $P_{k+1} = (P_k - \{\gamma\}) \cup \{\eta\}$.

The following example illustrates an application of Rule R3.

Example 2. Suppose that the following clause belongs to P_k:

$\gamma\colon\ h(X) \leftarrow X \geq 1 \wedge 2\,Y = 3\,X + 2 \wedge p(X, Y)$

and suppose that the following clause is a definition clause in $Defs_k$:

$\delta\colon\ new(V, Z) \leftarrow Z > 2 \wedge p(V, Z)$

We have that the substitution $\rho = \{V/X, Z/Y\}$ satisfies Conditions (i) and (ii) of the positive folding rule because $X \geq 1 \wedge 2\,Y = 3\,X + 2 \sqsubseteq_{\mathbb{R}} (Z > 2)\rho$. Thus, by folding clause γ using clause δ, we derive:

$\eta\colon\ h(X) \leftarrow X \geq 1 \wedge 2\,Y = 3\,X + 2 \wedge new(X, Y)$

R4. Negative Folding. Let $\gamma\colon H \leftarrow c \wedge G_L \wedge \neg A \wedge G_R$ be a clause in P_k and let $\delta\colon K \leftarrow d \wedge B$ be a clause in (a renamed apart variant of) $Defs_k$. Suppose that there exists a renaming substitution ρ such that: (i) $A = B\,\rho$, and (ii) $c \sqsubseteq_{\mathbb{R}} d\,\rho$. By *folding γ using δ* we derive the clause $\eta\colon H \leftarrow c \wedge G_L \wedge \neg K\rho \wedge G_R$ and from program P_k we derive the program $P_{k+1} = (P_k - \{\gamma\}) \cup \{\eta\}$.

The following notion will be used for introducing the *clause removal* rule. Given two clauses of the form $\gamma\colon H \leftarrow c \wedge B$ and $\delta\colon H \leftarrow d$, respectively, we say that γ is $\mathbb{Z}$-*subsumed* by δ, if $c \sqsubseteq_{\mathbb{Z}} d$. Similarly, we say that γ is $\mathbb{R}$-*subsumed* by δ, if $c \sqsubseteq_{\mathbb{R}} d$.

By Lemma 1, if γ is $\mathbb{R}$-subsumed by δ, then γ is $\mathbb{Z}$-subsumed by δ.

R5. Clause Removal. Let γ be a clause in P_k. By *clause removal* we derive the program $P_{k+1} = P_k - \{\gamma\}$ if clause γ is $\mathbb{R}$-subsumed by a clause occurring in $P_k - \{\gamma\}$.

The following Theorem 1 states that the transformation rules R1–R5 preserve the perfect $\mathbb{Z}$-model semantics.

Theorem 1 (Correctness of the Transformation Rules). *Let P_0 be a locally stratified program and let $P_0, \ldots, P_n$ be a transformation sequence obtained by applying rules R1–R5. Let us assume that for every k, with $0 < k < n-1$, if P_{k+1} is derived by applying positive folding to a clause in P_k using a clause δ in $Defs_k$, then there exists j, with $0 < j < n-1$, such that: (i) δ belongs to P_j, and (ii) P_{j+1} is derived by unfolding δ w.r.t. the only atom in its body.*
Then P_n is locally stratified and for every ground atom A whose predicate occurs in P_0, we have that $A \in M_{\mathbb{Z}}(P_0)$ iff $A \in M_{\mathbb{Z}}(P_n)$.

Proof. (Sketch) Let us consider variants of Rules R1–R5 where the applicability conditions are obtained from those for R1–R5 by replacing $\mathbb{R}$ by $\mathbb{Z}$. Let us denote R1$_\mathbb{Z}$–R5$_\mathbb{Z}$ these variants of the rules. Rules R1$_\mathbb{Z}$, R2$_\mathbb{Z}$, R3$_\mathbb{Z}$, R4$_\mathbb{Z}$, and R5$_\mathbb{Z}$ can be viewed as instances (for $\mathcal{D} = \mathbb{Z}$) of the rules R1, R2p, R3(P), R3(N), and R4s, respectively, for specializing CLP($\mathcal{D}$) programs presented in [15]. By Theorem 3.3.10 of [15] we have that P_n is locally stratified and for every ground atom A whose predicate occurs in P_0, we have that $A \in M_\mathbb{Z}(P_0)$ iff $A \in M_\mathbb{Z}(P_n)$. Since, by Lemma 1 we have that the applicability conditions of R1–R5 imply the applicability conditions of R1$_\mathbb{Z}$–R5$_\mathbb{Z}$, we get the thesis.   □

4   The Specialization Strategy

Now we present a strategy for specializing a program P_0 with respect to a query of the form $c \wedge p(X_1, \ldots, X_h)$, where c is a constraint and p is a predicate occurring in P_0. Our strategy constructs a transformation sequence $P_0, \ldots, P_n$ by using the rules R1–R5 defined in Section 3. The last program P_n is the specialized version of P_0 with respect to $c \wedge p(X_1, \ldots, X_h)$. P_n is the output program P_{sp} of the specialization strategy presented below.

The Specialization Strategy makes use of two auxiliary operators: an *unfolding operator* and a *generalization operator* that tell us how to apply the unfolding rule R2 and the constrained atomic definition rule R1, respectively. The problem of designing suitable unfolding and generalization operators has been addressed in many papers and various solutions have been proposed in the literature (see, for instance, [16,19,31] and [28] for a survey in the case of logic programs). In this paper we will not focus on this aspect and we will simply assume that we are given: (i) an operator *Unfold*(δ, P) which, for every clause δ occurring in a program P, returns a set of clauses derived from δ by applying $n (\geq 1)$ times the unfolding rule R2, and (ii) an operator *Gen*($c \wedge A$, *Defs*) which, for every constraint c, atom A with *vars*(c) $\subseteq$ *vars*(A), and set *Defs* of the definition clauses introduced so far by Rule R1 during the Specialization Strategy, returns a constraint g that is *more general than* c, that is: (i) *vars*(g) $\subseteq$ *vars*(c) and (ii) $c \sqsubseteq_\mathbb{R} g$. An example of the generalization operator *Gen* will be presented in Section 5.

The Specialization Strategy

Input: A program P_0 and a query $c \wedge p(X_1, \ldots, X_h)$ where: (i) c is a constraint with *vars*(c) $\subseteq \{X_1, \ldots, X_h\}$, and (ii) p occurs in P_0.

Output: A program P_{sp} such that for every tuple $\langle n_1, \ldots, n_h \rangle \in \mathbb{Z}^h$,

$$p_{sp}(n_1, \ldots, n_h) \in M_\mathbb{Z}(P_0 \cup \{\delta_0\}) \text{ iff } p_{sp}(n_1, \ldots, n_h) \in M_\mathbb{Z}(P_{sp}),$$

where: (i) δ_0 is the definition clause $p_{sp}(X_1, \ldots, X_h) \leftarrow c \wedge p(X_1, \ldots, X_h)$ and (ii) p_{sp} is a predicate not occurring in P_0.

INITIALIZATION:

$P_{sp} := P_0 \cup \{\delta_0\}$; *Defs* $:= \{\delta_0\}$;

while there exists a clause δ in $P_{sp} \cap Defs$ do

UNFOLDING: $\Gamma := Unfold(\delta, P_{sp})$;

CLAUSE REMOVAL:
while in Γ there exist two distinct clauses γ_1 and γ_2 such that γ_1 is $\mathbb{R}$-subsumed by γ_2 do $\Gamma := \Gamma - \{\gamma_1\}$ *end-while*;

DEFINITION & FOLDING:
while in Γ there exists a clause γ: $H \leftarrow c \wedge G_1 \wedge L \wedge G_2$, where L is a literal whose predicate occurs in P_0 *do*

let c_p be the $\mathbb{R}$-projection of c on $vars(L)$ and let A be the atom such that L is either A or $\neg A$;

if in $Defs$ there exists a clause $K \leftarrow d \wedge B$ and a renaming substitution ρ such that: (i) $A = B\rho$ and (ii) $c_p \sqsubseteq_{\mathbb{R}} d\rho$

then $\Gamma := (\Gamma - \{\gamma\}) \cup \{H \leftarrow c \wedge G_1 \wedge M \wedge G_2\}$
where M is $K\rho$ if L is A, and M is $\neg K\rho$ if L is $\neg A$;

else $P_{sp} := P_{sp} \cup \{K \leftarrow g \wedge A\}$; $Defs := Defs \cup \{K \leftarrow g \wedge A\}$
where: (i) $K = newp(Y_1, \ldots, Y_m)$, (ii) $newp$ is a predicate symbol not occurring in $P_0 \cup Defs$, (iii) $\{Y_1, \ldots, Y_m\} = vars(A)$, and (iv) $g = Gen(c_p \wedge A, Defs)$;

$\Gamma := (\Gamma - \{\gamma\}) \cup \{H \leftarrow c \wedge G_1 \wedge M \wedge G_2\}$
where M is K if L is A, and M is $\neg K$ if L is $\neg A$;
end-while;

$P_{sp} := (P_{sp} - \{\delta\}) \cup \Gamma$;

end-while

In the Specialization Strategy we use real relaxations at several points: (i) when we apply the *Unfold* operator (because for applying rule R2 we check $\mathbb{R}$-satisfiability of constraints); (ii) when we check $\mathbb{R}$-subsumption during clause removal; (iii) when we compute the $\mathbb{R}$-projection c_p of the constraint c, (iv) when we check whether or not $c_p \sqsubseteq_{\mathbb{R}} d\rho$, and (v) when we compute the constraint $g = Gen(c_p \wedge A, Defs)$ such that $c_p \sqsubseteq_{\mathbb{R}} g$. Note that the condition $c_p \sqsubseteq_{\mathbb{R}} g$ ensures that clause γ can be folded using the new clause $K \leftarrow g \wedge A$, as it can be checked by inspecting Rules R3 and R4 and recalling that, by Lemma 1, $c \sqsubseteq_{\mathbb{R}} c_p$.

The correctness of the Specialization Strategy derives from the correctness of the transformation rules (see Theorem 1). Indeed, the sequence of values assigned to P_{sp} during the strategy can be viewed as (a subsequence of) a transformation sequence satisfying the hypotheses of Theorem 1.

We assume that the unfolding and the generalization operators guarantee that the Specialization Strategy terminates. In particular, we assume that: (i) the *Unfold* operator performs a finite number of unfolding steps, and (ii) the set *Defs* stabilizes after a finite number of applications of the *Gen* operator, that is, there exist two consecutive values, say $Defs_k$ and $Defs_{k+1}$, of *Defs* such that $Defs_k = Defs_{k+1}$. This stabilization property can be enforced by defining the generalization operator similarly to the *widening* operator on polyhedra, which is often used in the static analysis of programs [10].

5   Application to the Verification of Reactive Systems

In this section we show how our Specialization Strategy based on real relaxations can be used for the verification of properties of infinite state reactive systems.

Suppose that we are given an infinite state reactive system such that: (i) the set of *states* is a subset of $\mathbb{Z}^k$, and (ii) the state *transition relation* is a binary relation on $\mathbb{Z}^k$ specified as a set of constraints over $\mathbb{Z}^k \times \mathbb{Z}^k$. In this section we will take into consideration *safety* properties, but our technique can be applied to more complex properties, such as CTL temporal properties [9,19]. A reactive system is said to be *safe* if from every *initial state* it is not possible to reach, by zero or more applications of the transition relation, a state, called an *unsafe state*, satisfying an undesired property. Let *Unsafe* be the set of all unsafe states. A standard method to verify whether or not the system is safe consists in: (i) computing (backwards from *Unsafe*) the set BR of the states from which it is possible to reach an unsafe state, and (ii) checking whether or not $BR \cap Init = \emptyset$, where *Init* denotes the set of initial states.

In order to compute the set BR of backward reachable states, we introduce a CLP($\mathbb{Z}$) program P_{BR} defining a predicate br such that $\langle n_1, \ldots, n_k \rangle \in BR$ iff $br(n_1, \ldots, n_k) \in M_{\mathbb{Z}}(P_{BR})$. Then we can show that the reactive system is safe, by showing that there is no atom $br(n_1, \ldots, n_k) \in M_{\mathbb{Z}}(P_{BR})$ such that $init(n_1, \ldots, n_k)$ holds, where $init(X_1, \ldots, X_k)$ is a constraint that represents the set *Init* of states. Unfortunately, the computation of the perfect $\mathbb{Z}$-model $M_{\mathbb{Z}}(P_{BR})$ by a bottom-up evaluation of the immediate consequence operator may not terminate, and in that case we are unable to check whether or not the system is safe.

It has been shown in [19] that the termination of the bottom-up construction of the perfect model of a program can be improved by first specializing the program with respect to the query of interest. In this paper, we use a variant of the specialization-based method presented in [19] which is tailored to the verification of safety properties.

Our specialization-based method for verification consists of two steps. In Step 1 we apply the Specialization Strategy of Section 4 and we specialize program P_{BR} with respect to the initial states of the system, that is, w.r.t. the query $init(X_1, \ldots, X_k) \wedge br(X_1, \ldots, X_k)$. In Step 2 we compute the perfect $\mathbb{Z}$-model of the specialized program by a bottom-up evaluation of the immediate consequence operator associated with the program.

Before presenting an example of application of our verification method, let us introduce the generalization operator we will use in the Specialization Strategy. We will define our generalization operator by using the widening operator [10], but we could have made other choices by using suitable combinations of the widening operator, the *convex hull* operator, and *thin well-quasi orderings* based on the coefficients of the polynomials (see [11,19,31] for details).

First, we need to structure the set *Defs* of definition clauses as a tree, also called *Defs* (a similar approach is followed in [19]): (i) the root clause of that tree is δ_0, and (ii) the children of a definition clause δ are the new definition clauses added to *Defs* (see the *else* branch in the body of the inner while-loop of the Specialization Strategy) during the execution relative to δ (see the test 'δ in $P_{sp} \cap Defs$') of the body of the outer while-loop of the Specialization Strategy.

Given a constraint c_p and an atom A obtained from a clause δ as described in the Specialization Strategy, $Gen(c_p \wedge A, Defs)$ is the constraint g defined as follows. If in *Defs* there exists a (most recent) ancestor clause $K \leftarrow d \wedge B$ of δ (possibly δ itself) such that: (i) $A = B\rho$ for some renaming substitution ρ, and (ii) $d\rho = a_1 \wedge \ldots \wedge a_m$ then $g = \bigwedge_{i=1}^{m}\{a_i \mid c_p \sqsubseteq_{\mathbb{R}} a_i\}$. Otherwise, if no such ancestor of δ exists in *Defs*, then $g = c_p$.

Now let us present an example of application of our verification technique based on the Specialization Strategy of Section 4. The states of the infinite state reactive system we consider are pairs of integers and the transitions from states to states, denoted by $\longrightarrow$, are the following ones: for all $X, Y \in \mathbb{Z}$,

(1) $\langle X, Y \rangle \longrightarrow \langle X, Y-1 \rangle$ if $X \geq 1$
(2) $\langle X, Y \rangle \longrightarrow \langle X, Y+2 \rangle$ if $X \leq 2$
(3) $\langle X, Y \rangle \longrightarrow \langle X, -1 \rangle$ if $\exists Z \in \mathbb{Z}\ (Y = 2Z+1)$

(Thus, transition (3) is applicable only if Y is a positive or negative odd number.) The initial state is $\langle 0, 0 \rangle$ and, thus, *Init* is the singleton $\{\langle 0, 0 \rangle\}$. We want to prove that the system is safe in the sense that from the initial state we cannot reach any state $\langle X, Y \rangle$ with $Y < 0$. As mentioned above, we define the set $BR = \{\langle m, n \rangle \in \mathbb{Z}^2 \mid \exists \langle x, y \rangle \in \mathbb{Z}^2\ (\langle m, n \rangle \longrightarrow^* \langle x, y \rangle \wedge y < 0)\}$, where $\longrightarrow^*$ is the reflexive, transitive closure of the transition relation $\longrightarrow$. Thus, BR is the set of states from which an unsafe state is reachable. We have to prove that $Init \cap BR = \emptyset$.

We proceed as follows. First, we introduce the following program P_{BR}:

1. $br(X, Y) \leftarrow X \geq 1 \wedge X'=X \wedge Y'=Y-1 \wedge br(X', Y')$
2. $br(X, Y) \leftarrow X \leq 2 \wedge X'=X \wedge Y'=Y+2 \wedge br(X', Y')$
3. $br(X, Y) \leftarrow Y = 2Z+1 \wedge X'=X \wedge Y'=-1 \wedge br(X', Y')$
4. $br(X, Y) \leftarrow Y < 0$

The predicate br computes the set BR of states, in the sense that: for all $\langle m, n \rangle \in \mathbb{Z}^2$, $\langle m, n \rangle \in BR$ iff $br(m, n) \in M_{\mathbb{Z}}(P_{BR})$. Thus, in order to prove the safety of the system it is enough to show that $br(0, 0) \notin M_{\mathbb{Z}}(P_{BR})$. Unfortunately, the construction of $M_{\mathbb{Z}}(P_{BR})$ performed by means of the bottom-up evaluation of the immediate consequence operator does not terminate.

Note that the use of a *tabled* logic programming system [8], augmented with a solver for constraints on the integers, would not overcome this difficulty. Indeed, a top-down evaluation of the query $br(0, 0)$ generates infinitely many calls of the form $br(0, 2n)$, for $n \geq 1$.

Now we show that our two step verification method successfully terminates. *Step* 1. We apply the Specialization Strategy which takes as input the program P_{BR} and the query $X=0 \wedge Y=0 \wedge br(X, Y)$. Thus, the clause δ_0 is:

$\delta_0.\;\; br_{sp}(X,Y) \leftarrow X{=}0 \wedge Y{=}0 \wedge br(X,Y)$

By applying the *Unfold* operator we obtain the two clauses:

5. $br_{sp}(X,Y) \leftarrow X{=}0 \wedge Y{=}0 \wedge X'{=}0 \wedge Y'{=}2 \wedge br(X',Y')$
6. $br_{sp}(X,Y) \leftarrow X{=}0 \wedge Y{=}0 \wedge Y{=}2\,Z{+}1 \wedge X'{=}0 \wedge Y'{=}{-}1 \wedge br(X',Y')$

Since clause δ_0 cannot be used for folding clause 5, we apply the generalization operator and we compute $Gen((X'{=}0 \wedge Y'{=}2 \wedge br(X',Y')),\ \{\delta_0\})$ as follows. We consider the definition clause δ_0 to be an ancestor clause of itself. Then, we consider its constraint, rewritten as $d \equiv (X \geq 0 \wedge X \leq 0 \wedge Y \geq 0 \wedge Y \leq 0)$, and we generalize the constraint $d\rho \equiv (X' \geq 0 \wedge X' \leq 0 \wedge Y' \geq 0 \wedge Y' \leq 0)$, using $c_p \equiv (X'{=}0 \wedge Y'{=}2)$, thereby introducing the following definition (modulo variable renaming):

$\delta_1.\;\; new1(X,Y) \leftarrow X{=}0 \wedge Y \geq 0 \wedge br(X,Y)$

Similarly, in order to fold clause 6, we introduce the following definition:

$\delta_2.\;\; new2(X,Y) \leftarrow X{=}0 \wedge Y \leq 0 \wedge br(X,Y)$

By folding clauses 5 and 6 by using definitions δ_1 and δ_2, respectively, we derive the following clauses:

7. $br_{sp}(X,Y) \leftarrow X{=}0 \wedge Y{=}0 \wedge X'{=}0 \wedge Y'{=}2 \wedge new1(X',Y')$
8. $br_{sp}(X,Y) \leftarrow X{=}0 \wedge Y{=}0 \wedge Y{=}2\,Z{+}1 \wedge X'{=}0 \wedge Y'{=}{-}1 \wedge$
    $\qquad new2(X',Y')$

Then, we proceed with the next iterations of the body of the outermost while-loop of the Specialization Strategy, and we process first clause δ_1 and then clause δ_2. By using clauses δ_0, δ_1, and δ_2, we cannot fold all the clauses which are obtained by unfolding δ_1 and δ_2 w.r.t. the atom $br(X,Y)$. Thus, we apply again the generalization operator and we introduce the following definition (modulo variable renaming):

$\delta_3.\;\; new3(X,Y) \leftarrow X{=}0 \wedge br(X,Y)$

After processing also this clause δ_3 and performing the unfolding and folding steps as indicated by the Specialization Strategy, we obtain the clauses:

9. $new1(X,Y) \leftarrow X{=}0 \wedge Y \geq 0 \wedge X'{=}0 \wedge Y'{=}Y{+}2 \wedge new1(X',Y')$
10. $new1(X,Y) \leftarrow X{=}0 \wedge Y \geq 0 \wedge Y{=}2\,Z{+}1 \wedge X'{=}0 \wedge Y'{=}{-}1 \wedge$
    $\qquad new2(X',Y')$
11. $new2(X,Y) \leftarrow X{=}0 \wedge Y \leq 0 \wedge X'{=}0 \wedge Y'{=}Y{+}2 \wedge new3(X',Y')$
12. $new2(X,Y) \leftarrow X{=}0 \wedge Y \leq 0 \wedge Y{=}2\,Z{+}1 \wedge X'{=}0 \wedge Y'{=}{-}1 \wedge$
    $\qquad new2(X',Y')$
13. $new2(X,Y) \leftarrow X{=}0 \wedge Y < 0$
14. $new3(X,Y) \leftarrow X{=}0 \wedge X'{=}0 \wedge Y'{=}Y{+}2 \wedge new3(X',Y')$
15. $new3(X,Y) \leftarrow X{=}0 \wedge Y{=}2\,Z{+}1 \wedge X'{=}0 \wedge Y'{=}{-}1 \wedge new3(X',Y')$
16. $new3(X,Y) \leftarrow X{=}0 \wedge Y < 0$

The final program P_{sp} consists of clauses 7–16.

Step 2. Now we construct the perfect $\mathbb{Z}$-model of P_{sp} by computing the least fixpoint of the immediate consequence operator associated with P_{sp} (note that in our case the least fixpoint exists, because the program is definite, and is reached after a finite number of iterations) and we have that:

$$M_{\mathbb{Z}}(P_{sp}) = \{new1(X,Y) \mid X=0 \wedge Y \geq 0 \wedge Y = 2\,Z+1\} \quad \cup$$
$$\{new2(X,Y) \mid X=0 \wedge (Y<0 \vee Y = 2\,Z+1)\} \cup$$
$$\{new3(X,Y) \mid X=0 \wedge (Y<0 \vee Y = 2\,Z+1)\}.$$

By inspection, we immediately get that $br_{sp}(0,0) \notin M_{\mathbb{Z}}(P_{sp})$ and, thus, the safety property has been proved.

Our Specialization Strategy has been implemented on the MAP transformation system (available at http://www.iasi.cnr.it/~proietti/system.html) by suitably modifying the specialization strategy presented in [19], so as to use the transformation rules based on real relaxations we have presented in this paper. We have tested our implementation on the set of infinite state systems used for the experimental evaluation in [19] and we managed to prove the same properties. However, the technique proposed in [19] encodes the temporal properties of the reactive systems we consider as CLP($\mathbb{Q}$) programs, where $\mathbb{Q}$ is the set of rational numbers. Thus, a proof of correctness of the encoding is needed for each system, to show that the properties of interest hold in the CLP($\mathbb{Q}$) encoding iff they hold in the CLP($\mathbb{Z}$) one. In contrast, the method presented in this paper makes use of constraint solvers over the real numbers, but it preserves *equivalence* with respect to the perfect $\mathbb{Z}$-model, thereby avoiding the need for *ad hoc* proofs of the correctness of the encoding.

Finally, note that the example presented in this section cannot be worked out by first applying the relaxation from integers to reals to the initial program and then applying polyhedral approximations, such as those considered in static program analysis [10]. Indeed, we have that $br(0,0) \notin M_{\mathbb{Z}}(P_{BR})$, but if the system is interpreted over the reals, instead of the integers, we have that $br(0,0) \in M_{\mathbb{R}}(P_{BR})$ (where $M_{\mathbb{R}}$ denotes the perfect model constructed over $\mathbb{R}$). This is due to the fact that $\exists Z(0=2\,Z+1)$ holds on the reals (but it does not hold on the integers) and, hence, we derive $br(0,0)$ from clauses 3 and 4 of program P_{BR}. Thus, $br(0,0)$ is a member of every over-approximation of $M_{\mathbb{R}}(P_{BR})$ and the safety property cannot be proved.

6   Related Work and Conclusions

We have presented a technique for specializing a CLP($\mathbb{Z}$) program with respect to a query of interest. Our technique is based on the unfold/fold transformation rules and its main novelty is that it makes use of the relaxation from the integers $\mathbb{Z}$ to the reals $\mathbb{R}$, that is, during specialization the constraints are interpreted over the set of the real numbers, instead of the integer numbers. The most interesting feature of our specialization technique is that, despite the relaxation, the initial program and the derived, specialized program are equivalent with respect to the perfect model constructed over $\mathbb{Z}$ (restricted to the query of interest). In essence, the reason for this equivalence is that, if the unsatisfiability of constraints or the entailment between constraints that occur in the applicability conditions of the transformation rules hold in $\mathbb{R}$, then they hold also in $\mathbb{Z}$.

The main practical advantage of our specialization technique is that, during transformation, we can use tools for manipulating constraints over the reals, such as the libraries for constraint solving, usually available within CLP($\mathbb{R}$) systems and, in particular, the Parma Polyhedral Library [3]. These tools are significantly more efficient than constraint solvers over the integers and, moreover, they implement operators which are often used during program specialization and program analysis, such as, the widening and convex hull operators. The price we pay, at least in principle, for the efficiency improvement, is that the result of program specialization may be sub-optimal with respect to the one which can be achieved by manipulating integer constraints. Indeed, our specialization strategy might fail to exploit properties which hold for the integers and not for the reals, while transforming the input program. For example, it may be unable to detect that a clause could be removed because it contains constraints which are unsatisfiable on the integers. However, we have checked that, for the significant set of examples taken from [19], this sub-optimality never occurs.

The main application of our specialization technique is the verification of infinite state reactive systems by following the approach presented in [19]. Those systems are often specified by using constraints over integer variables, and their properties (for instance, reachability, safety and liveness) can be specified by using CLP($\mathbb{Z}$) programs [20,21]. It has been shown in [19] that properties of infinite state reactive systems can be verified by first (1) specializing the program that encodes the properties of the system with respect to the property of interest, and then (2) constructing the perfect model of the specialized program by the standard bottom-up procedure based on the evaluation of the immediate consequence operator. However, in [19] the reactive systems and their properties were encoded by using CLP programs over the rational numbers (or, equivalently in the case of linear constraints, over the real numbers), instead of integer numbers. Thus, a proof of correctness of the encoding is needed for each system (or for some classes of systems, as in [7,13]). In contrast, our specialization technique makes use of constraint solvers over the real numbers, but preserves equivalence with respect to the perfect model constructed over the integer numbers, thereby avoiding the need for *ad hoc* proofs of the correctness of the encoding.

Specialization techniques for constraint logic programs have been presented in several papers [12,16,23,31,35]. However, those techniques consider CLP($\mathcal{D}$) programs, where $\mathcal{D}$ is either a generic domain, or the domain of the rational numbers, or the domain of the real numbers. None of those papers proposes techniques for specializing CLP($\mathbb{Z}$) programs by manipulating constraints interpreted over the real numbers, as we do here.

Also the application of program specialization to the verification of infinite state systems is not a novel idea [17,19,29,30,31] and, indeed, the technique outlined in Section 5 is a variant of the one proposed in [19]. The partial deduction techniques presented in [29,30] do not make use of constraints. The papers [17,19,31] propose verification techniques for reactive systems which are based on the specialization of constraint logic programs, where the constraints are linear equalities and inequalities over the rational or real numbers. When applying these specialization techniques to reactive systems whose native specifications are given by using constraints over the integers, we need to prove the correctness of the encoding. Indeed, as shown by our example in Section 3, if we specify a system by using constraints over the integers and then we interpret those constraints over the reals (or the rationals), we may get an incorrect result. The approach investigated in this paper avoids extra correctness proofs, and allows us to do the specialization by interpreting constraints over the reals.

The verification of program properties based on real convex polyhedral approximations (that is, linear inequalities over the reals) has been first proposed in the field of static program analysis [10,11] and then applied in many contexts. In particular, [4,5,13] consider CLP($\mathbb{R}$) encodings of infinite state reactive systems. In the case where a reactive system is specified by constraints over the integers and we want to prove a property of a set of reachable states, these encodings determine an over-approximation of that set. Thus, by static analysis a further over-approximation is computed, besides the one due the interpretation over the reals, instead of the integers, and the property of interest is checked on the approximated set of reachable states. (Clearly this method can only be applied to prove that certain states are *not* reachable.)

A relevant difference between our approach and the program analysis techniques based on polyhedral approximations is that we apply equivalence preserving transformations and, therefore, the property to be verified holds in the initial CLP($\mathbb{Z}$) program *if and only if* it holds in the specialized CLP($\mathbb{Z}$) program. In some cases this equivalence preservation is an advantage of the specialization-based verification techniques over the approximation-based techniques. For instance, if we want to prove that a given state is not reachable and this property does not hold in the CLP($\mathbb{R}$) encoding (even if it holds in the CLP($\mathbb{Z}$) encoding), then we will not be able to prove the unreachability property of interest by computing any further approximation (see our example in Section 5).

Another difference between specialization-based verification techniques and static program analysis techniques is that program specialization allows *polyvariance* [24], that is, it can produce several specialized versions for the same predicate (see our example in Section 5), while static program analysis produces one approximation for each predicate. Polyvariance is a potential advantage, as it could be exploited for a more precise analysis, but, at the same time, it requires a suitable control to avoid the explosion of the size of the specialized program. The issue of controlling polyvariance is left for future work.

In this paper we have considered constraints consisting of conjunctions of linear inequalities. In the case of non-linear inequalities the relaxation from integer to real numbers is even more advantageous, as the satisfiability of non-linear inequalities is undecidable on the integers and decidable on the reals. Our techniques smoothly extend to the non-linear case which, for reasons of simplicity, we have not considered here.

Finally, in this paper we have considered transformation rules and strategies for program specialization. An issue for future research is the extension of relaxation techniques to more general unfold/fold rules, including, for instance:

(i) negative unfolding, and (ii) folding using multiple clauses with multiple literals in their bodies (see, for instance, [18]).

Acknowledgements. This work has been partially supported by PRIN-MIUR. The last author has been supported by an ERCIM grant during his visit at LORIA-INRIA, France.

References

1. Annichini, A., Bouajjani, A., Sighireanu, M.: TREX: A Tool for Reachability Analysis of Complex Systems. In: Berry, G., Comon, H., Finkel, A. (eds.) CAV 2001. LNCS, vol. 2102, pp. 368–372. Springer, Heidelberg (2001)
2. Apt, K.R., Bol, R.N.: Logic programming and negation: A survey. Journal of Logic Programming 19/20, 9–71 (1994)
3. Bagnara, R., Hill, P.M., Zaffanella, E.: The Parma Polyhedra Library: Toward a complete set of numerical abstractions for the analysis and verification of hardware and software systems. Science of Computer Programming 72(1-2), 3–21 (2008)
4. Banda, G., Gallagher, J.P.: Analysis of Linear Hybrid Systems in CLP. In: Hanus, M. (ed.) LOPSTR 2008. LNCS, vol. 5438, pp. 55–70. Springer, Heidelberg (2009)
5. Banda, G., Gallagher, J.P.: Constraint-Based Abstract Semantics for Temporal Logic: A Direct Approach to Design and Implementation. In: Clarke, E.M., Voronkov, A. (eds.) LPAR-16 2010. LNCS(LNAI), vol. 6355, pp. 27–45. Springer, Heidelberg (2010)
6. Bardin, S., Finkel, A., Leroux, J., Petrucci, L.: FAST: Acceleration from theory to practice. International Journal on Software Tools for Technology Transfer 10(5), 401–424 (2008)
7. Bérard, B., Fribourg, L.: Reachability Analysis of (Timed) Petri Nets Using Real Arithmetic. In: Baeten, J.C.M., Mauw, S. (eds.) CONCUR 1999. LNCS, vol. 1664, pp. 178–193. Springer, Heidelberg (1999)
8. Chen, W., Warren, D.S.: Tabled evaluation with delaying for general logic programs. JACM 43(1) (1996)
9. Clarke, E.M., Grumberg, O., Peled, D.: Model Checking. MIT Press (1999)
10. Cousot, P., Cousot, R.: Abstract interpretation: A unified lattice model for static analysis of programs by construction of approximation of fixpoints. In: Proceedings of the 4th ACM-SIGPLAN POPL 1977, pp. 238–252. ACM Press (1977)
11. Cousot, P., Halbwachs, N.: Automatic discovery of linear restraints among variables of a program. In: Proceedings of the Fifth ACM Symposium on Principles of Programming Languages (POPL 1978), pp. 84–96. ACM Press (1978)
12. Craig, S.-J., Leuschel, M.: A Compiler Generator for Constraint Logic Programs. In: Broy, M., Zamulin, A.V. (eds.) PSI 2003. LNCS, vol. 2890, pp. 148–161. Springer, Heidelberg (2003)
13. Delzanno, G., Podelski, A.: Constraint-based deductive model checking. International Journal on Software Tools for Technology Transfer 3(3), 250–270 (2001)
14. Etalle, S., Gabbrielli, M.: Transformations of CLP modules. Theoretical Computer Science 166, 101–146 (1996)
15. Fioravanti, F.: Transformation of Constraint Logic Programs for Software Specialization and Verification. PhD thesis, Università di Roma "La Sapienza", Italy (2002)

16. Fioravanti, F., Pettorossi, A., Proietti, M.: Automated Strategies for Specializing Constraint Logic Programs. In: Lau, K.-K. (ed.) LOPSTR 2000. LNCS, vol. 2042, pp. 125–146. Springer, Heidelberg (2001)

17. Fioravanti, F., Pettorossi, A., Proietti, M.: Verifying CTL properties of infinite state systems by specializing constraint logic programs. In: Proceedings of the ACM SIGPLAN Workshop on Verification and Computational Logic, VCL 2001, Florence, Italy, Technical Report DSSE-TR-2001-3, pp. 85–96. University of Southampton, UK (2001)

18. Fioravanti, F., Pettorossi, A., Proietti, M.: Transformation Rules for Locally Stratified Constraint Logic Programs. In: Bruynooghe, M., Lau, K.-K. (eds.) Program Development in CL. LNCS, vol. 3049, pp. 291–339. Springer, Heidelberg (2004)

19. Fioravanti, F., Pettorossi, A., Proietti, M., Senni, V.: Program Specialization for Verifying Infinite State Systems: An Experimental Evaluation. In: Alpuente, M. (ed.) LOPSTR 2010. LNCS, vol. 6564, pp. 164–183. Springer, Heidelberg (2011)

20. Fribourg, L.: Constraint Logic Programming Applied to Model Checking. In: Bossi, A. (ed.) LOPSTR 1999. LNCS, vol. 1817, pp. 30–41. Springer, Heidelberg (2000)

21. Fribourg, L., Olsén, H.: A decompositional approach for computing least fixedpoints of Datalog programs with Z-counters. Constraints 2(3/4), 305–335 (1997)

22. Henriksen, J.G., Jensen, J.L., Jørgensen, M.E., Klarlund, N., Paige, R., Rauhe, T., Sandholm, A.: Mona: Monadic Second-Order Logic in Practice. In: Brinksma, E., Steffen, B., Cleaveland, W.R., Larsen, K.G., Margaria, T. (eds.) TACAS 1995. LNCS, vol. 1019, pp. 89–110. Springer, Heidelberg (1995)

23. Hickey, T.J., Smith, D.A.: Towards the partial evaluation of CLP languages. In: Proceedings of the 1991 ACM Symposium PEPM 1991, New Haven, CT, USA. SIGPLAN Notices, vol. 26(9), pp. 43–51. ACM Press (1991)

24. Jones, N.D., Gomard, C.K., Sestoft, P.: Partial Evaluation and Automatic Program Generation. Prentice Hall (1993)

25. LASH, homepage: http://www.montefiore.ulg.ac.be/~boigelot/research/lash

26. Leroux, J.: Vector addition system reachability problem: a short self-contained proof. In: Proceedings of the 38th ACM SIGPLAN-SIGACT Symposium on Principles of Programming Languages, POPL 2011, Austin, TX, USA, January 26-28, pp. 307–316. ACM (2011)

27. Leroux, J., Sutre, G.: Flat Counter Automata Almost Everywhere! In: Peled, D.A., Tsay, Y.-K. (eds.) ATVA 2005. LNCS, vol. 3707, pp. 489–503. Springer, Heidelberg (2005)

28. Leuschel, M., Bruynooghe, M.: Logic program specialisation through partial deduction: Control issues. Theory and Practice of Logic Programming 2(4&5), 461–515 (2002)

29. Leuschel, M., Lehmann, H.: Coverability of Reset Petri Nets and Other Well-Structured Transition Systems by Partial Deduction. In: Palamidessi, C., Moniz Pereira, L., Lloyd, J.W., Dahl, V., Furbach, U., Kerber, M., Lau, K.-K., Sagiv, Y., Stuckey, P.J. (eds.) CL 2000. LNCS (LNAI), vol. 1861, pp. 101–115. Springer, Heidelberg (2000)

30. Leuschel, M., Massart, T.: Infinite State Model Checking by Abstract Interpretation and Program Specialization. In: Bossi, A. (ed.) LOPSTR 1999. LNCS, vol. 1817, pp. 62–81. Springer, Heidelberg (2000)

31. Peralta, J.C., Gallagher, J.P.: Convex Hull Abstractions in Specialization of CLP Programs. In: Leuschel, M. (ed.) LOPSTR 2002. LNCS, vol. 2664, pp. 90–108. Springer, Heidelberg (2003)

32. Pugh, W.: A practical algorithm for exact array dependence analysis. Communications of the ACM 35(8), 102–114 (1992)
33. Schrijver, A.: Theory of Linear and Integer Programming. John Wiley & Sons (1986)
34. Seki, H.: On Negative Unfolding in the Answer Set Semantics. In: Hanus, M. (ed.) LOPSTR 2008. LNCS, vol. 5438, pp. 168–184. Springer, Heidelberg (2009)
35. Wrzos-Kaminska, A.: Partial Evaluation in Constraint Logic Programming. In: Michalewicz, M., Raś, Z.W. (eds.) ISMIS 1996. LNCS, vol. 1079, pp. 98–107. Springer, Heidelberg (1996)
36. Yavuz-Kahveci, T., Bultan, T.: Action Language Verifier: An infinite-state model checker for reactive software specifications. Formal Methods in System Design 35(3), 325–367 (2009)

Marker-Directed Optimization of UnCAL Graph Transformations

Soichiro Hidaka[1], Zhenjiang Hu[1], Kazuhiro Inaba[1], Hiroyuki Kato[1],
Kazutaka Matsuda[3], Keisuke Nakano[2], and Isao Sasano[4]

[1] National Institute of Informatics, Japan
{hidaka,hu,kinaba,kato}@nii.ac.jp
[2] The University of Electro-Communications, Japan
ksk@cs.uec.ac.jp
[3] Tohoku University, Japan
kztk@kb.ecei.tohoku.ac.jp
[4] Shibaura Institute of Technology, Japan
sasano@sic.shibaura-it.ac.jp

Abstract. Buneman et al. proposed a graph algebra called UnCAL (Unstructured CALculus) for compositional graph transformations based on
structural recursion, and we have recently applied to model transformations. The compositional nature of the algebra greatly enhances the
modularity of transformations. However, intermediate results generated
between composed transformations cause overhead. Buneman et al. proposed fusion rules that eliminate the intermediate results, but auxiliary
rewriting rules that enable the actual application of the fusion rules are
not apparent so far. UnCAL graph model includes the concept of markers, which correspond to recursive function call in the structural recursion. We have found that there are many optimization opportunities at
rewriting level based on static analysis, especially focusing on markers.
The analysis can safely eliminate redundant function calls. Performance
evaluation shows its practical effectiveness for non-trivial examples in
model transformations.

Keywords: program transformations, graph transformations, UnCAL.

1 Introduction

Graph transformation has been an active research topic [9] and plays an important role in model-driven engineering [5,11]; models such as UML dia-
grams are represented as graphs, and model transformation is essentially graph
transformation. We have recently proposed a bidirectional graph transformation
framework [6] based on providing bidirectional semantics to an existing graph
transformation language UnCAL [4], and applied it to bidirectional model
transformation by translating from existing model transformation language to
UnCAL [10]. Our success in providing well-behaved bidirectional transformation
framework was due to structural recursion in UnCAL, which is a powerful mechanism of visiting and transforming graphs. Transformation based on structural

G. Vidal (Ed.): LOPSTR 2011, LNCS 7225, pp. 123–138, 2012.

recursion is inherently compositional, thus facilitates modular model transformation programming.

However, compositional programming may lead to many unnecessary intermediate results, which would make a graph transformation program terribly inefficient. As actively studied in programming language community, optimization like fusion transformation [12] is desired to make it practically useful. Despite a lot of work being devoted to fusion transformation of programs manipulating lists and trees, little work has been done on fusion on programs manipulating graphs. Although the original UnCAL has provided some fusion rules and rewriting rules to optimize graph transformations [4], we believe that further work and enhancement on fusion and rewriting are required.

The key idea presented in this paper is to analyze input/output markers, which are sort of labels on specific set of nodes in the UnCAL graph model and are used to compose graphs by connecting nodes with matching input and output markers. By statically analyzing connectivity of UnCAL by our marker analysis, we can simplify existing fusion rule. Consider, for instance, the following existing generic fusion rule of the structural recursion operator in UnCAL:

$$\mathbf{rec}(\lambda(\$l_2, \$t_2).e_2)(\mathbf{rec}(\lambda(\$l_1, \$t_1).e_1)(e_0))$$
$$= \mathbf{rec}(\lambda(\$l_1, \$t_1).\, \mathbf{rec}(\lambda(\$l_2, \$t_2).e_2)(e_1 \, @ \, \mathbf{rec}(\lambda(\$l_1, \$t_1).e_1)(\$t_1)))(e_0)$$

where $\mathbf{rec}(\lambda(\$l, \$t).e)$ applies transformation e on each edge (whose label is bound to $\$l$ and subgraph pointed by the edge is bound to $\$g$) of the input graph, and combine the results of e to produce the output graph. $\mathbf{rec}$ encodes a structural recursive function which is an important computation pattern and explained later. The graph constructor @ connects two graphs by matching markers on nodes, and in this case, result of transformation e_1 is combined to another structural recursion $\mathbf{rec}(\lambda(\$l_1, \$t_1).e_1)$. If we know by static analysis that e_1 creates no output markers, or equivalently, $\mathbf{rec}(\lambda(\$l_1, \$t_1).e_1)$ makes no recursive function call, then we can eliminate $@ \, \mathbf{rec}(\lambda(\$l_1, \$t_1).e_1)(\$t_1)$ and further simplify the fusion rule. Our preliminary performance analysis reports relatively good evidence of usefulness of this optimization.

The main technical contributions of this paper are two folds:

- A sound static inference of markers that is refined over that in [4] (Section 3). In the prior inference, the set of output markers was inferred using subtyping rule, which could lead to a set that is unnecessarily larger than actually produced at run time. For example, the set of output markers of the body of **rec** was treated as identical to the set of input markers. This over-approximation missed the chance of expression simplification exemplified above. Our inference can avoid this over-approximation by avoiding subtyping rule and computing the sets in a bottom-up manner.
- A set of rewriting rules for optimization using inferred markers (Section 4), that is more powerful than that in [4] in the sense that more expressions are simplified as exemplified above.

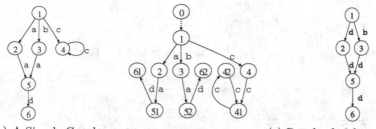

(a) A Simple Graph   (b) An Equivalent Graph   (c) Result of $a2d\_xc$ on Fig. 1(a)

Fig. 1. Graph Equivalence Based on Bisimulation

All have been implemented and tested with graph transformations widely recognized in software engineering research. The source code of the implementation can be downloaded via our project web site at www.biglab.org.

The rest of this paper is organized as follows. Section 2 reviews UnCAL graph model, graph transformation language and existing optimizations. Section 3 proposes enhanced static analysis of markers. In Section 4, we build enhanced rewriting optimization algorithm based on the static analysis. Section 5 reports preliminary performance results. Section 6 reviews related work, and Section 7 concludes this paper.

2   UnCAL Graph Algebra and Prior Optimizations

In this section, we review the UnCAL graph algebra [3,4], in which our graph transformation is specified.

2.1   Graph Data Model

We deal with rooted, directed, and edge-labeled graphs with no order on outgoing edges. UnCAL graph data model has two prominent features, *markers* and *ε-edges*. Nodes may be marked with *input* and *output markers*, which are used as an interface to connect them to other graphs. An ε-edge represents a shortcut of two nodes, working like the ε-transition in an automaton. We use *Label* to denote the set of labels and $\mathcal{M}$ to denote the set of markers.

Formally, a graph G is a quadruple (V, E, I, O), where V is a set of nodes, $E \subseteq V \times (Label \cup \{\varepsilon\}) \times V$ is a set of edges, $I \subseteq \mathcal{M} \times V$ is a set of pairs of an input marker and the corresponding node, and $O \subseteq V \times \mathcal{M}$ is a set of pairs of nodes and associated output markers. For each marker $\&x \in \mathcal{M}$, there is at most one node v such that $(\&x, v) \in I$. The node v is called an *input node* with marker $\&x$ and is denoted by $I(\&x)$. Unlike input markers, more than one node can be marked with an identical output marker. They are called *output nodes*. Intuitively, input nodes are root nodes of the graph (we allow a graph to have multiple root nodes, and for singly rooted graphs, we often use default marker

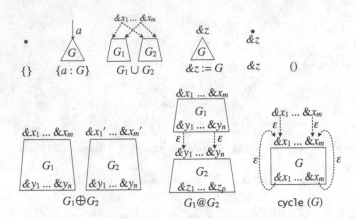

Fig. 2. Graph Constructors

& to indicate the root), while an output node can be seen as a "context-hole" of graphs where an input node with the same marker will be plugged later. We write inMarker(G) to denote the set of input markers and outMarker(G) to denote the set of output markers in a graph G.

Note that multiple-marker graphs are meant to be an internal data structure for graph composition. In fact, the initial source graphs of our transformation have one input marker (single-rooted) and no output markers (no holes). For instance, the graph in Fig. 1(a) is denoted by (V, E, I, O) where $V = \{1, 2, 3, 4, 5, 6\}$, $E = \{(1, \mathsf{a}, 2), (1, \mathsf{b}, 3), (1, \mathsf{c}, 4), (2, \mathsf{a}, 5), (3, \mathsf{a}, 5),$ $(5, \mathsf{d}, 6)\}$, $I = \{(\&, 1)\}$, and $O = \{\}$. $DB_{\mathcal{Y}}^{\mathcal{X}}$ denotes graphs with sets of input markers $\mathcal{X}$ and output markers $\mathcal{Y}$. $DB_{\mathcal{Y}}^{\{\&\}}$ is abbreviated to $DB_{\mathcal{Y}}$.

2.2 Notion of Graph Equivalence

Two graphs are value equivalent if they are bisimilar. Please refer to [4] for the complete definition. For instance, the graph in Fig. 1(b) is value equivalent to the graph in Fig. 1(a); the new graph has an additional ε-edge (denoted by the dotted line), duplicates the graph rooted at node 5, and unfolds and splits the cycle at node 4. Unreachable parts are also disregarded, i.e., two bisimilar graphs are still bisimilar if one adds subgraphs unreachable from input nodes.

This value equivalence provides optimization opportunities because we can rewrite transformation so that transformation before and after rewriting produce results that are bisimilar to each other [4]. For example, optimizer can freely cut off expressions that is statically determined to produce unreachable parts.

2.3 Graph Constructors

Figure 2 summarizes the nine graph constructors that are powerful enough to describe arbitrary (directed, edge-labeled, and rooted) graphs [4]. Here, {} constructs a root-only graph, $\{a : G\}$ constructs a graph by adding an edge with

$$
\begin{aligned}
e ::= {}\ &|\ \{l : e\}\ |\ e \cup e\ |\ \&x := e\ |\ \&y\ |\ () \\
&|\ e \oplus e\ |\ e @ e\ |\ \mathbf{cycle}(e) && \{\ \text{constructor}\ \} \\
&|\ \$g && \{\ \text{graph variable}\ \} \\
&|\ \mathbf{let}\ \$g = e\ \mathbf{in}\ e && \{\ \text{variable binding}\ \} \\
&|\ \mathbf{if}\ l = l\ \mathbf{then}\ e\ \mathbf{else}\ e && \{\ \text{conditional}\ \} \\
&|\ \mathbf{rec}(\lambda(\$l, \$g).e)(e) && \{\ \text{structural recursion application}\ \} \\
l ::= a\ &|\ \$l && \{\ \text{label}\ (a \in Label)\ \text{and label variable}\ \}
\end{aligned}
$$

Fig. 3. Core UnCAL Language

label $a \in Label \cup \{\varepsilon\}$ pointing to the root of graph G, and $G_1 \cup G_2$ adds two ε-edges from the new root to the roots of G_1 and G_2. Also, $\&x := G$ associates an input marker, $\&x$, to the root node of G, $\&y$ constructs a graph with a single node marked with one output marker $\&y$, and () constructs an empty graph that has neither a node nor an edge. Further, $G_1 \oplus G_2$ constructs a graph by using a componentwise $(V, E, I$ and $O)$ union. $\cup$ differs from $\oplus$ in that $\cup$ unifies input nodes while $\oplus$ does not. $\oplus$ requires input markers of operands to be disjoint, while $\cup$ requires them to be identical. $G_1 @ G_2$ composes two graphs vertically by connecting the output nodes of G_1 with the corresponding input nodes of G_2 with ε-edges, and $\mathbf{cycle}(G)$ connects the output nodes with the input nodes of G to form cycles. Formal definitions can be found in the full version of [6]. These graph constructors are, together with other operators, bisimulation generic [4], i.e., bisimilar result is obtained for bisimilar operands.

Example 1. The graph equivalent to that in Fig. 1(a) can be constructed as follows (though not uniquely).

$$
\begin{aligned}
\&z @ \mathbf{cycle}((\&z := \{a : \{a : \&z_1\}\} &\cup \{b : \{a : \&z_1\}\} \cup \{c : \&z_2\}) \\
&\oplus (\&z_1 := \{d : \{\}\}) \\
&\oplus (\&z_2 := \{c : \&z_2\}))
\end{aligned}
$$
□

For simplicity, we often write $\{a_1 : G_1, \ldots, a_n : G_n\}$ to denote $\{a_1 : G_1\} \cup \cdots \cup \{a_n : G_n\}$, and $(G_1, \ldots, G_n)$ to denote $(G_1 \oplus \cdots \oplus G_n)$.

2.4 UnCAL Syntax

UnCAL (Unstructured CALculus) is an internal graph algebra for the graph query language UnQL, and its core syntax is depicted in Fig. 3. It consists of the graph constructors, variables, variable bindings (**let** is our extension and is used for rewriting), conditionals, and structural recursion. We have already detailed the data constructors, while variables, variable bindings and conditionals are self explanatory. Therefore, we will focus on *structural recursion*, which is a powerful mechanism in UnCAL to describe graph transformations.

A function f on graphs is called a *structural recursion* if it is defined by the following equations

$$f(\{\}) \quad\;\; = \{\}$$
$$f(\{\$l : \$g\}) = e \mathbin{@} f(\$g)$$
$$f(\$g_1 \cup \$g_2) = f(\$g_1) \cup f(\$g_2),$$

and f can be encoded by $\mathbf{rec}(\lambda(\$l, \$g).e)$. Despite its simplicity, the core UnCAL is powerful enough to describe interesting graph transformation including all graph queries (in UnQL) [4], and nontrivial model transformations [8].

Example 2. The following structural recursion $a2d\_xc$ replaces all labels a with d and removes edges labeled c.

$$a2d\_xc(\$db) = \mathbf{rec}(\lambda(\$l, \$g).\, \mathbf{if}\; \$l = \mathrm{a}\; \mathbf{then} \qquad \{\mathrm{d} : \&\}$$
$$\mathbf{else\; if}\; \$l = \mathrm{c}\; \mathbf{then}\; \{\varepsilon : \&\}$$
$$\mathbf{else} \qquad\qquad \{\$l : \&\})\, (\$db)$$

The outer **if** of the nested **if**s corresponds to e in the above equations. Applying the function $a2d\_xc$ to the graph in Fig. 1(a) yields the graph in Fig. 1(c).   □

2.5   Revisiting Original Marker Analysis

There were actually previous work on marker analysis by original authors of UnCAL. The original typing rules appeared in the technical report version of [2]. Note that we call type to denote sets of input and output markers. Compared to our analysis, these rules were provided declaratively. For example, the rule for **if** says that if sets of output markers in both branches are equal, then the result have that set of output markers. It is not apparent how we obtain the output marker of **if** if the branches have different sets of output markers.

Buneman et al. [4] did mention optimization based on marker analysis, to avoid evaluating unnecessary subexpressions. But it was mainly based on *runtime* analysis. As we propose in the following sections, we can *statically* compute the set of markers and further simplify the transformation itself.

2.6   Fusion Rules and Output Marker Analysis

Buneman et al. [3,4] proposed the following fusion rules that aim to remove intermediate results in successive applications of structural recursion **rec**.

$$\mathbf{rec}(\lambda(\$l_2, \$t_2).e_2)(\mathbf{rec}(\lambda(\$l_1, \$t_1).e_1)(e_0))$$
$$= \begin{cases} \mathbf{rec}(\lambda(\$l_1, \$t_1).\, \mathbf{rec}(\lambda(\$l_2, \$t_2).e_2)(e_1))(e_0) & \text{if } \$t_2 \text{ does not appear free in } e_2 \\ \mathbf{rec}(\lambda(\$l_1, \$t_1).\, \mathbf{rec}(\lambda(\$l_2, \$t_2).e_2) \\ \quad (e_1 \mathbin{@} \mathbf{rec}(\lambda(\$l_1, \$t_1).e_1)(\$t_1)))(e_0) & \text{for arbitrary } e_2 \end{cases}$$

$$(1)$$

If you can statically guarantee that e_1 does not produce any output marker, which means the **rec** is "non-recursive", then the second rule is promoted to the first rule, opening another optimization opportunities.

$$\&x := (\&z := e) \longrightarrow \&x.\&z := e \quad \&x := (e_1 \oplus e_2) \longrightarrow (\&x := e_1) \oplus (\&x := e_2)$$

$$e \cup \{\} \longrightarrow e \quad \{\} \cup e \longrightarrow e \quad e \oplus () \longrightarrow e \quad () \oplus e \longrightarrow e$$

$$() @ e \longrightarrow () \qquad \frac{e :: DB_{\mathcal{Y}}^{\mathcal{X}} \quad \mathcal{X} \cap \mathcal{Y} = \phi}{\mathbf{cycle}(e) \longrightarrow e}$$

Fig. 4. Auxiliary Rewriting Rules

Non-recursive Query. Now questions that might be asked would be how often do such kind of "non-recursive" queries appear. Actually it frequently appears as *extraction* or *join*. Extraction transformation is a transformation in which some subgraphs are simply extracted. It is achieved by direct reference of the bound graph variable in the body of **rec**. Join is achieved by nesting of these extraction transformations. Finite steps of edge traversals are expressed by this nesting.

Example 3. The following structural recursion *consecutive* extracts subgraphs that can be accessible by traversing two connected edges of the same label.

$$consecutive(\$db) = \mathbf{rec}(\lambda(\$l, \$g).\, \mathbf{rec}(\lambda(\$l', \$g').$$
$$\qquad \mathbf{if}\ \$l = \$l'\ \mathbf{then}\ \{\mathtt{result} : \$g'\}$$
$$\qquad \mathbf{else} \qquad\qquad \{\} \qquad\qquad)(\$g))(\$db)$$

For example, we have *consecutive* $\left(\begin{smallmatrix}a & a & X \\ \circ & \bullet\to\bullet\to\bullet \\ b & a & Y \\ & \bullet\to\bullet\to\bullet\end{smallmatrix}\right) = \circ \xrightarrow{\text{result}} \bullet \xrightarrow{X} \bullet$.

If this transformation is followed by $\mathbf{rec}(\lambda(\$l_2, \$t_2).e_2)$ where e_2 refers to $\$t_2$, the second condition of fusion rule applies, but it will be promoted to the first, since the body of **rec** in *consecutive*, which corresponds to e_1 in the fusion rule, does not have output markers. We revisit this case in Example 4 in Section 4.

2.7 Other Prior Rewriting Rules

Apart from fusion rules, the following rewriting rules for **rec** are proposed in [4] for optimizations. Type of e is assumed to be $DB_{\mathcal{Z}}^{\mathcal{Z}}$. They simplify the argument of **rec** and increase chances of fusions. Some of them are recapped below.

$$\mathbf{rec}(\lambda(\$l, \$t).e)(\{\}) \quad =^1 \bigoplus_{\&z \in \mathcal{Z}} \&z := \{\}$$
$$\mathbf{rec}(\lambda(\$l, \$t).e)(\{l : d\}) = e[^l\!/\$l][^d\!/\$t] @ \mathbf{rec}(\lambda(\$l, \$t).e)(d)$$

The first rule eliminates **rec**, while the second rule eliminates an edge from the argument.

Additional rules proposed by (full version of) Hidaka et al. [8] to further simplify the body of **rec** are given in Fig. 4. The rules in the last line in Fig. 4 can be generalized by static analysis of the marker in the following section. And given the static analysis, we can optimize further as described in Section 4.

---

[1] Original right hand side was $\{\}$ in [4], but we corrected here.

$$(\&x \cdot \&y) \cdot \&z = \&x \cdot (\&y \cdot \&z) \quad \& \cdot \&x = \&x \cdot \& = \&x \quad \mathcal{X} \cdot \mathcal{Y} \overset{\text{def}}{=} \{\&x \cdot \&y \mid \&x \in \mathcal{X}, \&y \in \mathcal{Y}\}$$

$$\frac{}{\Gamma \vdash \{\} :: DB_\emptyset} \quad \frac{\Gamma \vdash l :: Label \qquad \Gamma \vdash e :: DB_\mathcal{Y}}{\Gamma \vdash \{l : e\} :: DB_\mathcal{Y}} \quad \frac{\Gamma \vdash e_1 :: DB_{\mathcal{Y}_1}^{\mathcal{X}} \qquad \Gamma \vdash e_2 :: DB_{\mathcal{Y}_2}^{\mathcal{X}}}{\Gamma \vdash e_1 \cup e_2 :: DB_{\mathcal{Y}_1 \cup \mathcal{Y}_2}^{\mathcal{X}}}$$

$$\frac{}{\Gamma \vdash () :: DB_\emptyset^\emptyset} \quad \frac{\Gamma \vdash e :: DB_\mathcal{Y}^\mathcal{Z}}{\Gamma \vdash \&x := e :: DB_\mathcal{Y}^{\{\&x\} \cdot \mathcal{Z}}} \quad \frac{}{\Gamma \vdash \&y :: DB_{\{\&y\}}}$$

$$\frac{\Gamma \vdash e_1 :: DB_{\mathcal{Y}_1}^{\mathcal{X}_1} \quad \Gamma \vdash e_2 :: DB_{\mathcal{Y}_2}^{\mathcal{X}_2}}{\mathcal{X}_1 \cap \mathcal{X}_2 = \emptyset} \quad \frac{\Gamma \vdash e_1 :: DB_{\mathcal{Y}_1}^{\mathcal{X}_1}}{\Gamma \vdash e_2 :: DB_{\mathcal{Y}_2}^{\mathcal{X}_2}} \, 2 \quad \frac{\Gamma \vdash e :: DB_\mathcal{Y}^\mathcal{X}}{\Gamma \vdash \mathbf{cycle}(e) :: DB_{\mathcal{Y} \setminus \mathcal{X}}^{\mathcal{X}}}$$
$$\frac{}{\Gamma \vdash e_1 \oplus e_2 :: DB_{\mathcal{Y}_1 \cup \mathcal{Y}_2}^{\mathcal{X}_1 \cup \mathcal{X}_2}} \quad \frac{}{\Gamma \vdash e_1 @ e_2 :: DB_{\mathcal{Y}_2}^{\mathcal{X}_1}}$$

$$\frac{\Gamma(\$g) = DB_\mathcal{Y}^\mathcal{X}}{\Gamma \vdash \$g :: DB_\mathcal{Y}^\mathcal{X}} \quad \frac{\Gamma \vdash e_a :: DB_\mathcal{Y}^\mathcal{X} \qquad \Gamma\{\$l \mapsto Label, \$g \mapsto DB_\mathcal{Y}\} \vdash e_b :: DB_{\mathcal{Z}_o}^{\mathcal{Z}_i} \quad \mathcal{Z} = \mathcal{Z}_i \cup \mathcal{Z}_o}{\Gamma \vdash \mathbf{rec}(\lambda(\$l, \$g).e_b)(e_a) :: DB_{\mathcal{Y} \cdot \mathcal{Z}}^{\mathcal{X} \cdot \mathcal{Z}}}$$

$$\frac{\Gamma \vdash l_1 :: Label \quad \Gamma \vdash l_2 :: Label}{\Gamma \vdash e_t :: DB_{\mathcal{Y}_t}^{\mathcal{X}} \quad \Gamma \vdash e_f :: DB_{\mathcal{Y}_f}^{\mathcal{X}}}{\Gamma \vdash \mathbf{if}\, l_1 = l_2\, \mathbf{then}\, e_t\, \mathbf{else}\, e_f :: DB_{\mathcal{Y}_t \cup \mathcal{Y}_f}^{\mathcal{X}}} \quad \frac{\Gamma \vdash e_1 :: DB_{\mathcal{Y}_1}^{\mathcal{X}_1}}{\Gamma\{\$g \mapsto DB_{\mathcal{Y}_1}^{\mathcal{X}_1}\} \vdash e_2 :: DB_{\mathcal{Y}_2}^{\mathcal{X}_2}}{\Gamma \vdash \mathbf{let}\, \$g = e_1\, \mathbf{in}\, e_2 :: DB_{\mathcal{Y}_2}^{\mathcal{X}_2}}$$

Fig. 5. UnCAL Static Typing (Marker Inference) Rules: Rules for *Label* are Omitted

3   Enhanced Static Analysis

This section proposes our enhanced marker analysis. Figure 5 shows the proposed marker *inference* rules for UnCAL. Dot notation ($\cdot$) between markers and sets of markers represents "concatenation" of markers that satisfies the properties at the top of the figure. Static environment Γ denotes mapping from variables to their types. We assume that the types of free variables are given. Since we focus on graph values, we omit rules for labels. Roughly speaking, $DB_\mathcal{Y}^\mathcal{X}$ is a type for graphs that have $\mathcal{X}$ input markers exactly and have at most $\mathcal{Y}$ output markers, which will be shown formally by Lemma 1.

The original typing rules were provided based on the subtyping rule

$$\frac{\Gamma \vdash e :: DB_\mathcal{Y}^\mathcal{X} \qquad \mathcal{Y} \subseteq \mathcal{Y}'}{\Gamma \vdash e :: DB_{\mathcal{Y}'}^\mathcal{X}}$$

and required the arguments of $\cup$, $\oplus$, **if** to have identical sets of output markers. Unlike the original rules, the proposed type system does not use the subtyping

---

[2] Original rule (let's say $@_o$) which requires $\mathcal{Y}_1 = \mathcal{X}_2$ is relaxed here. Our $@$ can be defined by $g_1 @ g_2 = (g_1 @_o \mathsf{Id}_{\mathcal{X}_2 \setminus \mathcal{Y}_1}) @_o (\mathsf{Bot}_{\mathcal{Y}_1 \setminus \mathcal{X}_2} \oplus g_2)$, where Bot and Id are defined in Section 4.2.1. This particular definition in which markers $\mathcal{Y}_1 \setminus \mathcal{X}_2$ are peeled off is close to the original semantics because final output markers coincide. Extension in which these excess output markers remain would be possible, allowing the markers to be used later to connect to other graphs.

rule directly for inference. Combined with the forward evaluation semantics $\mathcal{F}[\![]\!]$ that is summarized in [6], we have the following type safety property.

Lemma 1 (Type Safety). *Assume that g is the graph obtained by $g = \mathcal{F}[\![e]\!]$ for an expression e. Then, $\vdash e :: DB_{\mathcal{Y}}^{\mathcal{X}}$ implies both* $\mathsf{inMarker}(g) = \mathcal{X}$ *and* $\mathsf{outMarker}(g) \subseteq \mathcal{Y}$.

Lemma 1 guarantees that the set of input markers estimated by the type inference is exact in the sense that the set of input markers generated by evaluation exactly coincides with that of the inferred type. For the output markers, the type system provides an over-approximation in the sense that the set of output markers generated by evaluation is a subset of the inferred set of output markers. Since the treatment of the input markers are identical to that in [4], we focus that on the output markers and prove it. The proof, which is based on induction on the structure of UnCAL expressions, is in the full version [7] of this paper.

Between the original typing rules in [4], the following property holds: for all $\mathcal{X}$ and $\mathcal{Y}$, $e :: DB_{\mathcal{Y}}^{\mathcal{X}}$ for some $\mathcal{Y}' \supseteq \mathcal{Y}$ if and only if e has a type $DB_{\mathcal{Y}'}^{\mathcal{X}}$ in the original type system. The proof appears in the full version [7].

4 Enhanced Rewiring Optimization

This section proposes enhanced rewriting optimization rules based on the static analysis shown in the previous section.

4.1 Rule for @ and Revised Fusion Rule

Statically-inferred markers enable us to optimize expressions much more. We can generalize, for example, the rewriting rule $() @ e \longrightarrow ()$ in the last row of Fig. 4 to the following, by not just referring to the pattern of subexpressions but its estimated markers.

$$\frac{e_1 :: DB_{\emptyset}^{\mathcal{X}}}{e_1 @ c_2 \longrightarrow e_1} \tag{2}$$

As we have seen in Sect. 2, we have two fusion rules (1) for **rec**. Although the first rule can be used to gain performance, the second rule is more complex so less performance gain is expected. Using (2), we can relax the first condition of the fusion rule (1) to increase chances to apply the first rule as follows.

$$\mathbf{rec}(\lambda(\$l_2, \$t_2).e_2)(\mathbf{rec}(\lambda(\$l_1, \$t_1).e_1)(e_0))$$
$$= \mathbf{rec}(\lambda(\$l_1, \$t_1).\, \mathbf{rec}(\lambda(\$l_2, \$t_2).e_2)(e_1))(e_0)$$
$$\text{if } \$t_2 \text{ does not appear free in } e_2, \text{ or } \underline{e_1 :: DB_{\emptyset}^{\mathcal{X}}}$$

Here, the underlined part is added to relax the entire condition.

4.2 Further Optimization with Static Marker Information

In this section, general rules for $e_1 @ e_2$ is investigated. First to eliminate @ e_2, and then to statically compute @ by plugging e_2 into e_1.

4.2.1 Static Output Marker Removal Algorithm and Soundness

For more general cases of @ where connections by ε do not happen, we have the following rule.

$$\frac{e_1 :: DB_{\mathcal{Y}_1}^{\mathcal{X}} \quad e_2 :: DB_{\mathcal{Z}}^{\mathcal{Y}_2} \quad \mathcal{Y}_1 \cap \mathcal{Y}_2 = \emptyset \quad \mathsf{Rm}_{\mathcal{Y}_1}\langle\!\langle e_1 \rangle\!\rangle = e}{e_1 @ e_2 \longrightarrow e}$$

$\mathsf{Rm}_{\mathcal{Y}}\langle\!\langle e \rangle\!\rangle$ denotes static removal of the set of output markers, i.e., if $\vdash e :: DB_{\mathcal{Y}}^{\mathcal{X}}$, then $\vdash \mathsf{Rm}_{\mathcal{W}}\langle\!\langle e \rangle\!\rangle :: DB_{\mathcal{Y}\backslash\mathcal{W}}^{\mathcal{X}}$. Without this, rewriting result in spurious output markers from e_1 remained in the final result. The formal definition of $\mathsf{Rm}_{\mathcal{Y}}\langle\!\langle e \rangle\!\rangle$ is shown below.

$$\mathsf{Rm}_{\emptyset}\langle\!\langle e \rangle\!\rangle = e \quad \mathsf{Rm}_{\mathcal{X}\cup\mathcal{Y}}\langle\!\langle e \rangle\!\rangle = \mathsf{Rm}_{\mathcal{Y}}\langle\!\langle \mathsf{Rm}_{\mathcal{X}}\langle\!\langle e \rangle\!\rangle \rangle\!\rangle \quad \mathsf{Rm}_{\mathcal{Y}}\langle\!\langle \{\} \rangle\!\rangle = \{\}$$
$$\mathsf{Rm}_{\mathcal{Y}}\langle\!\langle () \rangle\!\rangle = () \quad \mathsf{Rm}_{\{\&y\}}\langle\!\langle \&y \rangle\!\rangle = \{\} \quad \mathsf{Rm}_{\{\&y\}}\langle\!\langle \&x \rangle\!\rangle = \&x$$

$$\mathsf{Rm}_{\mathcal{Y}}\langle\!\langle e_1 \odot e_2 \rangle\!\rangle = \mathsf{Rm}_{\mathcal{Y}}\langle\!\langle e_1 \rangle\!\rangle \odot \mathsf{Rm}_{\mathcal{Y}}\langle\!\langle e_2 \rangle\!\rangle \quad (\odot \in \{\cup, \oplus\})$$
$$\mathsf{Rm}_{\mathcal{Y}}\langle\!\langle \&x := e \rangle\!\rangle = (\&x := \mathsf{Rm}_{\mathcal{Y}}\langle\!\langle e \rangle\!\rangle)$$
$$\mathsf{Rm}_{\mathcal{Y}}\langle\!\langle \{l : e\} \rangle\!\rangle = \{l : \mathsf{Rm}_{\mathcal{Y}}\langle\!\langle e \rangle\!\rangle\}$$
$$\mathsf{Rm}_{\mathcal{Y}}\langle\!\langle e_1 @ e_2 \rangle\!\rangle = e_1 @ \mathsf{Rm}_{\mathcal{Y}}\langle\!\langle e_2 \rangle\!\rangle$$
$$\mathsf{Rm}_{\mathcal{Y}}\langle\!\langle \mathbf{if}\ b\ \mathbf{then}\ e_1\ \mathbf{else}\ e_2 \rangle\!\rangle = \mathbf{if}\ b\ \mathbf{then}\ \mathsf{Rm}_{\mathcal{Y}}\langle\!\langle e_1 \rangle\!\rangle\ \mathbf{else}\ \mathsf{Rm}_{\mathcal{Y}}\langle\!\langle e_2 \rangle\!\rangle$$

Since the output markers of the result of $e_1 @ e_2$ are not affected by those of e_1, e_1 is not visited in the rule of @. In the following, $\mathsf{Id}_{\mathcal{Y}}$ and $\mathsf{Bot}_{\mathcal{Y}}$ are respectively defined as $\bigoplus_{\&z \in \mathcal{Y}} \&z := \&z$ and $\bigoplus_{\&z \in \mathcal{Y}} \&z := \{\}$.

$$\frac{e :: DB_{\mathcal{Y}}^{\mathcal{X}} \quad \&y \in (\mathcal{Y} \backslash \mathcal{X}) \quad \mathsf{Rm}_{\{\&y\}}\langle\!\langle e \rangle\!\rangle = e'}{\mathsf{Rm}_{\{\&y\}}\langle\!\langle \mathbf{cycle}(e) \rangle\!\rangle = \mathbf{cycle}(e')} \quad \frac{e :: DB_{\mathcal{Y}}^{\mathcal{X}} \quad \&y \notin (\mathcal{Y} \backslash \mathcal{X})}{\mathsf{Rm}_{\{\&y\}}\langle\!\langle \mathbf{cycle}(e) \rangle\!\rangle = \mathbf{cycle}(e)}$$

$$\frac{\$v :: DB_{\mathcal{Y}}^{\mathcal{X}} \quad \&y \notin \mathcal{Y}}{\mathsf{Rm}_{\{\&y\}}\langle\!\langle \$v \rangle\!\rangle = \$v} \quad \frac{\$v :: DB_{\mathcal{Y}}^{\mathcal{X}} \quad \&y \in \mathcal{Y}}{\mathsf{Rm}_{\{\&y\}}\langle\!\langle \$v \rangle\!\rangle = \$v @ (\mathsf{Bot}_{\{\&y\}} \oplus \mathsf{Id}_{\mathcal{Y}\backslash\{\&y\}})}$$

The first rule of $\$v$ says that according to the safety of type inference, $\&y$ is guaranteed not to result at run-time, so the expression $\$v$ remains unchanged. The second rule actually removes the output marker $\&y_j$, but static removal is impossible. So the removal is deferred till run-time. The output node marked $\&y_j$ is connected to node produced by $\&y := \{\}$. Since the latter node has no output marker, the original output marker disappears from the graph produced by the evaluation. The rest of the $\&y_k := \&y_k$ does no operation on the marker. Since estimation $\mathcal{Y}$ is the upper bound, the output maker may not be produced at run-time. If it is the case, connection with ε-edge by @ does not occur, and the nodes produced by the := expressions are left unreachable, so the transformation is still valid. As another side effect, @ may connect identically marked output nodes to single node. However, the graph before and after this "funneling" connection are bisimilar, since every leaf node with identical output markers are bisimilar by definition. Should the output nodes are to be further connected to other input nodes, the target node is always single, because more than one node with identical input marker is disallowed by the data model. So this connection does no harm. Note that the second rule increases the size of the expression, so it may increase the cost of evaluation.

$$\frac{\mathbf{rec}(\lambda(\$l,\$t).e_b)(e_a) :: DB_{\mathcal{Y}.\mathcal{Z}}^{\mathcal{X}.\mathcal{Z}} \quad \&y \in \mathcal{Y} \quad \mathsf{Rm}_{\{\&y\}} \langle\!\langle e_a \rangle\!\rangle = e_a'}{\mathsf{Rm}_{\{\&y.\&z | \&z \in \mathcal{Z}\}} \langle\!\langle \mathbf{rec}(\lambda(\$l,\$t).e_b)(e_a) \rangle\!\rangle = \mathbf{rec}(\lambda(\$l,\$t).e_b)(e_a')}$$

For **rec**, one output marker $\&y$ in e_a corresponds to $\{\&y\} \cdot \mathcal{Z} = \{\&y.\&z \mid \&z \in \mathcal{Z}\}$ in the result. So removal of $\&y$ from e_a results in removal of all of the $\{\&y\} \cdot \mathcal{Z}$. So only removal of all of $\{\&y.\&z \mid \&z \in \mathcal{Z}\}$ at a time is allowed.

Lemma 2 (Soundness of Static Output-Marker Removal Algorithm).
Assume that $G = (V, E, I, O)$ is a graph obtained by $G = \mathcal{F}[\![e]\!]$ for an expression e, and e' is the expression obtained by $\mathsf{Rm}_{\mathcal{Y}} \langle\!\langle e \rangle\!\rangle$. Then, we have $\mathcal{F}[\![e']\!] = (V, E, I, \{(v, \&y) \in O \mid \&y \notin \mathcal{Y}\})$.

Lemma 2 guarantees that no output marker in $\mathcal{Y}$ appears at run-time if $\mathsf{Rm}_{\mathcal{Y}} \langle\!\langle e \rangle\!\rangle$ is evaluated.

4.2.2 Plugging Expression to Output Marker Expression
The following rewriting rule is to plug an expression into another through correspondingly marked node.

$$\{l : \&y\} @ (\&y := e) \longrightarrow \{l : e\}$$

This kind of rewriting was actually implicitly used in the exemplification of optimization in [4], but was not generalized. We can generalize this rewriting as

$$e @ (\&y := e') \longrightarrow \begin{cases} \mathsf{Rm}_{\mathcal{Y} \setminus \{\&y\}} \langle\!\langle e \rangle\!\rangle [e'/\&y] & \text{if } \&y \in \mathcal{Y} \text{ where } e :: DB_{\mathcal{Y}}^{\mathcal{X}} \\ \mathsf{Rm}_{\mathcal{Y}} \langle\!\langle e \rangle\!\rangle & \text{otherwise.} \end{cases}$$

where $e[e'/\&y]$ denotes substitution of $\&y$ by e' in e. Since nullrary constructors $\{\}$, (), and $\&x \neq \&y$ do not produce output marker $\&y$, the substitution takes no effect and the rule in the latter case apply. So we focus on the former case in the sequel. For most of the constructors the substitution rules are rather straightforward:

$$\&y[e/\&y] = e$$
$$(e_1 \odot e_2)[e/\&y] = (e_1[e/\&y]) \odot (e_2[e/\&y]) \quad (\odot \in \{\cup, \oplus\})$$
$$(\&x := e)[e/\&y] = (\&x := (e[e/\&y]))$$
$$\{l : e\}[e/\&y] = \{l : (e[e/\&y])\}$$
$$(e_1 @ e_2)[e/\&y] = e_1 @ (e_2[e/\&y])$$
$$(\mathbf{if}\ b\ \mathbf{then}\ e_1\ \mathbf{else}\ e_2)[e/\&y] = \mathbf{if}\ b\ \mathbf{then}\ (e_1[e/\&y])\ \mathbf{else}\ (e_2[e/\&y])$$

Since the final output marker for @ is not affected by that of e_1, e_1 is not visited in the rule of @. For **cycle**, we should be careful to avoid capturing of marker.

$$\mathbf{cycle}(e)[e'/\&y] = \begin{cases} \mathbf{cycle}(e[e'/\&y]) & \text{if } (\mathcal{Y}' \cap \mathcal{X}) = \emptyset \text{ where } e :: DB_{\mathcal{Y}}^{\mathcal{X}} \quad e' :: DB_{\mathcal{Y}'} \\ \overline{\mathbf{cycle}(e)}[e'/\&y] & \text{otherwise.} \end{cases}$$

The above rule says that if $\mathcal{Y}'$ will be "free" markers in e, that is, the output markers in e', namely $\mathcal{Y}'$ will not be captured by **cycle**, then we can plug e' into output marker expression in e. If some of the output markers in $\mathcal{Y}'$ are included in $\mathcal{X}$, then the renaming is necessary. As suggested in the full version of [3], markers in $\mathcal{X}$ instead of those in $\mathcal{Y}'$ should be renamed. And that renaming can be compensated outside of **cycle** as follows:

$$\overline{\mathbf{cycle}}(e) \stackrel{\text{def}}{=} (\bigoplus_{\&x \in \mathcal{X}} \&x := \&tmp_x) \,@\, \mathbf{cycle}(e[\&tmp_{x_1}/\&x_1] \dots [\&tmp_{x_M}/\&x_M])$$

where $\&x_1, \dots, \&x_M = \mathcal{X}$ are the markers to be renamed, and $\mathcal{X}$ of $e :: DB_{\mathcal{Y}}^{\mathcal{X}}$ is used. Note that in the renaming, not only output markers, but also input markers are renamed. $\&tmp_{x_1}, \dots, \&tmp_{x_M}$ are corresponding fresh (temporary) markers. The left hand side of @ recovers the original name of the markers. After renaming by $\overline{\mathbf{cycle}}$, no marker is captured anymore, so substitution is guaranteed to succeed. For variable reference and **rec**, static substitution is impossible. So we resort to the following generic "fall back" rule.

$$\frac{e \in \{\$v, \mathbf{rec}(\_)(\_)\} \quad e :: DB_{\mathcal{Y}}^{\mathcal{X}} \quad \mathcal{Y} = \{\&y_1, \dots, \&y_j, \dots, \&y_n\}}{e[e'/\&y_j] = e \,@\, \begin{pmatrix} \&y_1 := \&y_1, \dots, \&y_{j-1} := \&y_{j-1}, \&y_j := e', \\ \&y_{j-1} := \&y_{j-1}, \dots, \&y_n := \&y_n \end{pmatrix}}$$

The "fall back" rule is used for **rec** because unlike output marker removal algorithm, we can not just plug e into e_a since that will not plug e but $\mathbf{rec}(\lambda(\$l, \$t).e_b)(e)$ in the result. We could have used the inverse $\mathbf{rec}(\lambda(\$l, \$t).e_b)^{-1}$ to plug $\mathbf{rec}(\lambda(\$l, \$t).e_b)^{-1}(e')$ instead, but the inverse does not always exist in general.

The overall rewriting is conducted by two mutually recursive functions as follows: a driver function first applies itself to subexpressions recursively, and then applies a function that implements $\longrightarrow$ and other rewriting rules recursively such as fusions described in this paper, on the result of the driver function. The implemented rewriting system is deterministic by imposing consistent order of rule applications by these functions.

With respect to proposed rewriting rules in this section, the following theorem holds.

Theorem 1 (Soundness of Rewriting). *If $e \longrightarrow e'$, then $\mathcal{F}[\![e]\!]$ is bisimilar to $\mathcal{F}[\![e']\!]$.*

It can be proved by simple induction on the structure of UnCAL expressions, and omitted here.

Example 4. The following transformation that apply selection after *consecutive* in Example 3

$$\mathbf{rec}(\lambda(\$l_1, \$g_1).\, \mathbf{if}\ \$l_1 = \mathbf{a}\ \mathbf{then}\ \{\$l_1 : \$g_1\}\ \mathbf{else}\ \{\})(consecutive(\$db))$$

is rewritten as follows:

$=$ { expand definition of *consecutive* and apply 2nd fusion rule }

$$\mathbf{rec}(\lambda(\$l, \$g).\, \mathbf{rec}(\lambda(\$l_1, \$g_1).\, \mathbf{if}\ \$l_1 = \mathbf{a}\ \mathbf{then}\ \{\$l_1 : \$g_1\}\ \mathbf{else}\ \{\})$$
$$(\mathbf{rec}(\lambda(\$l', \$g').\, \mathbf{if}\ \$l = \$l'\ \mathbf{then}\ \{result : \$g'\}\ \mathbf{else}\ \{\})(\$g)$$
$$@\, \mathbf{rec}(\lambda(\$l, \$g).\, \mathbf{rec}(\lambda(\$l', \$g').$$
$$\mathbf{if}\ \$l = \$l'\ \mathbf{then}\ \{result : \$g'\}\ \mathbf{else}\ \{\})(\$g))(\$g)))(\$db)$$

$=$ { (2) }

$$\mathbf{rec}(\lambda(\$l, \$g).\, \mathbf{rec}(\lambda(\$l_1, \$g_1).\, \mathbf{if}\ \$l_1 = \mathbf{a}\ \mathbf{then}\ \{\$l_1 : \$g_1\}\ \mathbf{else}\ \{\})$$
$$(\mathbf{rec}(\lambda(\$l', \$g').\, \mathbf{if}\ \$l = \$l'\ \mathbf{then}\ \{result : \$g'\}\ \mathbf{else}\ \{\})(\$g)))(\$db)$$

$=$ { 2nd fusion rule, (2), **rec** rule for if and $\{l : d\}$, static label comparison }

$$\mathbf{rec}(\lambda(\$l, \$g).\, \mathbf{rec}(\lambda(\$l', \$g').\{\})(\$g))(\$db)$$

This example demonstrates the second fusion rule promotes to the first. The top level edges of the result of *consecutive* are always labeled **result** while the selection selects subgraphs under edges labeled **a**. So the result will always be empty, and correspondingly the body of **rec** in the final result is {}.   □

More examples can be found in the full version [7] of this paper.

The following remark summarizes how far can we remove intermediate graphs. Proof can be found in the full version [7].

Remark 1 (Removal of Intermediage Graph). Suppose we have a composition of the form

$$\mathbf{rec}(\lambda(\$l_2, \$t_2).e_2)(C[\mathbf{rec}(\lambda(\$l_1, \$t_1).e_1)(e_0)])$$

where $C[]$ denotes context using constructors and if expressions. Then, (i) if $\$t_2$ does not appear free in e_2, then the composition of the above form, including the ones that are generated during fusion, are removed. (ii) if $\$t_2$ appears free in e_2 but $e_1 :: DB_{\bar{\emptyset}}$, and e_1 consists of nested **rec** with context not using @ or **cycle** with body of type $DB_{\bar{\emptyset}}$, then the composition, including the ones that are generated during the fusion rule application, are removed.

5   Implementation and Performance Evaluation

This section reports preliminary performance evaluations. All of the transformations in the paper are implemented in GRoundTram, or Graph Roundtrip Transformation for Models, which is a system to build a bidirectional transformation between two models (graphs). All the source codes are available online at www.biglab.org. The following experimental results are obtained by the system.

Performance evaluation was conducted on GRoundTram, running on MacOSX over MacBookPro 17 inch, with 3.06 GHz Intel Core 2 Duo CPU. An UnCAL transformation runs in time exponential to the size (number of compositions

Table 1. Summary of Experiments (running time is in CPU seconds)

	direction	no rewriting	previous [4,8]	ours
Class2RDB	forward	1.18	0.68	0.68
	backward	14.5	7.99	7.89
PIM2PSM	forward	0.08	0.77 (2*3)	0.07 (2*13)
	backward	1.62	3.64	0.75
C2Osel	forward	0.04	0.04 (2*1)	0.05 (2*11)
	backward	2.26	0.26	0.27
C2Osel'	forward	0.05	0.06 (2*1)	0.04 (2*11)
	backward	2.53	2.58	1.26
Ex1 [4]	forward	0.022	0.016 (1*1)	0.010 (1*1)
	backward	0.85	0.30	0.15

or nesting of **rec**s) of the transformation (and polynomial to the size of input graph [4]). Thus, the proposed rewriting, which can reduce the size of transformation, may change the elapsed time drastically even for the small graphs (up to a hundred of nodes) used in the experiments.

Table 1 shows the experimental results. Each running time includes time for forward and backward[3] transformations [6], and for backward transformations, algorithm for edge-renaming is used, and no modification on the target is actually given. However, we suppose presence of modification would not make much difference in the running time. Running time of forward transformation in which rewriting is applied (last two columns) includes time for rewriting. Rewriting took 0.006 CPU seconds at the worst case (*PIM2PSM*, ours). *Class2RDB* stands for class diagram to table diagram transformation, *PIM2PSM* for platform independent model to platform specific model transformation, *C2Osel* is for transformation of customer oriented database into order oriented database, followed by a simple selection, and *Ex1* is the example that is extracted from our previous paper [8], which was borrowed from [4]. It is a composition of two **rec**s.

The numbers in parentheses show how often the fusion transformation happened. For example, *PIM2PSM* led to 3 fusions based on the second rule, and further enhanced rewriting led to 10 more fusion rule applications, all of which promoted to the first rule via proposed rewriting rule (2). Same promotions happened to *C2Osel*. Except for *C2Osel'*, a run-time optimization in which unreachable parts are removed after every application of **rec** is applied. Enhanced rewriting led to performance improvements in both forward and backward evaluations, except *C2Osel*. Comparing "previous" with "no rewriting", *PIM2PSM* and *C2Osel'* led to slowdown. This slowdown is explained as follows. The fusion turns composition of **rec**s to their nesting. In the presence of the run-time optimization, composition is more advantageous than nesting when only small part of the result is passed to the subsequent **rec**s, which will run faster than when passed entire results (including unreachable parts). Once nested, interme-

---

[3] Since we are conducting research on bidirectional transformations, we are not only interested in the performance of forward transformations, but also that of backward transformations.

diate result is not produced, but the run-time optimization is suppressed because every execution of the inner **rec** traverses the input graph. *C2Osel'* in which run-time optimization is turned off, shows that the enhanced rewriting itself lead to performance improvements.

6 Related Work

Although some of our optimization rules were mentioned in [8], the relationship with static marker analysis was not covered in depth. Our optimization, based on the enhanced marker analysis in Sect. 3 , generalizes all the rules in [8] uniformly.

In our previous paper [6], an implementation of rewriting optimizations was reported, but concrete strategies were not included in the paper.

Plugging constructor-only expressions into output marker expressions was discussed in the full (technical report) version of [3]. Their motivation was to express semantics of @ at the constructor expression level and not graph data level as in [4]. It also mentioned renaming of markers to avoid capture of the output markers in **cycle** expressions. We do attempt the same thing at the expression level but we argue here more formally.

Our rewriting rules are inspired by the technical report but the idea there is not yet exploited fully. They discussed the semantics of **rec** on the **cycle** expressions, even when the body refered to graph variables, although marker environment that maps markers to connected subgraphs introduced there makes the semantics complex. But we could use the semantics to enhance rewriting rules for **rec** with **cycle** arguments.

The journal version [4] mentioned run-time optimization in which, assuming top-down evaluation, only necessary components of structural recursion are executed. For example, only $\&z_1$ component of **rec** in $\&z_1$ @ **rec**(_)(_) is evaluated. It is not applicable to our bidirectional settings which rely on bulk semantics [6].

A static analysis of UnCAL was described in [1], but the main motivation was to analyze the structure of graphs using graph schema, whereas our analysis focus on the connectivity of graphs.

7 Conclusion

In this paper, under the context of graph transformation using UnCAL graph algebra, enhanced static marker inference is first formalized. Fusion rule becomes more powerful thanks to the static marker analysis. Further rewriting rules based on this inference are also explored. Marker renaming for capture avoidance is formalized to support the rewriting rules. Under the context of bidirectional graph transformations [6], one of the advantage of static analysis is that we can keep implementation of bidirectional interpreter intact. The marker analysis and rewriting proposed can be considered as dead-code detection and elimination. We believe this technique can be used for other graph languages that based on graph model that have named connecting points like input/output nodes. Preliminary performance evaluation shows the usefulness of the optimization for various non-trivial transformations in the field of software engineering research.

Future work under this context includes reasoning about effects on the backward updatability. Although rewriting is sound with respect to well-behavedness of bidirectional transformations, backward transformation before and after rewriting may accept different update operations. Our conjecture is that simplified transformation accepts more updates, but this argument requires further discussions.

Acknowledgments. We thank reviewers and Kazuyuki Asada for their thorough comments on the earlier versions of the paper. The research was supported in part by the Grand-Challenging Project on "Linguistic Foundation for Bidirectional Model Transformation" from the National Institute of Informatics, Grant-in-Aid for Scientific Research No. 20700035, Grant-in-Aid for Research Activity Start-up No. 22800003, and Kayamori Foundation of Informational Science Advancement.

References

1. Benczúr, A.A., Kósa, B.: Static Analysis of Structural Recursion in Semistructured Databases and Its Consequences. In: Benczúr, A.A., Demetrovics, J., Gottlob, G. (eds.) ADBIS 2004. LNCS, vol. 3255, pp. 189–203. Springer, Heidelberg (2004)
2. Buneman, P., Davidson, S., Fernandez, M., Suciu, D.: Adding Structure to Unstructured Data. In: Afrati, F.N., Kolaitis, P.G. (eds.) ICDT 1997. LNCS, vol. 1186, pp. 336–350. Springer, Heidelberg (1996)
3. Buneman, P., Davidson, S., Hillebrand, G., Suciu, D.: A query language and optimization techniques for unstructured data. In: SIGMOD, pp. 505–516 (1996); long version appears as U.Penn TR MS-CIS-96-09
4. Buneman, P., Fernandez, M.F., Suciu, D.: UnQL: a query language and algebra for semistructured data based on structural recursion. VLDB J. 9(1), 76–110 (2000)
5. Ehrig, K., Guerra, E., de Lara, J., Lengyel, L., Levendovszky, T., Prange, U., Taentzer, G., Varró, D., Varró-Gyapay, S.: Model transformation by graph transformation: A comparative study. Presented at MTiP (2005), http://www.inf.mit.bme.hu/FTSRG/Publications/varro/2005/mtip05.pdf
6. Hidaka, S., Hu, Z., Inaba, K., Kato, H., Matsuda, K., Nakano, K.: Bidirectionalizing graph transformations. In: ACM SIGPLAN International Conference on Functional Programming, pp. 205–216. ACM (2010)
7. Hidaka, S., Hu, Z., Inaba, K., Kato, H., Matsuda, K., Nakano, K., Sasano, I.: Marker-directed Optimization of UnCAL Graph Transformations (revised version). Technical Report GRACE-TR-2011-06, GRACE Center, National Institute of Informatics (November 2011)
8. Hidaka, S., Hu, Z., Kato, H., Nakano, K.: Towards a compositional approach to model transformation for software development. In: SAC 2009, pp. 468–475 (2009)
9. Rozenberg, G. (ed.): Handbook of Graph Grammars and Computing by Graph Transformations. Volume 1: Foundations. World Scientific (1997)
10. Sasano, I., Hu, Z., Hidaka, S., Inaba, K., Kato, H., Nakano, K.: Toward Bidirectionalization of ATL with GRoundTram. In: Cabot, J., Visser, E. (eds.) ICMT 2011. LNCS, vol. 6707, pp. 138–151. Springer, Heidelberg (2011)
11. Stevens, P.: Bidirectional Model Transformations in QVT: Semantic Issues and Open Questions. In: Engels, G., Opdyke, B., Schmidt, D.C., Weil, F. (eds.) MoDELS 2007. LNCS, vol. 4735, pp. 1–15. Springer, Heidelberg (2007)
12. Wadler, P.: Deforestation: Transforming Programs to Eliminate Trees. In: Ganzinger, H. (ed.) ESOP 1988. LNCS, vol. 300, pp. 344–358. Springer, Heidelberg (1988)

Modular Extensions for Modular (Logic) Languages

José F. Morales[1], Manuel V. Hermenegildo[1,2], and Rémy Haemmerlé[2]

[1] IMDEA Software Research Institute, Madrid, Spain
[2] School of Computer Science, T. U. Madrid (UPM), Spain

Abstract. We address the problem of developing mechanisms for easily implementing modular extensions to modular (logic) languages. By
(language) extensions we refer to different groups of syntactic definitions
and translation rules that extend a language. Our use of the concept of
modularity in this context is twofold. We would like these extensions to
be modular, in the sense above, i.e., we should be able to develop different extensions mostly separately. At the same time, the sources and
targets for the extensions are modular languages, i.e., such extensions
may take as input separate pieces of code and also produce separate
pieces of code. Dealing with this double requirement involves interesting
challenges to ensure that modularity is not broken: first, combinations
of extensions (as if they were a single extension) must be given a precise
meaning. Also, the separate translation of multiple sources (as if they
were a single source) must be feasible. We present a detailed description of a code expansion-based framework that proposes novel solutions
for these problems. We argue that the approach, while implemented for
Ciao, can be adapted for other Prolog-based systems and languages.

Keywords: Compilation, Modules, Modular Program Processing, Separate Compilation, Prolog, Ciao, Language Extensions, Domain Specific
Languages.

1 Introduction

The choice of a good notation and adequate semantics when encoding a particular problem can dramatically affect the final outcome. Extreme examples are
programming pearls, whose beauty is often completely lost when translated to a
distant language. In practice, large projects are bigger than pearls and often no
single language fulfills all expectations (which can include many aspects, such
as development time or execution performance). The programmer is forced to
make a commitment to one language —and accept sub-optimal encoding— or
more than one language —at the expense of interoperability costs.

An alternative is to provide new features and idioms as syntactic and semantic
extensions of a language, thus achieving notational convenience while avoiding
inter-language communication costs. In the case of Prolog, language extension
through term-expansion systems (combined with operator definitions) has traditionally offered a quick way to develop variants of logic languages and semantics

G. Vidal (Ed.): LOPSTR 2011, LNCS 7225, pp. 139–154, 2012.
© Springer-Verlag Berlin Heidelberg 2012

(e.g., experimental domain-specific languages, constraint systems, optimizations, debugging tools, etc.). Some systems, and in particular Ciao [10], have placed special attention on these capabilities, extending them [1] and exploiting them as the base for many language extensions.

Once a good mechanism is available for writing extensions and a number of them are available, it is natural to consider whether combining a number of them following modular design principles is feasible. For example, consider embedding a simple handy *functional notation* [3] (syntactic sugar to write goals, marked with ~, in term positions), into a more complex extension, such as the Prolog-based implementation of CHR [8]. In this new dialect, the CHR rule (see Sect. 6.2.1 in Frühwirth's book [8]):

```
T eq and(T1, T2), T1 eq 1, T2 eq X <=> T eq X.
```

can be written more concisely as:

```
T eq and(~eq(1), ~eq(X)) <=> T eq X.
```

Intuitively, expansions are applied one after the other. This already points out that at least a mechanism to determine application order is needed. This is already undesirable because it requires users to be aware of the valid orderings. Furthermore, just ordering may not be enough. In our example, if functional syntax is applied first, it must normalize the ~eq(_) terms before CHR translation happens, but there is no simple way to indicate to the functional expansion that the CHR constraints have to be treated syntactically as goals. If CHR translation is done first, it will not recognize that ~eq(_) corresponds to a constraint, and incorrect code will be generated before the functional expansion takes place. Thus, the second rule cannot be translated into the first one by simply composing the two expansions, without tweaking the translation code, which is undesirable.

Moreover, current extension mechanisms have difficulties dealing with the module system. An example is the Typed Prolog extension of [13], which elegantly implements gradually typed Prolog in the style of Hindley-Milner, but needs to treat programs as monolitic, non-modular, units. Even if extensions are made module-aware, the *dynamic* features of traditional Prolog systems present an additional hurdle: programs can change dynamically, and modules may be loaded at runtime, with no clear distinction between program code and translation code, and with no limits on what data is available to the expansion (e.g., consulting the predicates compiled in other arbitrary modules). In the worst case, this leads to a chaotic scenario, where reasoning about language translations is an impossible task.

The previous examples illustrate the limitations of the current extension mechanisms for Prolog and motivate the goals of this work:

- **Predictable Combination of Fine-Grained Extensions:** The extension mechanisms must be fine-grained enough to allow rich combinations, but also provide a simple interface for users of extensions. Namely, programmers should be able to write modules using several extensions (e.g., functional notation, DCGs, and profiling), without being required to know the application order of rules or the compatibility of extensions. Obviously, the result of the combination of such extensions must be predictable. That indirectly leads us to the necessity of describing a precise compilation model that includes compilation and loading of the extension code.

- **Integration with Module Systems:** It is thus necessary to make the extensions module-aware, while at the same time constraining them to respect the module system. For example, it must be possible to determine during expansion to what module a goal being expanded belongs, if that information is available, or to export new declarations. It is well known that modularity, if not designed carefully, can make static analysis impossible [2]. A flexible extension system that however allows breaking modularity renders any efforts towards static transformations useless.

This paper presents a number of novel contributions aimed at addressing these problems. We take as starting point the module and extension system implemented in Ciao [1,10], which is more elaborate than the one offered by traditional Prolog systems. We provide in this paper a refined formal description of the compilation model, and use it to propose and justify the definition of a number of distinct translation phases as well as the information that is available at each one. Then, we propose a set of rule-based extension mechanisms that we argue generalize previous approaches and allows us to provide better solutions for a good number of the problems mentioned earlier.

The paper is structured as follows. Section 2 gives a detailed description of the core translation process for extensions. Section 3 defines a compilation model that integrates the extensions. Section 4 and Section 5 illustrate the rules defined in the previous section by defining several (real-life) language features and extensions. We close with a discussion of related and future work in Section 6, and conclusions in Section 7.

2   Language Extensions as Translation Rules

By language extensions we refer to translations that manipulate a symbolic representation of a given program. For simplicity we will use *terms* representing abstract syntax trees, denoted by $\mathcal{T}$, following the usual definition of *ground* terms in *first order logic*. To simplify notation, we include sequences of terms $(Seq(\mathcal{T}))$ as part of $\mathcal{T}$.[1] We also assume some standard definitions and operations on terms: $\mathsf{termFn}(x)$ denotes the $(name, arity)$ of the term, $\mathsf{args} : \mathcal{T} \to Seq(\mathcal{T})$ obtains the term arguments (i.e., the sequence of children), and $\mathsf{setArgs} : \mathcal{T} \times Seq(\mathcal{T}) \to \mathcal{T}$ replaces the arguments of a term.

We use a homogeneous term representation for the program, but terms may represent a variety of language elements. The meaning of each term is often given by its surrounding context. In order to reflect this, each input term is labeled with a symbolic *kind* annotation. That annotation determines which transformation to apply to each term.

We define the main transformation algorithm $\mathbf{tr}[\![x : \kappa]\!] = x'$ in Fig. 1. Given a term x of *kind* κ, it obtains a term x' by applying the available rules. Translation

---

[1] We will assume –for simplicity and contrary to common practice– that when compiling a program variables are read as special ground terms.

$$\mathbf{tr}[\![x : final]\!] = x$$
$$\mathbf{tr}[\![x : \kappa]\!] = \mathbf{tr}[\![x : \kappa']\!] \quad (\text{if } \kappa \succ \kappa')$$
$$\mathbf{tr}[\![x : \kappa]\!] = \mathbf{tr}[\![x' : \kappa']\!] \quad (\text{if } x : \kappa \Longrightarrow x' : \kappa')$$
$$\mathbf{tr}[\![x : \kappa_x]\!] = \mathbf{tr}[\![x' : \kappa'_x]\!] \quad (\text{if } x : \kappa_x \overset{decons}{\Longrightarrow} \vec{a} : \vec{\kappa})$$
where
$$a'_i = \mathbf{tr}[\![a_i : \kappa_i]\!] \quad \forall i.1 < i < |\vec{a}|$$
$$(x : \kappa_x, \vec{a'}) \overset{recons}{\Longrightarrow} x' : \kappa'_x$$
$$\mathbf{tr}[\![x : try(t, \kappa_1, \kappa_2)]\!] = \begin{cases} \mathbf{tr}[\![x' : \kappa_1]\!] & \text{if } t(x, x') \\ \mathbf{tr}[\![x : \kappa_2]\!] & \text{otherwise} \end{cases}$$
$$\mathbf{tr}[\![x_1 \ldots x_n : seq(\kappa)]\!] = (\mathbf{tr}[\![x_1 : \kappa]\!] \ldots \mathbf{tr}[\![x_n : \kappa]\!])$$

Fig. 1. The Transformation Algorithm

ends for a term when the *final* kind is found. The transformation is driven by rules (defined in *compilation modules*). Note that the rules may contain guards in order to make them conditional on the term. Rule $x : \kappa \Longrightarrow x' : \kappa'$ denotes that when a term x of kind κ is found, it is replaced by x' of kind κ'. Rule $\kappa \succ \kappa'$ is the same, but the term is unmodified. Finally, rules $x : \kappa_x \overset{decons}{\Longrightarrow} \vec{a} : \vec{\kappa}$ and $(x : \kappa_x, \vec{a'}) \overset{recons}{\Longrightarrow} x' : \kappa'_x$ allow the deconstruction (*decons*) of a term into smaller parts, which are translated and then put together by reconstruction of the term (*recons*). Intuitively, this pair of rules allows performing a complex expansion that reuses other rules (which may be defined elsewhere). We will see examples of all these rules later. We divided expansions into finer-grained translations because we want to be able to combine them and to allow them to be interleaved with other rules in such combinations. Monolithic expansions would render their combination infeasible in many cases.

Additionally, there are some rules for *special kinds*, which are provided here for programmer convenience, even if they can be defined in terms of the previous rules. Their meaning is the following: the $try(t, \kappa_1, \kappa_2)$ kind tries to transform the input with the relation t. If it is possible, the resulting term is transformed with kind κ_1. Otherwise, the untransformed input is retried with kind κ_2. This is useful to compose translations. The $seq(\kappa)$ kind indicates that the input term is a sequence of elements of kind κ.[2]

Composition of Transformations. Note that the transformation algorithm does not make any assumption regarding the order in which rules are defined in the program, given that the rules define a fixed order relation between kinds. We will see in Section 5 how to give an unambiguous meaning to conflicting rules targeting the same kind.

---

[2] In the Prolog implementation sequences are replaced by lists.

Example 1 (Combining Transformations). Consider the example about merging CHR and *functional syntax* presented in the introduction. It can be solved in our framework by introducing rules such as:

$$(a \setminus b <=> c) : chrclause_1 \overset{\text{decons}}{\Longrightarrow} (a\ b\ c) : (goal_1\ goal_1\ goal_1)$$

$$(\_ : chrclause_1, (a'\ b'\ c')) \overset{\text{recons}}{\Longrightarrow} (a' \setminus b' <=> c') : chrclause_2$$

Those rules expose the internal structure of some constructs to allow the cooperation between translations. That is, those rules mean that in the middle of the translation from the kinds $chrclause_1$ and $chrclause_2$ we allow treatment of a kind $goal_1$, which could be treated, e.g., by the functional syntax package. Note that neither the CHR nor the functional package are required to know about the existence of each other.

3   Integration in the Compilation Model

In our compilation model programs are separated into modules. Modules can import and use code from other modules. Additionally, modules may load language extensions through special *compilation modules*. For the sake of simplicity, we will show here the compilation passes required for a simplified language with exported symbols (e.g., predicates) and imported modules. Extending it to support more features is straightforward.

We assume that compilation is performed on a single module at a time, in two phases.[3] Let us also assume that each phase reads and writes a series of initial (sources), temporal, and final compilation results (linkable/executable binaries). We will call those elements *nodes*, since clearly there is a dependency relation between them. In practice, nodes are read and written from a (persistent) memory, that we will abstract here as a mapping V. We denote as $V(n)$ the value of a node. We denote as $V(n) \leftarrow v$ the destructive assignment of v to n in V, and $V(n)$ the value of n in V.

Given a module i, the first phase (strPass) performs the source code (denoted for conciseness as $src[i]$) parsing and additional processing to obtain the *abstract syntax tree* ($ast[i]$) and module interface ($itf[i]$). In order to extend the compilation, we introduce a call to strTr. This will be defined herein in terms of the translation algorithm $\mathbf{tr}[\![\ \cdot : \tau]\!]$ (Fig. 1), working on the program definitions. We will see actual example definitions for them in Section 5. We call this the *structural* pass, since we can change arbitrarily the structure of the syntax tree, but we are not yet able to fully provide a semantics for the program, which may depend on external definitions from imported modules.[4] Indeed, the information about local definitions (e.g., defined predicates) and the module *interface*

---

[3] This is common in many languages, since at least two passes are required to allow identifiers in arbitrary order in the program text.

[4] It is important not to confuse *importing* a module with *including* a file. The latter is purely syntactic and can be performed during program reading. For the sake of clarity, we omit dependencies to included files in further sections.

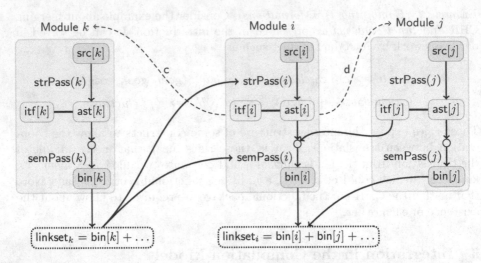

Fig. 2. Example of compilation dependencies for module i, which imports module j (d arrow), and requires compilation module k (c arrow)

(defined below) is available only at the end of this pass. Note that we load compilation modules dynamically during compilation. We will show later how this is done.

Once the first phase finishes, the module interfaces are available. In the second phase (semPass), the interface of the imported modules are collected and processed alongside with the module abstract syntax tree ($ast[i]$) and interface ($itf[i]$). The output of this phase (denoted as $bin[i]$), can be either in an executable form (e.g., bytecode), or in a suitable *kernel* language treatable by other tools (e.g., like program analysis). As in the previous phase, we introduce an extensible translation pass called semTr, similar to strTr. However, this time it can access the interface of imported modules. We name this the *semantic* pass. For loading compilation modules (or any other module) dynamically, we need to compute the *link-set* (the reflexive transitive closure of the *import* relation between modules), or the minimum set of modules required during execution of a module.

Compilation Order. In general, determining the order in which compilation must occur, and which recompilations have to take place whenever some source changes is not a straightforward task. For example, see Figure 2 which shows the dependencies for the incremental compilation of a module i depending on module j and compilation module k. We need an algorithm that automatically schedules compilation of modules (both program and translation modules) and which is incremental in order to reduce compilation times. Both of these requirements are necessary in a scalable, useful, dynamic environment. I.e., when developing, the user should not have to do anything special in order to ensure that all modules are compiled and up to date. However, since dependencies are dynamic we cannot (and would not want to) rely on traditional tools like `Makefile`s.

3.1 Incremental Compilation

We solve the problems of determining the compilation order, and making the algorithm incremental, with minor changes. The idea is to invoke the necessary compilation passes before each $V(n)$ is read, in order to access up-to-date values. For that, we define the UpdateNode(n) in Algorithm 1.

The algorithm works with *time-stamps*. We extend the V mapping with an (also persistent) mapping T between nodes and time-stamps, so that: $T(x) = \bot$ if the node does not have any value, and for each $V(n) \leftarrow v$, $T(n)$ is updated with a recent time-stamp. We need another mapping S, that relates a node with its *status*, and is non-persistent and monotonous during a *compilation session* (during which compilation of a set of modules takes place, with no allowed changes in source code). When a *compilation session* starts, we begin with the empty status for all nodes. Finally, we assume that for passes that produce more than one output node (e.g., the interface and the syntax tree), we can choose a fixed one of them as the *ruling node* (e.g., the interface). We denote by RulingNode(n) the ruling node of n.

---

data: mappings V, T, and S

```
1  def UpdateNode(n):
2      r = RulingNode(n) ; CheckNode(r)
3      if S(r) = invalid then
4          S(r) ← working
5          if GenNode(r) then S(r) ← valid(T(r)) else S(r) ← error

6  def CheckNode(r):
7      if S(r) = ⊥ then
8          if UpToDate(T(r), NodeInputs(r)) then S(r) ← valid(T(r))
9          else S(r) ← invalid

10 def UpToDate(t, oldin):
11     if t = ⊥ ∨ oldin = ⊥ then return False
12     foreach n_in ∈ oldin do
13         r_in = RulingNode(n_in) ; CheckNode(r_in)
14         if ¬((S(r_in) = valid(t_in)) ∧ (t_in ≤ t)) then return False
15     return True
```

Algorithm 1: UpdateNode

UpdateNode works by obtaining the ruling node, invoking CheckNode to update its status, and, depending on it, invoking GenNode to (re)generate the outputs if necessary. CheckNode (line 6) examines the node and updates its status.

If the node was visited before, the status will be different from $\perp$, and it will exit. If not, it will check that r is up to date (UpToDate($t, oldin$), line 10) w.r.t. all the dynamic input dependencies ($oldin$ = NodeInputs(r)). In our case, for strPass(i) the input nodes are src[i] and the *link-set* of all compilation modules specified in the (old) itf[i]. For semPass(i) the input nodes are itf[i] and itf[j], for each imported module j specified in itf[i], in addition to the nodes for compilation modules. The input nodes are $\perp$ if it was not possible to obtain them (i.e., no itf[i] is found). If the node is up-to-date, its status is marked as valid(t), indicating that it needs no recompilation. If not, it is marked as invalid. This may mark the status of other dependent nodes, but no action is performed on them.

For terminal nodes (e.g., source code src[i] for some module i), GenNode(r) will simply check that the node r exists, and NodeInputs(r) is empty. CheckNode will mark existing terminal nodes as *valid*. Non-existing nodes will be marked as *invalid*, and later UpdateNode will try to generate them. Since they do not exist, they will be marked as *error*. For computable nodes, GenNode(r) invokes the compilation pass that generates the corresponding output ruling node (based on static output dependencies, i.e., strPass(i) generates itf[i], semPass(i) generates bin[i]). If compilation was successful, the status is updated with valid($T(r)$) (indicating that it was successfully generated within this *compilation session*). On error, error is used as mark. An additional working status is used to mark nodes being computed and detect compilation loops (i.e., compilation modules depending on themselves). Note that for nodes whose value is assumed to be always up to date (*frozen nodes*, e.g., precompiled system libraries or static modules that cannot be updated) we make $S(n)$ = valid(0) by definition (denoting the oldest possible time-stamp).

Correctness and Optimality of Time-Stamp Approximation. The algorithm is based on, given a node, knowing if it needs to be recomputed. Based on the fact that each compilation pass only depends on its inputs, we can determine this by checking if the contents of a node have changed w.r.t. the contents used for the pass. For that, we could keep all the versions of each node, and number them in increasing order. Instead of local version numbers, we can use *time-stamps*, as a global version counter updated each time a node is written. This has the property that for each generated node n, $T(n) \geq T(m)$ for each m being an input dependency of n. If we can reason on time-stamps, then keeping the contents of each node version is unnecessary.[5] So if we find an input dependency with a time-stamp greater than $T(n)$, then it is possible that it may have changed. We may have false positives (time changed but the value is the same), which will result in more work than necessary, but not incorrect results. If the time-stamp is less or equal then we can be assured that it has not changed since n was generated. Unless time-stamps are artificially changed by hand, we will not have false negatives (whenever a node needs to be computed, it will be).

---

[5] When dealing with large dependencies, this seems inpractical, both in terms of time and space. We want this operation to be as fast as possible and not consume much additional space.

We only need to keep for each node its dependencies (the name of the nodes, not their value), or provide a way of inferring them from other stored values.[6]

Handling Compilation Module Loops. When a compilation module depends on modules that depend on it, a *deadlock* situation occurs. The compilation module cannot be compiled because it requires some modules that cannot be compiled yet. However, it is common to have languages that compile themselves. We solve the issue by distinguishing between normal and static modules. Static modules have been compiled previously and their bin[i] and itf[i] are kept for following compilations (say bin[i]^S and itf[i]^S respectively). In that case, (itf[i] = itf[i]^S ∧ bin[i] = bin[i]^S). The set of all static modules for the compiler constitutes the *bootstrap* system. Note that *self-compiling* modules require caution, since accidentally losing the bootstrap will make the source code useless (our source, only understood by our compilation module, may be written in a language for which there exists no working compiler).

Module Invariants and Extensions. Although the kernel language may provide low-level pathways if necessary (e.g., to implement debuggers, code inspection tools, or advanced transformation tools), it is important not to break the module *invariants*. One invariant is the module interface (itf[j]), which once computed cannot be changed without invalidating the compilation of any module i that imports it j. For this reason, a *semantic* expansion cannot modify the module interface.

4   Backward Compatibility

We now illustrate how the Ciao expansion primitives [1] can be easily emulated within the proposed approach. Ciao extensions are defined in special libraries called *packages*. They contain lexical and syntactic definitions (such as new operators), and hooks for language extension, defined in *compilation modules*. The available hooks can be seen as partial functional relations (or predicates that given an input have at least one solution) that translate different program parts: *term*, *sentence*, *clause*, and *goal* translations. For conciseness, we will denote them as $\mathcal{E}_t$, $\mathcal{E}_s$, $\mathcal{E}_c$, $\mathcal{E}_g \subseteq \mathcal{T} \times \mathcal{T}$, respectively. The transformations in a single package will be the tuple $\mathcal{E} = (\mathcal{E}_t, \mathcal{E}_s, \mathcal{E}_c, \mathcal{E}_g)$. We will denote with $\vec{\mathcal{E}} = (\mathcal{E}_1 \dots \mathcal{E}_n)$ all the transformation specifications that are local and used in module, and by $\vec{\mathcal{E}}_k$ the sequence of translations $(\mathcal{E}_{1_k} \dots \mathcal{E}_{n_k})$, for a particular $k \in \{t, s, c, g\}$. Fig. 3 shows the emulation of these translations.

The translations made during the strTr phase start with **tr**$[\![\cdot : sents]\!]$. A term of kind *sents* represents a sequence of sentences, that is translated as a *sents*$_{\vec{\mathcal{E}}_s}$. Subscripts are used here to represent families of *kinds*. The kind *sents*$_{ts}$ represents a sequence of sentences that require the translation *sent*$_{ts}$. The third rule

---

[6] That is of course not necessary for static dependencies (e.g., that each ast[i] depends on src[i]).

$$sents \succ sents_{\vec{\mathcal{E}}_s}$$

$$sents_{ts} \succ seq(sent_{ts})$$

$$sent_{[t|ts]} \succ try(t, sents_{ts}, sent_{ts})$$

$$sent_{[]} \succ term$$

$$term \succ term_{\vec{\mathcal{E}}_t}$$

$$term_{[t|ts]} \succ try(t, term_{ts}, term_{ts})$$

$$term_{[]} \succ rterm$$

$$x : rterm \overset{decons}{\Longrightarrow} args(x) : (term \ldots term)$$

$$(x : rterm, \vec{a}) \overset{recons}{\Longrightarrow} setArgs(x, \vec{a}) : final$$

$$clauses \succ seq(clause)$$

$$clause \succ clause_{\vec{\mathcal{E}}_c}$$

$$clause_{[t|ts]} \succ try(t, clause_{ts}, clause_{ts})$$

$$clause_{[]} \succ hb$$

$$(h\!:\!-b) : hb \overset{decons}{\Longrightarrow} (h\ b) : (head\ body)$$

$$(\_ : hb, (h\ b)) \overset{recons}{\Longrightarrow} (h\!:\!-b) : final$$

$$body \succ try(f, control, goal)$$

$$f(x, x) \equiv x \in \{ ,/2, ;/2, \ldots \}$$

$$x : control \overset{decons}{\Longrightarrow} args(x) : (body \ldots body)$$

$$(x : control, \vec{a}) \overset{recons}{\Longrightarrow} setArgs(x, \vec{a}) : final$$

$$goal \succ goal_{\vec{\mathcal{E}}_g}$$

$$goal_{[t|ts]} \succ try(t, body, goal_{ts})$$

$$goal_{[]} \succ resolv$$

Fig. 3. Emulating Ciao translation rules

indicates that a sentence of kind $sent_{[t|ts]}$ (we extract the first element of the list of transformations) will be transformed by t, yielding a term of kind $sents_{ts}$ (i.e., a sequence of sentences) on success.[7] In case of failure, the untransformed term will be treated as a $sent_{ts}$. In this way, all transformations in $\vec{\mathcal{E}}_s$ (i.e., all sentence_trans) will be applied. Once ts is empty, the result is translated as kind $term$, equivalent to $term_{\vec{\mathcal{E}}_t}$. Similarly to the previous case, all transformations in ts (i.e., all term_trans) are tried and removed from the list of pending transformations. When ts is empty, the datum is treated as an $rterm$, which divides the problem into the translation of arguments as kind $term$ and reuniting them as a final (non-suitable for further translations) result. Both transformations are applied in the same order as specified in Ciao.

The translations made during the semTr phase start with $\mathbf{tr}[\![\cdot : clauses]\!]$. Sequences of clauses are treated in a similar way as sentences, with the difference that the translation of a clause always returns one clause (not a sequence). When all translations in $\vec{\mathcal{E}}_c$ (all clause_trans) have been performed, the head and body are treated. In this figure, we do not show any successor for the $head$ kind, since this will be done in the following examples (we could add $head \succ final$ to mark the end of the translation). For $body$, we apply the same body translation on the arguments of control constructs (e.g. ,/2, ;/2, etc.). If we are not treating a control structure, the translations in $\vec{\mathcal{E}}_g$ are applied (all goal_trans). Note that the first kind in the try kind of goals is $goal$. In contrast with other translations,

---

[7] We assume that concatenation of sequences is implicit. We can adapt all the discussion to work with lists of sentences, but that would obscure the exposition.

when a goal translation is successfully applied, it is not removed from the list; all translations are tried again until none is applicable.[8] In such case, the term is translated as a *resolve* kind (for the same reason as for *head*, we leave it open for later translations).

Note the flexibility of the base framework: for instance, introducing changes in the expansion rules at fundamental levels can be done, even modularly.

Priority-Based Ordering of Transformations. The rules presented in this section establish a precise and fixed application order. However, when more than one sentence, term, clause, or goal translation is used in the same module the ordering among them also needs to be specified. The standard solution for this problem in Ciao is to use the order in which the packages which contain the expansion code are stated (e.g., in :- use_package([dcg, fsyntax]) the dcg transformations precede those of fsyntax. We propose an arguably better solution for this problem: to introduce a priority in each hook, so that all transformations in $\vec{\mathcal{E}}$ can be ordered beforehand. With this solution (now implemented in Ciao) directives such as :- use_package([dcg, fsyntax]) and :- use_package([fsyntax, dcg]) are fully equivalent, and both would apply the transformations in the right order. Of course, this moves the responsibility from the user of the extension to the extension developer. However, in practice this represents a huge advantage for users of packages.

5   Examples and Applications

We show the expressivity of the rules with fragments of two translations that deal with the module system and meta-predicates. Each of them is presented separately, but their combination results in a transformation equivalent to that hardwired in the Ciao compiler (and which was not expressible in the old transformation hooks). For the sake of clarity, we continue using the formal notation in all the following sections. Writing the Prolog equivalent of both the rules and the driver algorithm presented here is straightforward. As implemented in the Ciao compiler, Prolog terms can be used to represent the abstract syntax tree. The different stages of compilation can be kept in memory as facts in the dynamic database, with extra arguments to identify the module.

We indicate the current module as cm. We will assume that we have access to the information visible during the translation, such as parsed module code, declarations, interfaces, etc.

Example 2 (Predicate-based Module System). The following rules perform the module resolution and symbol replacement in all the clause goals to implement a predicate-based module system via a language extension. Instead of duplicating

---

[8] This preserves the semantics of the original translation hooks, where termination is up to the writer of the translation rules. Detecting those problems is out of the scope of this paper.

the logic to locate goal positions, the translations are *inserted* in the right place just after goal expansions are performed (Fig. 3).

We denote that a predicate symbol f is defined in the current module by $\mathsf{localdef}(f) \equiv \mathsf{defined}(f) \in \mathsf{ast}[\mathsf{cm}]$, and that f is exported by an imported module m by $\mathsf{importdef}(f, m) \equiv (\mathsf{exported}(f) \in \mathsf{itf}[m] \wedge \mathsf{imported}(m) \in \mathsf{ast}[\mathsf{cm}])$. Let $\mathsf{modlocal}(m, t)$ be a term operation that replaces the principal functor of term t by another one that is private to module m (e.g., by a special representation, not directly accessible from user code, that concatenates the name of the current module cm with the symbol). The translation of *head* transforms the term using that operation. The rule for *resolv* does the same, but uses the module obtained from lookup (that indicates where the predicate is defined).[9]

$$x : head \Longrightarrow \mathsf{modlocal}(\mathsf{cm}, x) : final$$

$$x : resolv \Longrightarrow \mathsf{modlocal}(m, x') : meta \qquad (\text{if } \mathsf{lookup}(x, m, x'))$$

$$x : resolv \Longrightarrow \mathsf{error}(\text{"module error"}) : final \qquad (\text{if } \neg\mathsf{lookup}(x, \_, \_))$$

$$\mathsf{lookup}(a, m, a') \equiv \begin{array}{l} (\neg\mathsf{qual}(a, \_, \_) \wedge a' = a \wedge \mathsf{localdef}(f) \wedge m = \mathsf{cm}) \vee \\ (\neg\mathsf{qual}(a, \_, \_) \wedge a' = a \wedge \mathsf{importdef}(f, m) \wedge m = \mathsf{cm}) \vee \\ (\mathsf{qual}(a, m, a') \wedge m = \mathsf{cm} \wedge \mathsf{localdef}(f)) \vee \\ (\mathsf{qual}(a, m, a') \wedge m \neq \mathsf{cm} \wedge \mathsf{importdef}(f, m)) \end{array}$$

$$\text{where } f = \mathsf{termFn}(a')$$

The complete specification is lengthy, but not more complicated. E.g., it would require more elaborate error handling, which checks for ambiguity on import (e.g., m in lookup must be unique, etc.).

Example 3 (Rules for Meta-predicates). Goals that call meta-predicates in Prolog require special handling of their arguments. We specify the translation of such goals though a kind *meta*. The translation rule decomposes the term into meta-arguments, each of kind $marg(\tau)$, where τ is the meta-type for the predicate, e.g., goal). Note that we assume that $\mathsf{ast}[\mathsf{cm}]$ includes a term $\mathsf{metaPred}(f, \vec{\tau})$ for each :- meta_predicate declaration. That relates the module-local symbol f of the predicate with each of the *meta-types* of the goal arguments. The translation of $marg(\tau)$ returns a pair of the transformed term and an optional goal. Then, the composition rule rebuilds the goal by placing the transformed terms as arguments, collecting the optional goals in front of it:

---

[9] $\mathsf{qual}(mg, m, g)$ is true iff the term mg is the qualification of term g with term m (e.g., lists:append([1],[2],[1,2])). We use it to avoid ambiguity with the colon symbol used elsewhere in rules.

$$g : meta \stackrel{decons}{\Longrightarrow} \vec{x} : \vec{\kappa} \quad \text{where}$$
$$\vec{x} = \text{args}(g)$$
$$\text{metaPred}(\text{termFn}(g), \vec{\tau}) \in \text{ast}[\text{cm}]$$
$$\kappa_i = marg(\tau_i) \quad \forall i.1 < i < |\vec{\tau}|$$
$$(g : meta, \vec{a}) \stackrel{recons}{\Longrightarrow} g' : final \quad \text{where}$$
$$a_i = (x_i, s_i) \quad \forall i.1 < i < l, \quad l = |\vec{a}|,$$
$$g' = \text{toConj}((s_1 \ s_2 \ \ldots \ s_l \ \text{setArgs}(g, \vec{x})))$$

The toConj function transforms the input sequence into a conjunction of literals. We list below the rules for arguments. In the cases where the arguments do not need any treatment, we use ϵ as the second element in the pair, which denotes the empty sequence. The case where the argument represents a goal, but is not known at compile time (e.g., x is a variable, or $x = qm : \_$, where qm is not an atom), is captured by needsRt(x). In such case the rule emits code that will perform an expansion at run time (which however may share code with those rules). Finally, if the argument represents a goal, we use a deconstruction rule to expose an argument of kind *body*, which once translated is put back in a pair, as required by $marg(\_)$.[10]

$$x : marg(\tau) \Longrightarrow (x, \epsilon) : final \qquad\qquad (\text{if } \tau \neq \text{goal})$$
$$x : marg(\tau) \Longrightarrow (x', (\text{rtexp}(x, \tau, \text{cm}, x'))) : final \quad (\text{if } \tau = \text{goal} \wedge \text{needsRt}(x))$$
$$\text{where } x' \text{ is a new variable}$$
$$x : marg(\tau) \stackrel{decons}{\Longrightarrow} x : body \qquad\qquad (\text{if } \tau = \text{goal} \wedge \neg\text{needsRt}(x))$$
$$(\_ : marg(\tau), \ x) \stackrel{recons}{\Longrightarrow} (x, \epsilon) : final$$

Example 4 (Combined Transformation). The previous transformations can be combined to translate goals involving meta-predicate calls into plain module-qualified goals. The rules defined in this section and in Section 4 can be used to transform the input goal:

$$G = \texttt{findall(X,member(X,[1,2,3]),Xs)}$$

as G' by evaluating $\mathbf{tr}[\![G : goal]\!]$ so that:

$$G' = \texttt{'aggregates:findall'(X,'lists:member'(X,[1,2,3]),Xs)}$$

assuming that `findall/3` is imported from the module `aggregates`, that `member/2` is imported from `lists`, and that the meta-predicate declaration of `findall/3` specifies that its second argument is a goal.

---

[10] This allows applying rules treating *bodies*, such as symbol renaming for the module system.

6   Related Work

In addition to the classic examples for imperative languages, such as the C pre-processor, or more semantic approaches like C++ *templates* and Java *generics*, much work has been carried out in the field of extensible syntax and semantics in the context of functional programming. Modern template systems such as the one implemented in the Glasgow Haskell compiler [14] generally provide syntax extension mechanisms in addition to static metaprogramming. The Objective Caml preprocessor, Camlp4 [4], provides similar features but focuses more on the syntax extension aspects. Both systems allow the combination of different syntax within the host language by using explicit mechanisms of quotations/antiquotations.

Another elegant approach consists on defining language extensions based on interpreters. In [11] a methodology for building domain-specific languages is shown, which combines the use of modular monad interpreters with a partial evaluation stage to reduce or eliminate the interpretation overhead. Although this approach provides a clean semantics for the extension, it has the disadvantage of requiring the (not always automatable) partial evaluation phase for efficiency, and its integration with the rest of the language and with the compilation architecture is more complex.

Another solution explored has been to expose the abstract syntax tree, through a reasonable interface, to the extensions. Racket (formerly PLT Scheme) [7] has an open macro system providing a flexible mechanism for writing language extensions. It allowed the design of domain-specific languages (including syntax), but also language features such as, e.g., the class and component systems, which in Racket are written using this framework. To the extent of our knowledge, there is no formal description of the framework nor whether and how multiple language extensions interact when specified simultaneously. However, it is interesting to note that despite growing independently, Ciao and Racket, both dynamic languages, have developed similar ideas, like separation of compile-time and run-time affairs and the necessity of expansions at different phases.

Finally, extensibility has also been achieved by making use of rewriting rules. For instance, by mixing such features with compilation inlining, the Glasgow Haskell compiler provides a powerful tool for purely functional code optimization [12]. It seems however that the result of the application of such rules can quickly become unpredictable [6]. In the context of constraint programming, a successful language transformation tool is Cadmium [5], which compiles solver-independent constraint models.

7   Conclusions

We have described an extensible compilation framework for dynamic programming languages that is amenable to performing separate, incremental compilation. Extensibility is ensured by a language of rewrite rules, defined in *pluggable* compilation modules. Although the work is mainly focused on Prolog-like languages, most of the presentation deals with common concepts (modules, interfaces, declarations, identifiers), and thus we believe that it can be adapted to other paradigms with minor effort.

In general, the availability of a rich and expressive extension system is a large asset for language design. One obvious advantage is that it helps accommodate the programmer's need for syntactic sugar, while keeping changes in the kernel language at a minimum. It also offers benefits for portability, since it makes it possible to keep a common front end (or a set of language features) and *plug in* different kernel engines (e.g., Prolog systems) at the back end, as long as they provide access to the same kernel language (or one that is rich enough) [15].

Beyond the obvious usefulness of the framework as a separation of concerns during the design of extensions (the support for extension composition and separate compilation, etc.), the translation rules can also be seen as a complementary specification mechanism for the language features designed. If such rules are succinct and clear enough, which is not that hard in practice, they can actually be exposed to programmers alongside standard documentation. We plan to modify the lpdoc tool [9] to provide support for this.

We believe that the model proposed makes it easier to provide unambiguous, composable specifications of language extensions, that should not only make reasoning about correctness easier, but also avoid causing and propagating erroneous language design decisions (such as, e.g., unintended compilation dependencies between modules that would ruin any *parallel* compilation or analysis efforts) that are normally hard to detect and correct. We also hope that our contribution will contribute, in the context of logic programming, towards setting a basis for interoperability and portability of language extensions among different systems.

Acknowledgments. This work was funded in part by EU project IST-215483 *SCUBE*, MICINN project TIN-2008-05624 *DOVES*, and CAM project S2009TIC-1465 *PROMETIDOS*.

References

1. Cabeza, D., Hermenegildo, M.V.: A New Module System for Prolog. In: Palamidessi, C., Moniz Pereira, L., Lloyd, J.W., Dahl, V., Furbach, U., Kerber, M., Lau, K.-K., Sagiv, Y., Stuckey, P.J. (eds.) CL 2000. LNCS (LNAI), vol. 1861, pp. 131–148. Springer, Heidelberg (2000)
2. Cardelli, L.: Program fragments, linking, and modularization. In: POPL, pp. 266–277 (1997)
3. Casas, A., Cabeza, D., Hermenegildo, M.V.: A Syntactic Approach to Combining Functional Notation, Lazy Evaluation, and Higher-Order in LP Systems. In: Hagiya, M. (ed.) FLOPS 2006. LNCS, vol. 3945, pp. 146–162. Springer, Heidelberg (2006)
4. de Rauglaudre, D., Pouillard, N.: Camlp4,
   http://brion.inria.fr/gallium/index.php/Camlp4
5. Duck, G.J., De Koninck, L., Stuckey, P.J.: Cadmium: An Implementation of ACD Term Rewriting. In: Garcia de la Banda, M., Pontelli, E. (eds.) ICLP 2008. LNCS, vol. 5366, pp. 531–545. Springer, Heidelberg (2008)

6. Elliott, C., Finne, S., de Moor, O.: Compiling embedded languages. J. Funct. Program. 13(3), 455–481 (2003)
7. Flatt, M.: PLT: Reference: Racket. Tech. Rep. PLT-TR-2010-1, PLT Inc. (2010), http://racket-lang.org/tr1/
8. Frühwirth, T.: Constraint Handling Rules. Cambridge University Press (August 2009)
9. Hermenegildo, M.V.: A Documentation Generator for (C)LP Systems. In: Palamidessi, C., Moniz Pereira, L., Lloyd, J.W., Dahl, V., Furbach, U., Kerber, M., Lau, K.-K., Sagiv, Y., Stuckey, P.J. (eds.) CL 2000. LNCS (LNAI), vol. 1861, pp. 1345–1361. Springer, Heidelberg (2000)
10. Hermenegildo, M.V., Bueno, F., Carro, M., López, P., Mera, E., Morales, J., Puebla, G.: An Overview of Ciao and its Design Philosophy. Theory and Practice of Logic Programming (2011), http://arxiv.org/abs/1102.5497
11. Hudak, P.: Modular domain specific languages and tools. In: Proceedings of Fifth International Conference on Software Reuse, pp. 134–142. IEEE Computer Society Press (1998)
12. Jones, S.P., Tolmach, A., Hoare, T.: Playing by the rules: rewriting as a practical optimisation technique in GHC. In: Haskell Workshop (2001)
13. Schrijvers, T., Santos Costa, V., Wielemaker, J., Demoen, B.: Towards Typed Prolog. In: Garcia de la Banda, M., Pontelli, E. (eds.) ICLP 2008. LNCS, vol. 5366, pp. 693–697. Springer, Heidelberg (2008)
14. Sheard, T., Jones, S.L.P.: Template meta-programming for haskell. In: Haskell Workshop, vol. 37(12), pp. 60–75 (2002)
15. Wielemaker, J., Santos-Costa, V.: On the Portability of Prolog Applications. In: Rocha, R., Launchbury, J. (eds.) PADL 2011. LNCS, vol. 6539, pp. 69–83. Springer, Heidelberg (2011)

Meta-predicate Semantics

Paulo Moura

Dep. of Computer Science, University of Beira Interior, Portugal
Center for Research in Advanced Computing Systems, INESC–TEC, Portugal
pmoura@di.ubi.pt

Abstract. We describe and compare design choices for meta-predicate semantics, as found in representative Prolog predicate-based module systems and in Logtalk. We look at the consequences of these design choices from a pragmatic perspective, discussing explicit qualification semantics, computational reflection support, expressiveness of meta-predicate directives, meta-predicate definitions safety, portability of meta-predicate definitions, and meta-predicate performance. We also describe how to extend the usefulness of meta-predicate definitions. Our aim is to provide useful insights to discuss meta-predicate semantics and portability issues based on actual implementations and common usage patterns.

Keywords: meta-predicates, predicate-based module systems, objects.

1 Introduction

Prolog and Logtalk [1,2] meta-predicates are predicates with one or more arguments that are either goals or closures[1] used for constructing goals, which are called in the body of a predicate clause. Common examples are all-solutions meta-predicates such as setof/3[2] and list mapping predicates. Prolog implementations may also classify predicates as meta-predicates whenever the predicate arguments need to be module-aware. Examples include built-in database predicates, such as assertz/1 and retract/1, and built-in reflection predicates, such as current_predicate/1 and predicate_property/2.

Meta-predicates provide a mechanism for reusing programming patterns. By encapsulating meta-predicate definitions in library modules or library objects, exported and public meta-predicates allow client modules or client objects to reuse these patterns, customized by calls to local predicates.

In order to compare meta-predicate semantics, as found in representative Prolog predicate-based module systems and in Logtalk, a number of design choices can be considered. These include explicit qualification semantics, computational reflection support, expressiveness of meta-predicate directives, safety of meta-predicate definitions, portability of meta-predicate definitions, and meta-predicate performance.

---

[1] In Prolog and Logtalk, a closure is defined as a callable term used to construct a goal by appending one or more additional arguments.

[2] Following common practice, predicates are referenced by their *predicate indicators*, i.e. by compound terms with the format *Functor/Arity*.

G. Vidal (Ed.): LOPSTR 2011, LNCS 7225, pp. 155–172, 2012.

When discussing meta-predicate semantics, it is useful to define the contexts where a meta-predicate is *defined*, *called*, and *executed*. The following definitions extend those found on [3] and will be used in this paper:

Definition Context. This is the object or module containing the meta-predicate definition.

Calling Context. This is the object or module from which a meta-predicate is called. This can be the object or module where the meta-predicate is defined in the case of a local call or another object or module assuming that the meta-predicate is within scope.

Execution Context. This includes both the calling context and the definition context. It is comprised by all the information required by the language runtime to correctly execute a meta-predicate call.

In this paper, we make use of an additional definition:

Lookup Context. This is the object or module where we start looking for the meta-predicate definition (note that the definition can always be reexported from another module or inherited from another object).

This paper is organized as follows. Section 2 describes meta-predicate directives. Section 3 discusses the consequences of using explicit qualified meta-predicate calls and the transparency of control constructs when using explicit qualification. Section 4 describes computational reflection support for meta-predicates. Section 5 describes a set of compilation rules aimed to prevent the use of meta-predicates to break module or object encapsulation. Section 6 discusses the portability of meta-predicate directives and meta-predicate definitions. Section 7 describes how lambda expressions can be used to extend the usefulness of meta-predicate definitions. Section 8 presents some remarks on meta-predicate performance. Section 9 summarizes our conclusions and discusses future work.

2   Meta-predicate Directives

Meta-predicate directives are required for proper compilation of meta-predicates in both Logtalk and Prolog predicate-based module systems in order to avoid forcing the programmer to explicitly qualify all meta-arguments. Meta-predicate directives are also useful for compilers to optimize meta-predicate calls (e.g. when using lambda expressions as meta-arguments) and to be able to check meta-predicate calls for errors (e.g. using a non-callable term in place of a meta-argument) and potential errors (e.g. arity mismatches when working with closures). The design choices behind the current variations of meta-predicate directives translate to different trade-offs between simplicity and expressiveness. The meta-predicate template information declared via meta-predicate directives can usually be programmatically retrieved using built-in reflection predicates such as `predicate_property/2`, as we will discuss in Section 4.

2.1   The ISO Prolog Standard metapredicate/1 Directive

The ISO Prolog standard for Modules [4] specifies a metapredicate/1 directive that allows us to describe which meta-predicate arguments are normal arguments and which are meta-arguments using a predicate template. In this template, the atom * represents a normal argument while the atom : represents a meta-argument. We are not aware of any Prolog module system implementing this directive. The standard does allow for alternative meta-predicate directives, providing solely as an example a meta/1 directive that takes a predicate indicator as argument. This alternative directive is similar to the tool/2 and module_transparent/1 directives discussed below. However, from the point-of-view of standardization and code portability, allowing for alternative directives is harmful, not helpful.

2.2   The Prolog meta_predicate/1 Directive

The ISO Prolog specification of a meta-predicate directive suffers from one major shortcoming [5]: it doesn't distinguish between goals and closures. The de facto standard solution for specifying closures is to use a non-negative integer representing the required number of additional arguments.[3] By interpreting a goal as a closure requiring zero additional arguments, we can reserve the atom : to represent arguments that need to be module-aware without necessarily referring to a predicate. This convention is found in recent B-Prolog, GNU Prolog, Qu-Prolog, SICStus Prolog, SWI-Prolog, and YAP versions and is being adopted by XSB and other Prolog compilers. In Prolog compilers without a module system, or with a module system where module expansion only needs to distinguish between normal arguments and meta-arguments, using an integer for representing closures can be useful for cross-reference tools and allows portable modularization extensions (such as Logtalk) to properly parse calls to proprietary built-in meta-predicates.

Despite being able to specify closure meta-arguments, there is still a known representation shortcoming. Some predicates accept a list of options where one or more options are module-aware. For example, the third argument of the predicate thread_create/3 [7] is a list of options that can include an at_exit/1 option. This option specifies a goal to be executed when a thread terminates. In this case, the argument is not a meta-argument but may *contain* a sub-term that will be used as a meta-argument. Although we could devise (a most likely cumbersome) syntax for these cases, the elegant solution for this representation problem is provided by the tool/2 and module_transparent/1 directives discussed below.

A minor limitation with the ISO Prolog metapredicate/1 directive, which is solved by the meta_predicate/1 directive, is the representation of the instantiation mode of the normal arguments. For representing the instantiation mode of

---

[3] This notation was first introduced on Quintus Prolog [6] in order to support meta-qualification and cross-referencing tools.

normal arguments, the atoms +, ?, @, and - are commonly used,[4] as specified in the ISO Prolog standard [8]. However, using mode indicators in `meta_predicate/1` directives is no replacement for a *mode* directive. Consider the following two `meta_predicate/1` directives for the standard `once/1` meta-predicate and the de facto standard `forall/2` meta-predicate:

```
:- meta_predicate(forall(0, 0)).
:- meta_predicate(once(0)).
```

For `forall/2`, 0 means @.[5] For `once/1`, 0 means +.[6] Thus, using mode indicators in meta-predicate directives is inherently ambiguous (but still common practice).

2.3   The Logtalk `meta_predicate/1` Directive

Logtalk uses a `meta_predicate/1` directive, based on the Prolog `meta_predicate/1` directive described above, extended with meta-predicate mode indicators for representing a predicate indicator, (/), a list of predicate indicators, [/], a list of goals, [0], and an existentially qualified goal, ^.[7] In addition, the atom : is replaced by :: for consistency with the message sending operator. Logtalk uses this information to verify meta-predicate definitions, as discussed in [3]. As Logtalk supports a `mode/2` predicate directive[8] for specifying the instantiation mode and the type of predicate arguments (plus the predicate determinism), the atom * is used to represent normal arguments in `meta_predicate/1` directives.

The extended set of meta-predicate mode indicators allows Logtalk to specify accurate meta-predicate templates for virtually all proprietary built-in meta-predicates found on all compatible Prolog compilers. This allows Logtalk to cope with the absence, limitations, differences, and sometimes ambiguity of meta-predicate templates in those Prolog compilers. Unfortunately, some Prolog compilers still don't implement the `meta_predicate/1` predicate property, while some other Prolog compilers return ambiguous meta-predicate templates due to the use of the : meta-predicate mode indicator for any kind of meta-argument.

2.4   The Ciao Prolog `meta_predicate/1` Directive

Ciao Prolog uses a `meta_predicate/1` directive that supports an extensive set of meta-predicate mode indicators [9] that, although apparently not adopted

---

[4] The meaning of these mode indicators atoms is as follows: + – argument must be instantiated; - – argument must be a variable; ? – argument can be either instantiated or a variable; @ – argument will not be (further) instantiated.

[5] The `forall/2` meta-predicate implements a generate and test loop using negation. Thus, no variable bindings are returned when calling it.

[6] The `once/1` meta-predicate proves its argument, possibly further instantiating it, and committing to the first solution found.

[7] This meta-predicate mode indicator was originally suggested by Jan Wielemaker and first implemented on SWI-Prolog 5.11.25. It is useful when defining wrappers for the `bagof/3` and `setof/3` built-in meta-predicates whenever the goal argument may use the ^/2 existential quantifier.

[8] http://logtalk.org/manuals/refman/directives/mode2.html

elsewhere, subsumes in expressive power the sets of meta-predicate mode indicators found on other Prolog compilers and in Logtalk. For example, it is possible to specify that a meta-argument should be a clause or, more specifically, a fact, using the mode indicators `clause` and `fact`. Moreover, a `list(Meta)` mode indicator, where `Meta` is itself a mode indicator, allows easy specification of lists of e.g. goals, predicate-indicators, or clauses.

2.5   The `tool/2` and `module_transparent/1` Directives

An alternative to the `meta_predicate/1` directive, found in ECLiPSe [10] and SWI-Prolog [11], is to declare meta-predicates as *module transparent*, forgoing the specification of which arguments are normal arguments and which arguments are meta-arguments. For this purpose, ECLiPSe provides a `tool/2` directive while SWI-Prolog provides a (apparently deprecated) `module_transparent/1` directive. These directives take predicate indicators as arguments and thus support a simpler, user-friendlier, solution when compared with the `meta_predicate/1` directive. More important, these directives allow the definition and encapsulation of meta-predicate definitions that cannot be (unambiguously) expressed using the alternative `meta_predicate/1` directive. Consider the following example, adapted from AutoBayes[9], an open-source NASA application:

```
cases(Pattern, [(Pattern -> Action)| _]) :-
    !,
    once(Action).
cases(Pattern, [_ | Cases]) :-
    cases(Pattern, Cases).
```

The `cases/2` meta-predicate is described as implementing a C-style pattern matching switch. The AutoBayes application is coded in plain Prolog. Thus, in the absence of an encapsulation mechanism, no meta-predicate directive is required. But if we attempt to modularize this code, using either a Prolog predicate-based module system or Logtalk, the `meta_predicate/1` directive can only express that somewhere in the second argument of the `cases/2` predicate there is a sub-term that is a meta-argument:

```
:- meta_predicate(cases(*, ::)).    % using Logtalk syntax
```

The inherent ambiguity is that, in general, the body of a library meta-predicate definition can contain both meta-calls to local predicates and meta-calls to client predicates. In the absence of explicit qualification, a system needs to know, for each meta-call, if it is going to executed in the context of the *caller* or in the context of the *callee*. The ECLiPSe `tool/2` directive and the SWI-Prolog `module_transparent/1` directive can be interpreted as assuming that all meta-calls are to client predicates.[10] This is a strong but fair assumption, which allows

---

[9] http://ti.arc.nasa.gov/tech/rse/synthesis-projects-applications/autobayes/
[10] Given that meta-calls and meta-arguments can always be explicitly qualified, this assumption does not prevent the definition of meta-predicates that perform local meta-calls.

e.g. the encapsulation and reuse of the `cases/2` meta-predicate above. In fact, this assumption is true for most common meta-predicate definitions. However, we have shown in [3] that distinguishing between goals and closures and specifying the exact number of closure additional arguments is necessary to avoid misusing meta-predicate definitions to break module and object encapsulation.

3   Explicit Qualification Semantics

The semantics of explicit qualification is perhaps the most significant design decision on meta-predicate semantics. This section compares two different semantics, found on actual implementations, for the explicit qualification of meta-predicates and control constructs.

3.1   Explicit Qualification of Meta-predicate Calls

Given an explicit qualified meta-predicate call, we have two choices for the corresponding semantics:

1. The explicit qualification sets only the initial lookup context for the meta-predicate definition. Therefore, all meta-arguments that are not explicitly-qualified are called in the meta-predicate calling context.
2. The explicit qualification sets both the initial lookup context for the meta-predicate definition and the meta-predicate calling context. Therefore, all meta-arguments that are not explicitly-qualified are called in the meta-predicate lookup context (usually the same as the meta-predicate definition context).

These two choices for explicit qualification semantics are also described in the ISO Prolog standard for modules. This standard specifies a read-only flag, `colon_sets_calling_context`, which would allow a programmer to query the semantics of a particular module implementation.[11]

Logtalk and the ECLiPSe module system implement the first choice. Prolog module systems derived from the Quintus Prolog module system [6], including those found on SICStus Prolog, SWI-Prolog, and YAP, implement the second choice (the native XSB module system is atom-based, not predicate-based).

In order to illustrate the differences between the two choices above, consider the following example, running on Prolog module systems implementing the second choice. First, we define a meta-predicate library module:

```
:- module(library, [my_call/1]).      % library exports my_call/1

:- meta_predicate(my_call(0)).         % my_call/1 takes a goal
```

---

[11] The `colon_sets_calling_context` read-only flag also means that two Prolog implementations could be fully compliant with ISO Prolog modules standard and still meta-predicate definitions written for one implementation would not be usable in the other implementation.

```
my_call(Goal) :-                        % as meta-argument
    write('Calling: '), writeq(Goal), nl,
    call(Goal).

me(library).                            % me/1 is a local predicate
```

The my_call/1 meta-predicate simply prints a message before calling its argument (which is a goal, as declared in its meta-predicate directive). Second, we define a simple client module that imports and calls our meta-predicate using a local predicate, me/1, as its argument:

```
:- module(client, [test/1]).           % client exports test/1

:- use_module(library, [my_call/1]).   % import the meta-predicate

test(Me) :-                            % call the meta-predicate
    my_call(me(Me)).                   % using implicit qualification

me(client).                            % me/1 is a local predicate
```

To test our code, we use the following query:

```
| ?- client:test(Me).
Calling: client:me(_)
Me = client
yes
```

This query provides the expected result: the meta-predicate argument is called in the context of the client, not in the context of the meta-predicate definition. But consider the following seemingly innocuous changes to the client module:

```
:- module(client, [test/1]).

test(Me) :-                            % call the meta-predicate
    library:my_call(me(Me)).           % using explicit qualification

me(client).
```

In this second version, instead of importing the my_goal/1 meta-predicate, we use explicit qualification in order to call it. Repeating our test query now gives:[12]

```
| ?- client:test(Me).
Calling: library:me(_)
Me = library
yes
```

---

[12] The test could not be performed using Ciao Prolog, which reports a bad module qualification error in the explicit qualified call, complaining that the meta-predicate is not imported, despite the library module being loaded. Importing the predicate eliminates the error but also makes the interpretation of the test result ambiguous.

In order for a programmer to understand this result, he/she needs to be aware that the :/2 operator both calls a predicate in another module and changes the calling context of the predicate to that module. The first use is expected. The second use is not intuitive, is not useful, and often not properly documented. First, in other programming languages, the choice between implicitly-qualified calls and explicitly-qualified calls is one of typing convenience to the programmer, not one of semantics. Second, in the most common case where a client is reusing a library meta-predicate, the client wants to customize the meta-predicate call with its own local predicate. Different clients will customize the call to the library meta-predicate using different local predicates. In those cases where the meta-predicate is defined and used locally, explicit qualification is seldom necessary. We can, however, conclude that the meta-predicate definition still works as expected as the calling context is set to the library module. If we still want the me/1 predicate to be called in the context of the client module instead, we need to explicitly qualify the meta-argument by writing:

```
test(Me) :-
    library:my_call(client:me(Me)).
```

This is an awkward solution but it works as expected in the cases where explicit qualification is required. It should be noted, however, that the idea of the meta_predicate/1 directive is to avoid the need for explicit qualifications in the first place. But that requires using use_module/1-2 directives for importing the meta-predicates and implicit qualification when calling them. This explicit qualification of meta-arguments is not necessary in Logtalk or in the ECLiPSe module system, where explicit qualification of a meta-predicate call sets where to start looking for the meta-predicate definition, not where to look for the meta-arguments definitions.

The semantics of the :/2 operator in Prolog module systems (derived from the Quintus Prolog module system) is rooted in optimization goals.[13] When a directive use_module/1 is used, most (if not all) Prolog compilers require the definition of the imported module to be available, thus resolving the call at compilation time. However, that does not seem to be required when compiling an explicitly qualified module call. For example, using recent versions of SICStus Prolog, SWI-Prolog, and YAP, the following code compiles without errors or warnings (despite the fact that the module fictitious does not exist):

```
:- module(client, [test/1]).

test(X) :-
    fictitious:predicate(X).
```

Thus, in this case, the fictitious:predicate/1 call is resolved at runtime. In our example above with the explicit call to the my_call/1 meta-predicate, the

---

[13] The goal of the original Quintus Prolog module system, according to former developers at Quintus, was to design a system with zero overhead over plain Prolog.

implementation of the :/2 operator propagates the module prefix to the meta-arguments that are not explicitly qualified at runtime. This runtime propagation results in a performance penalty. Therefore, and not surprisingly, the use of explicit qualification is discouraged by the Prolog implementers. In fact, until recently, most Prolog implementations provided poor performance for :/2 calls even when the necessary module information was available at compile time.

Logtalk and ECLiPSe illustrate the first choice for the semantics of explicitly-qualified meta-predicate calls. Consequently, both systems provide the same semantics for implicitly and explicitly qualified meta-predicate calls. Consider the following objects, corresponding to a Logtalk version[14] of the Prolog module example used in the previous section:

```
:- object(library).

    :- public(my_call/1).
    :- meta_predicate(my_call(0)).
    my_call(Goal) :-
        write('Calling: '), writeq(Goal), nl,
        call(Goal),
        sender(Sender), write('Sender: '), writeq(Sender).

    me(library).

:- end_object.


:- object(client).

    :- public(test/1).
    test(Me) :-                          % call the meta-predicate
        library::my_call(me(Me)).        % using explicit qualification

    me(client).

:- end_object.
```

Our test query becomes:

```
| ?- client::test(Me).
Calling: me(_)
Sender:  client
Me = client.
yes
```

That is, meta-arguments are always called in the context of the meta-predicate call. Logtalk also implements common built-in meta-predicates such as call/1-N,

---

[14] We extend the definition of the my_call/1 meta-predicate to also print the *sender* of the my_call/1 *message* by using Logtalk's built-in predicate sender/1.

`\+/1`, `findall/3`, and `phrase/3` with the same semantics as user-defined meta-predicates. In order to avoid misinterpretations, these built-in meta-predicates are implemented as private predicates.[15] Thus, the following call is illegal and results in a permission error:

```
| ?- an_object::findall(T, g(T), L).
error(permission_error(access, private_predicate, findall(T,g(T),L)),
      an_object::findall(T, g(T), L),
      user)
```

The correct call would be:

```
| ?- findall(T, an_object::g(T), L).
```

3.2   Transparency of Control Constructs

One of the design choices regarding meta-predicate semantics is the transparency of control constructs to explicit qualification. The relevance of this topic is that most control constructs can also be regarded as meta-predicates. In fact, there is a lack of agreement in the Prolog community on which language elements are control constructs and which language elements are predicates. For the purposes of our discussion, we use the classification found on the ISO Prolog standard, which specifies the following control constructs: `call/1`, *conjunction*, *disjunction*, *if-then*, *if-then-else*, and `catch/3`. The standard also specifies `true/0`, `fail/0`, `!/0`, and `throw/1` as control constructs but none of these can be interpreted as a meta-predicate.

When a control construct is transparent to explicit qualification, the qualification propagates to all the control constructs arguments that are not explicitly qualified. For example, the following equivalences hold for most Prolog module systems[16] (left column) and Logtalk (right column):[17]

```
M:(A, B)        ⇔ (M:A, M:B)         O::(A, B)        ⇔ (O::A, O::B)
M:(A; B)        ⇔ (M:A; M:B)         O::(A; B)        ⇔ (O::A; O::B)
M:(A -> B; C) ⇔ (M:A -> M:B; M:C) O::(A -> B; C) ⇔ (O::A -> O::B; O::C)
```

In Prolog module systems where the `:/1` operator sets both the meta-predicate lookup context and the meta-arguments calling context, the above equivalences are consistent with the explicit qualification semantics of meta-predicates described in the previous section. For example:

---

[15] Logtalk supports *private*, *protected*, and *public* predicates. A predicate may also be *local* if no scope directive is present, making the predicate invisible to the built-in reflection predicates (`current_predicate/1` and `predicate_property/2`).

[16] Note, however, that some Prolog compilers, such as Ciao and ECLiPSe, don't support explicit qualification of control constructs.

[17] Although both columns seem similar, the `::/2` Logtalk operator is a *message-sending* operator whose semantics differ from the module `:/2` *explicit-qualification* operator.

```
M:findall(T, G, L) ⇔ findall(T, M:G, L)
M:assertz(A)       ⇔ assertz(M:A)
```

This is also true for user-defined meta-predicates. For the example presented in the previous section, the following equivalence holds:

```
library:my_call(me(Me)) ⇔ my_call(library:me(Me))
```

Thus, the different semantics of implicitly and explicitly qualified meta-predicate calls allows the semantics of explicitly qualified control constructs to be consistent with the semantics of explicitly qualified meta-predicate calls.

In Logtalk, where explicit qualification of meta-predicates calls only sets the lookup context, the semantics of control constructs are different: the above equivalences are handy, supported, and can be interpreted as a shorthand notation for sending a set of messages to the same object. ECLiPSe implements a simpler design choice, disallowing the above shorthands, and thus treating control constructs and meta-predicates uniformly. We can conclude that ensuring the same semantics for implicitly and explicitly qualified meta-predicate calls requires either disallowing explicit qualification of control constructs (as found on e.g. Ciao and ECLiPSe) or different semantics for explicitly qualified control constructs, and thus a clear distinction between control constructs and predicates.

4 Computational Reflection Support

Computational reflection allows us to perform computations about the *structure* and the *behavior* of an application. For meta-predicates, structural reflection allows us to find where the meta-predicate is defined and about the meta-predicate template, while behavioral reflection allows us to access the meta-predicate execution context. As described in Section 1, a meta-predicate execution context includes information about from where the meta-predicate is called. This is only meaningful, however, in the presence of a predicate encapsulation mechanism such as modules or objects. Access to the execution-context is usually not required for common user-level meta-predicate definitions but can be necessary when meta-predicates are used to extend sytem meta-call features. In Logtalk, full access to predicate execution context is provided by the sender/1, self/1, this/1, and parameter/2 built-in predicates. For Prolog compilers supporting predicate-based module systems, the following table provides an overview of the available reflection built-in predicates:

Prolog compiler	Built-in reflection predicates
Ciao 1.10	predicate_property/2 (in library prolog_sys)
ECLiPSe 6.1	get_flag/3
SICStus Prolog 4.2	predicate_property/2
SWI-Prolog 5.10.4	context_module/1, predicate_property/2, strip_module/3
YAP 6.2	context_module/1, predicate_property/2, strip_module/3

From this table we conclude that the most common built-in predicate is `predicate_property/2`. Together with the ECLiPSe `get_flag/3` and the SWI-Prolog and YAP `context_module/1` predicates, these built-ins only provide *structural* reflection. Specifically, information about the meta-predicate template and the definition context of the meta-predicate. SWI-Prolog and YAP are the only systems that provide *built-in* access to the meta-predicate calling context using the predicate `strip_module/3`. As a simple example of using this predicate consider the following module:

```
:- module(m, [mp/2]).

:- meta_predicate(mp(0, -)).
mp(Goal, Caller) :-
    strip_module(Goal, Caller, _),
    call(Goal).
```

After compiling and loading this module, the following queries illustrate both the functionality of the `strip_module/3` predicate and the consequences of explicit qualification of the meta-predicate call:

```
| ?- mp(true, Caller).
Caller = user
yes

| ?- m:mp(true, Caller).
Caller = m
yes
```

For Prolog compiler module systems descending from the Quintus Prolog module system, it is possible to access the meta-predicate calling context by looking into the implicit qualification of a meta-argument:

```
:- module(m, [mp/2]).

:- meta_predicate(mp(0, -)).
mp(Goal, Caller) :-
    Goal = Caller:_,
    call(Goal).
```

After compiling and loading this module, we can reproduce the results illustrated by the queries above for the SWI-Prolog/YAP version of this module. One possible caveat would be if the Prolog compiler fails to ensure that there is always a single qualifier for a goal. That is, that terms such as `M1:(M2:(M3:G))` are never generated internally when propagating module qualifications.

In the case of ECLiPSe, a built-in predicate for accessing the meta-predicate calling context is not necessary. The `tool/2` directive works by connecting a meta-predicate interface with its implementation, which is extended with an extra argument that carries the meta-predicate calling context:

```
:- module(m).

:- export(mp/2).                    % due to the tool/2 directive, the
:- tool(mp/2, mp/3).                % ECLiPSe runtime system passes the
mp(Goal, Caller, Caller) :-         % calling context of mp/2 in the
    call(Goal).                     % third argument of mp/3
```

After loading this module, repeating the above queries illustrates the difference in explicit qualification semantics between ECLiPSe and the other compilers:

```
[eclipse 16]: mp(true, Caller).
Caller = eclipse
Yes (0.00s cpu)

[eclipse 17]: m:mp(true, Caller).
Caller = eclipse
Yes (0.00s cpu)
```

Note that the module eclipse is the equivalent of the module user in other Prolog compilers.

5   Secure Meta-predicate Definitions

Meta-predicate definitions should not provide a mechanism for calling client predicates other than the ones intended by the meta-predicate calls. This, however, is mostly meaningful for languages such as Logtalk and for Prolog module systems, such as ECLiPSe and Ciao [12], that aim to enforce object and module predicate scope rules. The following set of compilation rules, discussed and illustrated in detail in [3], contribute to make meta-predicate definitions secure:

1. The meta-arguments of a meta-predicate clause head must be variables.
2. Meta-calls whose arguments are not variables appearing in meta-argument positions in the clause head must be compiled as calls to local predicates.
3. Meta-predicate closures must be used within a call/2-N built-in predicate call that complies with the corresponding meta-predicate directive.

These rules are implemented in Logtalk. For Prolog module systems whose design allows any module predicate to be called using explicit module qualification, these rules may be regarded as best practice for writing meta-predicates and thus useful for checking meta-predicate definitions for possible errors (e.g. as part of lint checkers). Note that the third compilation rule above requires a meta-predicate directive capable of representing the number of additional arguments taken by a closure. The reader is invited to consult [3] for full details.

6   Portability of Meta-predicate Definitions

The portability of meta-predicate definitions depends on three main factors: the use of implicit qualification when calling meta-predicates in order to avoid the different semantics for explicitly qualified calls discussed in Section 3, the portability of the meta-predicate directives, and the portability of the meta-call primitives used when implementing the meta-predicates. Other factors that may impact portability are the preprocessing solutions for improving meta-predicate performance, described in Section 8, and the mechanisms for computational reflection about meta-predicate definition and execution, discussed in Section 4.

6.1   The call/1-N Control Constructs

The call/1 control construct is specified in the ISO Prolog standard [8]. This control construct is implemented by virtually all Prolog compilers. The call/2-N control constructs[18], whose use is strongly recommended for meta-predicates working with closures [3], is included in the latest revision of the ISO Prolog Core standard. A growing number of Prolog compilers implement these control constructs but with different maximum values for N, which can raise some portability problems. Ideally, the call/1-N control constructs would support N up to the maximum predicate arity. That depends, however, on the design decisions of a Prolog compiler implementation. For the Prolog systems listed in the table below, only five out of twelve systems support a value of N up to the maximum predicate arity. From a pragmatic point-of-view, it is not common that user written code (but not necessarily user *generated* code) would require a large upper limit of N. Despite some lack of agreement, the only portability issue is Prolog compilers only supporting an arguably small value of N. The following table summarizes the implementations of the call/2-N control construct on selected Prolog compilers:

System	N	Notes
B-Prolog 7.4	10/65535	(interpreted/compiled i.e. maximum arity)
Ciao 1.10	255	(maximum arity using the hiord library)
CxProlog 0.95.0	9	—
ECLiPSe 6.1#68	255	(maximum arity)
GNU Prolog 1.3.1	11	—
JIProlog 3.0.2	5	—
K-Prolog 6.0.4	9	—
Qu-Prolog 8.12	9	—
SICStus Prolog 4.2	255	(maximum arity)
SWI-Prolog 5.10.4	8/1024	(interpreted/compiled i.e. maximum arity)
XSB 3.3	11	—
YAP 6.2	12	—

This table only lists *built-in* support for call/2-N control construct. While this control construct can be defined by the programmer using the built-in pred-

---

[18] A call(Closure, Arg1, ...) goal is true iff call(Goal) is true where Goal is constructed by appending Arg1, ... additional arguments to the arguments (if any) of the callable term Closure.

icate =../2 and an append/3 predicate, such definitions provide relative poor performance due to the construction and appending of temporary lists.

6.2   Specification of Closures and Instantiation Modes in Meta-predicate Directives

The main portability issue of meta-predicate directives is the use of non-negative integers to specify closures and the atoms used to specify the instantiation mode of normal arguments. Although the use of non-negative integers comes from Quintus Prolog, it was historically regarded as a way to provide information to cross-reference and documentation tools, with Prolog compilers accepting this notation only for backward-compatibility with existing code. Other Prolog compilers such as Ciao define alternative but incompatible syntaxes for specifying closures. There is also some variation in the atoms used for representing the instantiation modes of normal arguments. Some Prolog compilers use an extended set of atoms for documenting argument instantiation modes compared to the basic set (+, ?, @, and -) found in the ISO Prolog standard. It is therefore tempting to use these extended sets in meta-predicate directives, which will likely raise portability issues. Hopefully, recent Prolog standardization initiatives, specially the development of portable libraries, will lead to a de facto standard meta-predicate directive derived from the extended directive described in Section 2.2.

7   Extending Meta-predicate Definitions Usefulness

The usefulness of meta-predicate definitions can be extended by adding support for lambda expressions. In the same way meta-predicates avoid the repeated coding of common programming patterns, lambda expressions avoid the definition of auxiliary predicates whose sole purpose is to be used as arguments in meta-predicate calls. Consider the following example (using Logtalk lambda expression syntax[19]) where we compute the distance to the origin for each point in a list using a mapping meta-predicate:

```
| ?- meta::map([(X,Y),Z]>>(Z is sqrt(X*X+Y*Y)),[(1,4),(2,5),(8,3)],Ds).
Ds = [4.1231056256176606,5.3851648071345037,8.5440037453175304]
yes
```

Without lambda expressions, it would be necessary to define an auxiliary predicate to compute the distance from a point to the origin:

```
distance((X, Y), Distance) :-
    Distance is sqrt(X*X+Y*Y).

| ?- meta::map(distance, [(1,4),(2,5),(8,3)], Ds).
Ds = [4.1231056256176606,5.3851648071345037,8.5440037453175304]
yes
```

---

[19] http://logtalk.org/manuals/refman/grammar.html#grammar_lambdas

This example also illustrates an additional issue when using meta-predicates: the map/3 list mapping meta-predicate accepts as first argument a closure that is extended by *appending* two arguments. Thus, an existing predicate for calculating the distance, e.g. distance(X, Y, Distance), cannot not be used without writing an auxiliary predicate for the sole purpose of packing the first two arguments.

Native support for lambda expressions can be found in e.g. λProlog [13], Qu-Prolog, and Logtalk. For Prolog compilers supporting a module system, a library is available [14] that adds lambda expressions support. There is, however, a lack of community agreement on lambda expression syntax. But the main issue of taking advantage of lambda expressions is the performance penalty resulting from the runtime processing of lambda parameters. In the case of Logtalk, recent releases include a preprocessor for both meta-predicates and lambda expressions that eliminate the performance penalty when compared with hand-coded and optimized (non-meta-predicate) alternative solutions.

8   Meta-predicate Performance

Considering that meta-programming is often touted as a major feature of Prolog, the relative poor performance of meta-calls often drive programmers to avoid using meta-predicates in production code where performance is crucial. A common solution is to interpret meta-predicate definitions as high-level macros and to preprocess meta-predicate calls in order to replace them with calls to automatically generated auxiliary predicates whose definitions that do not contain meta-calls. This preprocessing is usually only performed on stable code as the auxiliary predicates often complicate debugging. The preprocessing code is often implemented in optional libraries, which can be found on Logtalk and several Prolog compilers such as ECLiPSe, SWI-Prolog, and YAP. Consider as an example adding 1 to every integer in the list [1..100000] using a simple recursive predicate, a mapping meta-predicate using a closure, and a mapping meta-predicate using a lambda expression, with and without preprocessing.[20] Using Logtalk 2.43.2 with YAP 6.3.0 we get (times in seconds):

	Non-optimized	Optimized
Recursive predicate	0.002	0.002
Mapping predicate with a closure	0.072	0.008
Mapping predicate with a lambda expression	0.119	0.004

Equivalent results are obtained using Prolog implementations of this example with meta-predicate and preprocessing libraries. A performance penalty of one order of magnitude is commonly observed when comparing hand-optimized code with meta-predicate alternatives without any preprocessing. The use of lambda

---

[20] The full source code of this example is available at http://trac.logtalk.org/browser/trunk/examples/lambdas (no preprocessing) and http://trac.logtalk.org/browser/trunk/examples/lambdas_compiled (using preprocessing).

expressions adds another source of performance penalty. As illustrated above, preprocessing both the meta-predicate calls and the lambda expressions closes the performance gap. But the preprocessing libraries require custom code for each meta-predicate. Thus, user-defined meta-predicates will fail to match the performance of library-supported meta-predicates unless the user also writes its own custom preprocessing code. A more generic solution for preprocessing meta-predicate definitions, based on more powerful compile time code analysis and partial evaluation techniques, is needed to make meta-predicate programming patterns more appealing for applications where performance is crucial.

9   Conclusions and Future Work

We presented and discussed a comprehensive set of meta-predicate design decisions based on current practice in Logtalk and in Prolog predicate-based module systems. An interesting result is that none of the two commonly implemented semantics for explicitly qualified calls provides an ideal solution that both meets user expectations and allows the distinction between meta-predicates and control constructs to be waived. By describing the consequences of these design decisions we provided useful insight to discuss meta-predicate semantics, often a difficult subject for inexperienced programmers and a source of misunderstandings when porting applications and discussing Prolog standardization. From the point-of-view of writing portable code (including portable libraries), the current state of meta-predicate syntax and semantics across Prolog compilers is still a challenge, despite recent community efforts. We hope that this paper contributes to a convergence of meta-predicate directive syntax, meta-predicate semantics, and meta-predicate related reflection built-in predicates among Prolog compilers. But the main obstacle to improving the de facto standardization of meta-predicate syntax and semantics is backwards compatibility. Understandably, most Prolog implementers are wary of making changes that would break compatibility with existing applications and upset long time users. Nevertheless, we recommend that Prolog module systems make the semantics of implicitly- and explicitly-qualified meta-predicate calls the same (as found in Logtalk and ECLiPSe) and forbid the explicit qualification of control constructs (as found in ECLiPSe and Ciao) and built-in meta-predicates (as found in Logtalk). These changes would contribute to simpler and more uniform semantics while avoid programming constructs with unclear meaning for novice programmers.

Future work will include comparing Logtalk and Prolog predicate-based module systems with Prolog atom-based module system and derived object-oriented extensions. The most prominent example of a Prolog compiler featuring an atom-based module system is XSB [15] (which is used in the implementation of the object-oriented extension Flora [16]). XSB does not support a meta-predicate directive. Explicit-qualification of meta-arguments is used whenever the atom-based semantics fail to provide the desired behavior for the implementation of a specific meta-predicate. Interestingly, although atom-based module systems appear to solve or avoid some the issues discussed along this paper, predicate-based module systems are the most common implementation choice. This may be due to historical reasons but a deep understanding of the pros and cons of atom-based systems, at both the conceptual and implementation levels, will be required to perform a detailed comparison with the better known predicate-based systems.

Acknowledgements. We are grateful to Joachim Schimpf, Ulrich Neumerkel, Jan Wielemaker, and Richard O'Keefe for their feedback on explicitly-qualified meta-predicate call semantics in predicate-based module systems. We thank also the anonymous reviewers for their informative comments. This work was partially supported by the LEAP (PTDC/EIA-CCO/112158/2009) research project.

References

1. Moura, P.: Logtalk – Design of an Object-Oriented Logic Programming Language. PhD thesis, Department of Computer Science, University of Beira Interior, Portugal (September 2003)
2. Moura, P.: Logtalk 2.43.2 User and Reference Manuals (October 2011)
3. Moura, P.: Secure Implementation of Meta-predicates. In: Gill, A., Swift, T. (eds.) PADL 2009. LNCS, vol. 5418, pp. 269–283. Springer, Heidelberg (2009)
4. ISO/IEC: International Standard ISO/IEC 13211-2 Information Technology — Programming Languages — Prolog — Part II: Modules. ISO/IEC (2000)
5. O'Keefe, R.: An Elementary Prolog Library,
   http://www.cs.otago.ac.nz/staffpriv/ok/pllib.html
6. Swedish Institute for Computer Science: Quintus Prolog User's Manual (Release 3.5). Swedish Institute for Computer Science (December 2003)
7. Moura, P. (ed.): ISO/IEC DTR 13211-5:2007 Prolog Multi-threading predicates,
   http://logtalk.org/plstd/threads.pdf
8. ISO/IEC: International Standard ISO/IEC 13211-1 Information Technology — Programming Languages — Prolog — Part I: General core. ISO/IEC (1995)
9. Bueno, F., Cabeza, D., Carro, M., Hermenegildo, M.V., López, P., Puebla, G.: Ciao Prolog System Manual
10. Cheadle, A.M., Harvey, W., Sadler, A.J., Schimpf, J., Shen, K., Wallace, M.G.: ECLiPSe: A tutorial introduction. Technical Report IC-Parc-03-1, IC-Parc, Imperial College, London (2003)
11. Wielemaker, J.: An overview of the SWI-Prolog programming environment. In: Mesnard, F., Serebenik, A. (eds.) Proceedings of the 13th International Workshop on Logic Programming Environments, Heverlee, Belgium, pp. 1–16. Katholieke Universiteit Leuven (December 2003); CW 371
12. Cabeza, D., Hermenegildo, M.V.: A New Module System for Prolog. In: Palamidessi, C., Moniz Pereira, L., Lloyd, J.W., Dahl, V., Furbach, U., Kerber, M., Lau, K.-K., Sagiv, Y., Stuckey, P.J. (eds.) CL 2000. LNCS (LNAI), vol. 1861, pp. 131–148. Springer, Heidelberg (2000)
13. Nadathur, G., Miller, D.: An Overview of λProlog. In: Fifth International Logic Programming Conference, Seattle, pp. 810–827. MIT Press (August 1988)
14. Neumerkel, U.: Lambdas in ISO Prolog,
   http://www.complang.tuwien.ac.at/ulrich/Prolog-inedit/ISO-Hiord
15. Group, T.X.R.: The XSB Programmer's Manual: version 3.3 (April 2011)
16. Yang, G., Kifer, M.: Flora-2: User's manual (2001)

A Strategy Language for Graph Rewriting

Maribel Fernández[1], Hélène Kirchner[2], and Olivier Namet[1]

[1] King's College London, Department of Informatics, London WC2R 2LS, UK
{maribel.fernandez,olivier.namet}@kcl.ac.uk
[2] Inria, Domaine de Voluceau - Rocquencourt B.P. 105 - 78153 Le Chesnay France
helene.kirchner@inria.fr

Abstract. We give a formal semantics for a graph-based programming language, where a program consists of a collection of graph rewriting rules, a user-defined strategy to control the application of rules, and an initial graph to be rewritten. The traditional operators found in strategy languages for term rewriting have been adapted to deal with the more general setting of graph rewriting, and some new constructs have been included in the language to deal with graph traversal and management of rewriting positions in the graph. This language is part of the graph transformation and visualisation environment PORGY.

Keywords: port graph, graph rewriting, strategies, visual environment.

1 Introduction

To model complex systems, graphical formalisms are often preferred to textual ones, since they make it easier to visualise a system and convey intuitions about it. The dynamics of the system can then be specified using graph rewriting rules.

Graph rewriting has solid logic, algebraic and categorical foundations [15,18], and graph transformations have many applications in specification, programming, and simulation tools [18]. In this paper, we focus on *port graph rewriting systems* [3], a general class of graph rewriting systems that have been successfully used to model biochemical systems and interaction net systems [27].

PORGY [2] is a visual environment that allows users to define port graphs and port graph rewriting rules, and to experiment with a rewriting system in an interactive way. To control the application of rewriting rules, PORGY uses a *strategy language*, which is the main subject of this paper.

Reduction strategies define which (sub)expression(s) should be selected for evaluation and which rule(s) should be applied (see [26,12] for general definitions). These choices affect fundamental properties of computations such as laziness, strictness, completeness, termination and efficiency, to name a few (see, e.g.,[43,41,28]). Used for a long time in λ-calculus [8], strategies are present in programming languages such as Clean [33], Curry [24], and Haskell [25] and can be explicitly defined to rewrite terms in languages such as ELAN [11], Stratego [42], Maude [29] or Tom [7]. They are also present in graph transformation tools such as PROGRES [39], AGG [19], Fujaba [32], GROOVE [37], GrGen [21]

G. Vidal (Ed.): LOPSTR 2011, LNCS 7225, pp. 173–188, 2012.

and GP [36]. PORGY's strategy language draws inspiration from these previous works, but a distinctive feature of PORGY's language is that it allows users to define strategies using not only operators to combine graph rewriting rules but also operators to define the location in the target graph where rules should, or should not, apply.

The PORGY environment is described in [2,31], and we refer the reader to [20] for programming examples. In this paper, we focus on the design of the strategy language. Our main contribution is a formal semantics for the language, which allows users to analyse programs and reason about computations. We formalise the concept of graph program, and present semantic rules that associate to each graph program a result set (according to a given strategy), which is an abstraction of the derivation tree.

Strategies are used to control PORGY's rewrite engine: users can create graph rewriting derivations and specify graph traversals using the language primitives to select rewriting rules and the position where the rules apply. Subgraphs can be selected as focusing positions for rewriting interactively (in a visual way), or intensionally (using a focusing expression). Alternatively, rewrite positions could be encoded in the rewrite rules using markers or conditions, as done in other languages based on graph rewriting which do not have focusing primitives. We prefer to separate the two notions of positions and rules to make programs more readable (the rewrite rules are not cluttered with encodings), and easier to maintain and adapt. In this sense, the language follows the separation of concerns principle [17]. For example, to change a traversal algorithm, it is sufficient to change the strategy and not the whole rewriting system.

The paper is organised as follows. In Section 2, we recall the concept of port graph. In Section 3, we present the syntax of the strategy language and formally define its semantics. Section 4 illustrates the language with examples, and Section 5 states some properties. Section 6 discusses related languages before concluding and giving directions for future work.

2   Background: Port Graph Rewriting

Several definitions of graph rewriting are available, using different kinds of graphs and rewriting rules (see, for instance, [14,22,9,35,10,27]). In this paper we consider port graph rewriting systems [1,3,4], of which interaction nets [27] are a particular case.

Port graphs. Intuitively, a port graph is a graph where nodes have explicit connection points called *ports*; edges are attached to ports. Nodes, ports and edges are labelled and have attributes.

A port may be associated to a state (e.g., active/inactive or principal/auxiliary); this is formalised using a mapping from ports to port states. Similarly, nodes can also have associated properties such as colour, shape, root, leaf, etc. These attributes may be used for visualisation purposes and are also essential for the definition of strategy constructs, such as the `Property` and `NextSuc` operators introduced in Section 3; they are later illustrated in examples.

As shown in [1], a port graph can be considered as a labelled graph, and conversely, any labelled graph is a port graph in which each (port graph) node has a number of ports equal to the arity of its label. As a consequence, expressivity results, computational power, as well as correctness and completeness results on labelled graph rewriting can be translated to port graph rewriting. This is the approach taken here for defining rewriting on port graphs.

Let G and H be two port graphs. A *port graph morphism* $f : G \to H$ maps elements of G to elements of H preserving sources and targets of edges, constant node names and associated port name sets and states, up to variable renaming. We say that G and H are *isomorphic* if f is bijective.

A *port graph rewrite rule* $L \Rightarrow R$ is itself represented as a port graph consisting of two port graphs L and R called the *left-* and *right-hand side* respectively, and one special node $\Rightarrow$, called *arrow node*. The arrow node describes the interface of the rule, avoiding dangling edges [22,14] during rewriting as follows. For each port p in L, to which corresponds a non-empty set of ports $\{p_1, \ldots, p_n\}$ in R, the arrow node has a unique port r and the incident directed edges (p, r) and (r, p_i), for all $i = 1, \ldots, n$; all ports from L that are deleted in R are connected to the *black hole* port of the arrow node. We refer to [1] for full details. When the correspondence between ports in the left- and right-hand sides of the rule is obvious we omit the ports and edges involving the arrow node (as in Figure 2 in Section 4).

Port Graph Rewriting. Let $L \Rightarrow R$ be a port graph rewrite rule and G a port graph such that there is an injective port graph morphism g from L to G; hence $g(L)$ is a subgraph of G. A *rewriting step* on G using $L \Rightarrow R$, written $G \to_{L \Rightarrow R} G'$, transforms G into a new graph G' obtained from G by replacing the subgraph $g(L)$ of G by $g(R)$, and connecting $g(R)$ to the rest of the graph as specified in the arrow node of the rule. We call $g(L)$ a redex, and say that G rewrites to G' using $L \Rightarrow R$ at the position defined by $g(L)$, or that G' is a *result* of applying $L \Rightarrow R$ on G at $g(L)$. Several injective morphisms g from L to G may exist (leading to different rewriting steps); they are computed as solutions of a *matching* problem from L to (a subgraph of) G. If there is no such injective morphism, we say that G is *irreducible* by $L \Rightarrow R$. Given a finite set $\mathcal{R}$ of rules, a port graph G *rewrites* to G', denoted by $G \to_{\mathcal{R}} G'$, if there is a rule r in $\mathcal{R}$ such that $G \to_r G'$. This induces a transitive relation on port graphs, denoted by $\to_{\mathcal{R}}^*$. Each *rule application* is a rewriting step and a *derivation*, or computation, is a sequence of rewriting steps. A port graph on which no rule is applicable is in *normal form*. Rewriting is intrinsically non-deterministic since it may be possible to rewrite several subgraphs of a port graph with different rules or use the same one at different places, possibly getting different results.

3   Strategy Language

In this section we introduce the concept of graph program and give the syntax and semantics of the strategy language. In addition to the well-known constructs to select rewrite rules, the strategy language provides focusing primitives to select or ban specific positions in the graph for rewriting. The latter is useful to program graph traversals for instance, and is a distinctive feature of the language.

3.1 Graph Programs

Definition 1. *A located graph G_P^Q consists of a port graph G and two distinguished subgraphs P and Q of G, called respectively the* position subgraph, *or simply* position, *and the* banned subgraph.

In a located graph G_P^Q, P represents the subgraph of G where rewriting steps may take place (i.e., P is the focus of the rewriting) and Q represents the subgraph of G where rewriting steps are forbidden. We give a precise definition below; the intuition is that subgraphs of G that overlap with P may be rewritten, if they are outside Q.

When applying a port graph rewrite rule, not only the underlying graph G but also the position and banned subgraphs may change. A *located port graph rewrite rule*, defined below, specifies two disjoint subgraphs M and N of the right-hand side R that are used to update the position and banned subgraphs, respectively. If M (resp. N) is not specified, R (resp. the empty graph $\emptyset$) is used as default. Below, the set operators union, intersection and complement (denoted respectively $\cup, \cap, \backslash$) apply to port graphs considered as sets of nodes, ports and edges.

Definition 2. *A located port graph rewrite rule is given by a port graph rewrite rule $L \Rightarrow R$ and two disjoint subgraphs M and N of R. It is denoted $(L \Rightarrow R)_M^N$. We write $G_P^Q \rightarrow_{(L \Rightarrow R)_M^N}^g G'_{P'}^{Q'}$, and say that the located graph G_P^Q rewrites to $G'_{P'}^{Q'}$ using $(L \Rightarrow R)_M^N$ at position P avoiding Q, if $G \rightarrow_{L \Rightarrow R} G'$ with a morphism g such that $g(L) \cap P \neq \emptyset$ and $g(L) \cap Q = \emptyset$; the new position subgraph P' and banned subgraph Q' are defined as $P' = (P \backslash g(L)) \cup g(M)$, $Q' = Q \cup g(N)$.*

In general, for a given rule $(L \Rightarrow R)_M^N$ and located graph G_P^Q, more than one morphism g, such that $g(L) \cap P$ is not empty and $g(L) \cap Q$ is empty, might exist (i.e., several rewriting steps at P avoiding Q might be possible). Thus, the application of the rule at P avoiding Q produces a *set of located graphs*.

Definition 3. *A graph program consists of a set of located port graph rewrite rules $\mathcal{R}$, a strategy expression S (built from $\mathcal{R}$ using the grammar below) and a located graph G_P^Q. We denote it $\left[S_{\mathcal{R}}, G_P^Q \right]$, or simply $\left[S, G_P^Q \right]$ when $\mathcal{R}$ is obvious.*

The formal semantics of a graph program is given below, using an Abstract Reduction System [40,26,12]. The idea is to build a set of rewrite derivations out of $\left[S, G_P^Q \right]$ according to the strategy S (i.e., a *derivation tree*). A graph program $[S, G_P^Q]$ may define a non-terminating computation if there is an infinite derivation (written as an undefined result $[\bot, G_P^Q]$). All finite derivations produce results that are *values* of the form $[\mathsf{Id}, G_P^Q]$ or $[\mathsf{Fail}, G_P^Q]$ (see Property 2 in Section 5).

Let L, R be port graphs, M, N positions, ρ a property, m, n integers, $p_{i=1\ldots n} \in [0, 1]$

(Focusing) $F := \texttt{CrtGraph} \mid \texttt{CrtPos} \mid \texttt{CrtBan} \mid \texttt{AllSuc}(F) \mid \texttt{OneSuc}(F)$
$\qquad\qquad\quad \mid \texttt{NextSuc}(F) \mid \texttt{Property}(\rho, F) \mid F \cup F \mid F \cap F \mid F \setminus F \mid \emptyset$

(Transformations) $T := (L \Rightarrow R)_M^N \mid (T \parallel T)$

(Applications) $A := \texttt{Id} \mid \texttt{Fail} \mid T \mid \texttt{one}(T)$

(Strategies) $S := A \mid S; S \mid \texttt{ppick}(S_1, p_1, \ldots, S_n, p_n) \mid \texttt{while}(S)\texttt{do}(S)$
$\qquad\qquad\quad \mid (S)\texttt{orelse}(S) \mid \texttt{if}(S)\texttt{then}(S)\texttt{else}(S) \mid \texttt{isEmpty}(F)$
$\qquad\qquad\quad \mid \texttt{setPos}(F) \mid \texttt{setBan}(F)$

Fig. 1. Syntax of the strategy language

Definition 4. *Given a graph program* $[S, G_P^Q]$ *and its derivation tree, the* result set *is the multiset of values (*$[\texttt{Id}, G'^{Q'}_{P'}]$ *or* $[\texttt{Fail}, G'^{Q'}_{P'}]$*) in the tree, together with* $[\bot, G_P^Q]$ *if there is an infinite branch. We say that a result set is a* success *if there exists at least one graph program* $[\texttt{Id}, G'^{Q'}_{P'}]$ *(for some* $G'^{Q'}_{P'}$*) or is a* failure *otherwise.*

Definition 5. *A graph program* $[S, G_P^Q]$ *is* strongly terminating *if there are no infinite branches in its derivation tree (i.e., its result set contains only values* $[\texttt{Id}, G_P^Q]$ *or* $[\texttt{Fail}, G_P^Q]$*). It is* weakly terminating *if there is at least one finite branch (i.e., its result set contains at least one of* $[\texttt{Id}, G_P^Q]$ *or* $[\texttt{Fail}, G_P^Q]$*).*

3.2 Syntax and Informal Description

The syntax of the strategy language is given in Figure 1. The *strategy expressions* used in graph programs are generated by the grammar rules from the non-terminal S. A strategy expression combines applications, generated by A, and focusing operations, generated by F. The application constructs and some of the strategy constructs are strongly inspired from term rewriting languages such as ELAN [11], Stratego [42] and Tom [7]. The syntax presented here is a revised and simplified version of the one used in [2,20]; the main difference is that we now have an explicit notion of banned subgraph, a more concise syntax for iterative commands and a non-deterministic construct for rule applications.

Focusing. These constructs are used in strategy expressions to change the positions P and Q in the current located graph (e.g. to specify graph traversals).

- CrtGraph returns the whole current graph G. CrtPos and CrtBan return the current P and Q respectively in the located graph.
- AllSuc(F) returns the subgraph consisting of all immediate successors of the nodes in F, where an immediate successor of a node v is a node that has a port connected to a port of v. OneSuc(F) returns a subgraph consisting

of one immediate successor of a node in F, chosen non-deterministically. $\texttt{NextSuc}(F)$ computes successors of nodes in F using for each node only the subset of its ports that have the attribute "next"; we call the ports in this distinguished subset the *next* ports (so $\texttt{NextSuc}(F)$ returns a subset of the nodes in $\texttt{AllSuc}(F)$).

- $\texttt{Property}(\rho, F)$ is a filtering construct, that returns a subgraph of G containing only the nodes from F that satisfy the decidable property ρ. It typically tests a property on nodes or ports, allowing us for instance to select the subgraph of red nodes or nodes with active ports (as mentioned in Section 2, ports and nodes in port graphs may have associated properties).
- $\cup$, $\cap$ and $\setminus$ are the standard set theory operators; $\emptyset$ denotes the empty set.

Transformations. The focusing subgraphs P and Q in the target graph and the distinguished graphs M and N in a located port graph rewrite rule are original features of the language. $(L \Rightarrow R)_M^N$ represents the application of the rule $L \Rightarrow R$ at the current position P and avoiding Q in G_P^Q, according to Definition 2. The syntax $T \parallel T'$ represents simultaneous application of the transformations T and T' on disjoint subgraphs of G; it succeeds if both are possible *simultaneously*.

Applications. There are four kinds of applications according to the grammar: Id and Fail are two constant strategies that respectively denote success and failure. T denotes all possible applications of the transformation on the located graph at the current position, creating a new located graph for each application. In the derivation tree, this creates as many children as there are possible applications. $\texttt{one}(T)$ non-deterministically computes only one of the possible applications of the transformation and ignore the others.

Strategies. The expression $S;S'$ represents sequential application of S followed by S'. When probabilities $p_1, \ldots, p_n \in [0,1]$ are associated to strategies $S_1, \ldots, S_n$ such that $p_1 + \ldots + p_n = 1$, the strategy $\texttt{ppick}(S_1, p_1, \ldots, S_n, p_n)$ non-deterministically picks one of the strategies for application, according to the given probabilities. $\texttt{while}(S)\texttt{do}(S')$ keeps on sequentially applying S' while the expression S is successful; if S fails, then Id is returned. $(S)\texttt{orelse}(S')$ applies S if possible, otherwise applies S' and fails if both S and S' fail. $\texttt{if}(S)\texttt{then}(S')\texttt{else}(S'')$ checks if the application of S on (a copy of) G_P^Q returns Id, in which case S' is applied to (the original) G_P^Q, otherwise S'' is applied to the original G_P^Q. $\texttt{isEmpty}(F)$ behaves like Id if F returns an empty graph and Fail otherwise; this can be used for instance inside the condition of an if or while. $\texttt{setPos}(F)$ (resp. $\texttt{setBan}(F)$) sets the position subgraph P (resp. Q) to be the graph resulting from the focusing expression F.

3.3   Semantics

Focusing. The focusing operators defined by the grammar for F in Fig. 1 have a functional semantics. They apply to the current located graph, and compute a subgraph (i.e., they return a subgraph of G). We define below the result of focusing operations on a given located graph.

$$\texttt{CrtGraph}(G_P^Q) \quad = G \qquad\qquad \texttt{CrtPos}(G_P^Q) = P \qquad \texttt{CrtBan}(G_P^Q) = Q$$

$$\texttt{AllSuc}(F)(G_P^Q) \quad = G' \text{ where } G' \text{ consists of all immediate successors of}$$
$$\text{nodes in } F(G_P^Q)$$

$$\texttt{OneSuc}(F)(G_P^Q) \quad = G' \text{ where } G' \text{ consists of one immediate successor of}$$
$$\text{a node in } F(G_P^Q), \text{ chosen non-deterministically}$$

$$\texttt{NextSuc}(F)(G_P^Q) \quad = G' \text{ where } G' \text{ consists of the immediate successors,}$$
$$\text{via ports labelled "next", of nodes in } F(G_P^Q)$$

$$\texttt{Property}(\rho, F)(G_P^Q) = G' \text{ where } G' \text{ consists of all nodes in } F(G_P^Q) \text{ satisfying } \rho$$

$$(F_1 \; op \; F_2)(G_P^Q) \quad = F_1(G_P^Q) \; op \; F_2(G_P^Q) \text{ where } op \text{ is } \cup, \cap, \setminus$$

Transformations, Applications and Strategy Operators. The constructs in the grammars for T, A and S in Fig. 1 are defined by semantic rules given below. Thus, the semantics of the strategy language is also defined by rewriting, as done for example in [30]. Our semantic rules are applied to configurations containing graph programs, defining a *small step* operational semantics in the style of [34].

In order to deal with the non-determinism introduced by rewriting (see Definition 2), we introduce the notion of configuration.

Definition 6. *A* configuration *is a multiset* $\{O_1, \ldots, O_n\}$ *where each* O_i *is either a graph program or an intermediate object (built with auxiliary operators* $\texttt{if}_2, \texttt{orelse}_2$ *and* $;_2$ *on graph programs, strategy expressions and located graphs), denoted by angular brackets (e.g.* $\langle [S_1, G_P^Q];_2 S_2, G_P^Q \rangle$ *).*

In the semantic rules, we abuse notation and identify a singleton multiset with its element and work modulo the flattening of multisets, i.e., modulo associativity, commutativity and the axiom $\{\{X\}, Y\} = \{X, Y\}$; for instance $\{\{O_a, \ldots, O_b\}, O_y, \ldots, O_z\} = \{O_a, \ldots, O_b, O_y, \ldots, O_z\}$. We type variables in rules by naming them as the initial symbol of the corresponding grammar with an index number if needed (for example: F_2 represents a focusing expression; A_1 is a variable of type application; S_3 represents a strategy expression). The auxiliary function $isSuccess([S, G_P^Q])$ returns $True$ or $False$ depending on whether the result set associated to $[S, G_P^Q]$ is a success or a failure. This function terminates if the graph program is strongly terminating (in implementations, backtracking or breadth-first search will be used to ensure that even if a strategy is weakly terminating, a partial result set is computed).

– Graph rewrite rules are themselves strategy operators:

$$[(L \Rightarrow R)_M^N, G_P^Q] \to \{[\mathsf{Id}, G_{1}{}_{P_1}^{Q_1}], \ldots, [\mathsf{Id}, G_k{}_{P_k}^{Q_k}]\}$$
$$\text{if } G_P^Q \to_{(L \Rightarrow R)_M^N}^{g_i} G_i{}_{P_i}^{Q_i} (\forall i, 1 \le i \le k) \text{ with } g_1 \ldots g_k \text{ pairwise different.}$$
$$[(L \Rightarrow R)_M^N, G_P^Q] \to [\mathsf{Fail}, G_P^Q] \qquad\qquad \text{if the rule is } not \text{ applicable}$$

In the first rule, all possible applications of the rule are considered.

- Parallelism is allowed through the operator $||$ which works on rules only (not on general strategies). To define the semantics of $(L_1 \Rightarrow R_1)_{M_1}^{N_1} || \ldots || (L_k \Rightarrow R_k)_{M_k}^{N_k}$, we define a new rule $((L_1 \cup \ldots \cup L_k) \Rightarrow_{1\ldots k} (R_1 \cup \ldots \cup R_k))_{M_1 \cup \ldots \cup M_k}^{N_1 \cup \ldots \cup N_k}$, where $\Rightarrow_{1\ldots k}$ contains all the ports and edges of $\Rightarrow_i$ (for $1 \leq i \leq k$). It implements simultaneous application of rules at disjoint redexes (note that two nodes may have the same label, but if they are different nodes, then the union will contain both nodes).

$$[(L_1 \Rightarrow R_1)_{M_1}^{N_1} || \ldots || (L_k \Rightarrow R_k)_{M_k}^{N_k}, G_P^Q] \rightarrow$$
$$[((L_1 \cup \ldots \cup L_k) \Rightarrow_{1\ldots k} (R_1 \cup \ldots \cup R_k))_{M_1 \cup \ldots \cup M_k}^{N_1 \cup \ldots \cup N_k}, G_P^Q]$$
$$\text{if } \forall i, \ g_i(L_i) \cap P \neq \emptyset$$
$$[(L_1 \Rightarrow R_1)_{M_1}^{N_1} || \ldots || (L_k \Rightarrow R_k)_{M_k}^{N_k}, G_P^Q] \rightarrow [\mathsf{Fail}, G_P^Q] \qquad \text{otherwise}$$

- The non-deterministic `one()` operator takes as argument a rule or several rules in parallel (in the latter case, we create a new rule, as explained above). It selects only one of the reducts, non-deterministically.

$$[\mathbf{one}((L_1 \Rightarrow R_1)_{M_1}^{N_1} || \ldots || (L_k \Rightarrow R_k)_{M_k}^{N_k}), G_P^Q] \rightarrow$$
$$[\mathbf{one}(((L_1 \cup \ldots \cup L_k) \Rightarrow_{1\ldots k} (R_1 \cup \ldots \cup R_k))_{M_1 \cup \ldots \cup M_k}^{N_1 \cup \ldots \cup N_k}), G_P^Q]$$
$$\text{if } \forall i, \ g_i(L_i) \cap P \neq \emptyset$$
$$[\mathbf{one}((L_1 \Rightarrow R_1)_{M_1}^{N_1} || \ldots || (L_k \Rightarrow R_k)_{M_k}^{N_k}), G_P^Q] \rightarrow [\mathsf{Fail}, G_P^Q] \quad \text{otherwise}$$
$$[\mathbf{one}((L \Rightarrow R)_M^N), G_P^Q] \rightarrow [\mathsf{Id}, G'^{Q'}_{P'}] \qquad \qquad \text{if } G_P^Q \rightarrow^g_{(L \Rightarrow R)_M^N} G'^{Q'}_{P'}$$
$$\text{for a chosen } g$$
$$[\mathbf{one}((L \Rightarrow R)_M^N), G_P^Q] \rightarrow [\mathsf{Fail}, G_P^Q] \qquad \qquad \text{if the rule is } not \text{ applicable}$$

- Position definition:

$$[\mathbf{setPos}(F), G_P^Q] \rightarrow [\mathsf{Id}, G_{P'}^Q] \quad \text{where } P' \text{ is } F(G_P^Q)$$
$$[\mathbf{setBan}(F), G_P^Q] \rightarrow [\mathsf{Id}, G_P^{Q'}] \quad \text{where } Q' \text{ is } F(G_P^Q)$$

$$[\mathbf{isEmpty}(F), G_P^Q] \rightarrow [\mathsf{Id}, G_P^Q] \quad \text{if } F(G_P^Q) \text{ is empty}$$
$$[\mathbf{isEmpty}(F), G_P^Q] \rightarrow [\mathsf{Fail}, G_P^Q] \text{ if } F(G_P^Q) \text{ is not empty}$$

Note that with the semantics given above for `setPos()` and `setBan()`, it is possible for P and Q to have a non-empty intersection. Rules can still apply if the redex overlaps P but not Q.

- Sequential application: below E denotes Id or Fail.

$$[S_1; S_2, G_P^Q] \rightarrow \langle [S_1, G_P^Q];_2 S_2, G_P^Q \rangle$$
$$\text{if } S_1 \neq \mathsf{Id}, S_1 \neq \mathsf{Fail}$$
$$\langle \{[E, G_{0_{P_0}}^{Q_0}], [S_1^1, G_{1_{P_1}}^{Q_1}], \ldots, [S_1^k, G_{k_{P_k}}^{Q_k}]\};_2 S_2, G_P^Q \rangle \rightarrow \{[E; S_2, G_{0_{P_0}}^{Q_0}], \langle \{[S_1^1, G_{1_{P_1}}^{Q_1}],$$
$$\ldots, [S_1^k, G_{k_{P_k}}^{Q_k}]\}\};_2 S_2, G_P^Q \rangle \}$$
$$[\mathsf{Id}; S, G_P^Q] \rightarrow [S, G_P^Q]$$
$$[\mathsf{Fail}; S, G_P^Q] \rightarrow [\mathsf{Fail}, G_P^Q]$$

The first rule for sequences ensures that S_1 is applied first to G_P^Q: it builds the configuration $[S_1, G_P^Q]$ — we say that S_1 is *promoted* so that it can be applied. The second rule can then be applied when a value is obtained from S_1 (note that k could be zero, in which case there is only $[E; S_2, G_0 {}_{P_0}^{Q_0}]$ in the right hand side). Sequential application is strict: if the first strategy does not return a result, the final result is undefined.

- Conditional: the first rule promotes S_1, so that it is applied to G_P^Q and tested with the auxiliary function $isSuccess()$.

$$[\mathtt{if}(S_1)\mathtt{then}(S_2)\mathtt{else}(S_3), G_P^Q] \rightarrow$$
$$\langle \mathtt{if}_2(isSuccess([S_1, G_P]))\mathtt{then}(S_2)\mathtt{else}(S_3), G_P^Q\rangle$$
$$\langle \mathtt{if}_2(True)\mathtt{then}(S_2)\mathtt{else}(S_3), G_P^Q\rangle \rightarrow [S_2, G_P^Q]$$
$$\langle \mathtt{if}_2(False)\mathtt{then}(S_2)\mathtt{else}(S_3), G_P^Q\rangle \rightarrow [S_3, G_P^Q]$$

- Iteration:

$$[\mathtt{while}(S_1)\mathtt{do}(S_2), G_P^Q] \rightarrow$$
$$\langle \mathtt{if}_2(isSuccess([S_1, G_P^Q]))\mathtt{then}(S_2; \mathtt{while}(S_1)\mathtt{do}(S_2))\mathtt{else}(\mathsf{Id}), G_P^Q\rangle$$

- Priority choice:

$$[(S_1)\mathtt{orelse}(S_2), G_P^Q] \rightarrow$$
$$\langle ([S_1, G_P^Q])\mathtt{orelse}_2(S_2), G_P^Q\rangle$$
$$\langle (\{[\mathsf{Id}, G_0 {}_{P_0}^{Q_0}], [S_1^1, G_1 {}_{P_1}^{Q_1}], \ldots, [S_1^k, G_k {}_{P_k}^{Q_k}]\})\mathtt{orelse}_2(S_2), G_P^Q\rangle \rightarrow$$
$$\{[\mathsf{Id}, G_0 {}_{P_0}^{Q_0}], \langle (\{[S_1^1, G_1 {}_{P_1}^{Q_1}] \ldots, [S_1^k, G_k {}_{P_k}^{Q_k}]\})\mathtt{orelse}_2(S_2), G_P^Q\rangle\}$$
$$\langle (\{[\mathsf{Fail}, G_1 {}_{P_1}^{Q_1}], \ldots, [\mathsf{Fail}, G_k {}_{P_k}^{Q_k}]\})\mathtt{orelse}_2(S_2), G_P^Q\rangle \rightarrow [S_2, G_P^Q]$$

Here, S_1 is promoted so that it can be applied to G_P^Q. If it fails (i.e., all the derivations end with Fail) then S_2 is applied to the initial graph. Note that again k could be zero in the second rule, in which case the right hand side is just $[\mathsf{Id}, G_0 {}_{P_0}^{Q_0}]$. We chose to define $(S_1)\mathtt{orelse}(S_2)$ as a primitive operator instead of encoding it as $\mathtt{if}(S_1)\mathtt{then}(S_1)\mathtt{else}(S_2)$ since the language has non-deterministic operators: evaluating S_1 in the condition and in the "then" branch could yield different values.

- Probabilistic choice: we assume $prob(p_1, \ldots, p_n)$ returns the element $j \in \{1 \ldots n\}$ with probability p_j.

$$[\mathtt{ppick}(S_1, p_1, \ldots, S_n, p_n), G_P^Q] \rightarrow [S_j, G_P^Q] \quad \text{where } prob(p_1, \ldots, p_n) = j$$

4   Examples

In this section we give examples to illustrate the expressivity of the language. The **not** and **try** operators, well-known in strategy languages for term rewriting, are not primitive in our language but can be derived, as well as $\mathtt{repeat}(S)$, and bounded iteration; $\|\|$ is a weaker version of $\|$.

- $\mathtt{not}(S) \triangleq \mathtt{if}(S)\mathtt{then}(\mathsf{Fail})\mathtt{else}(\mathsf{Id})$ fails if S succeeds and succeeds if S fails.
- $\mathtt{try}(S) \triangleq (S)\mathtt{orelse}(\mathsf{Id})$ is a strategy that behaves like S if S succeeds, but if S fails then it behaves like Id.
- $\mathtt{repeat}(S) \triangleq \mathtt{while}(S)\mathtt{do}(S)$ applies S as long as possible.
- $\mathtt{while}(S_1)\mathtt{do}(S_2)\mathtt{max}(n) \triangleq$
  $\mathtt{if}(S_1)\mathtt{then}(S_2; \mathtt{if}(S_1)\mathtt{then}(S_2;\ldots)\mathtt{else}(\mathsf{Id}))\mathtt{else}(\mathsf{Id})$ representing a series of $\mathtt{if}()$s of the same form, with exactly n occurrences of S_2.
- $\mathtt{for}(n)\mathtt{do}(S) \triangleq S;\ldots; S$ where S is repeated n times.
- $A \parallel\!\!\mid A'$ is similar to $A\|A'$ except that it returns Id if at least one application of A or A' is possible (it can be generalised to n applications in parallel):
  $A_1\parallel\!\!\mid A_2 \triangleq \mathtt{if}(A_1)\mathtt{then}(\mathtt{if}(A_1\|A_2)\mathtt{then}(A_1\|A_2)\mathtt{else}(A_1))\mathtt{else}(A_2)$

Using focusing (specifically the Property construct), we can create concise strategies that perform traversals. In this way, we can define outermost or innermost term rewriting (on trees) without needing to change the rewrite rules. This is standard in term-based languages such as ELAN [11] or Stratego [42][13]; here we can also define traversals in graphs that are not trees.

Outermost rewriting on trees: We define the abbreviation $start \triangleq$ Property($root$, CrtGraph), which selects the subgraph containing just the root of the tree. The $next$ ports (see definition of the NextSuc(F) operator in Section 3.3) for each node in the tree are defined to be the ones that connect with their children. The strategy for outermost rewriting with a rule R is:

  setPos($start$);
  while(not(isEmpty(CrtPos)))do
        (if(R)then(R; setPos($start$))else(setPos(NextSuc(CrtPos))))

Thus, if R can be applied then we apply it and set the position back to the root of the tree. Otherwise, setPos(NextSuc(CrtPos)) makes all children of all elements in the current position the new current position, thus descending one step into the tree.

Innermost rewriting on trees: We define the abbreviations $start \triangleq$ Property($leaf$, CrtGraph), which selects the leaves of the tree, and $rest \triangleq$ CrtGraph \ $start$. For each node, the $next$ port connects with the parent node.

  setPos($start$); setBan($rest$);
  while(not(isEmpty(CrtPos)))do(
    if(R)then(R; setPos($start$); setBan($rest$))
    else(setPos(NextSuc(CrtPos)); setBan(CrtBan \ CrtPos)))

Thus, if R can be applied then we apply it and set the position back to the leaves of the tree and put all the other elements of the tree into the banned subgraph. Otherwise, we move up the tree one level with setPos(NextSuc(CrtPos)) and the banned subgraph is updated again to all the remaining elements of the tree (with setBan(CrtBan \ CrtPos)).

Sorting: The following example shows how a non-ordered list of three colours (Blue, Red and White) can be sorted to represent the French flag (Blue first, then White and finally Red). We have three port nodes representing each colour (shown in Figure 2) that have two ports each: a *previous* port (to the left of the

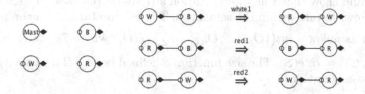

Fig. 2. The four port node types and the three flag sorting rules

node) and a *next* port (to the right of the node). We also have a *Mast* port node at the beginning of the list. Using the three rules in Figure 2, we can swap two colours if they are in the wrong order. Using the ||| operator we apply as many of these rules as we can in parallel. Our overall strategy would then be:

$$\texttt{repeat}(\texttt{setPos}(\texttt{CrtGraph}); ((\textit{white1}|||\textit{red1})|||\textit{red2}))$$

We can also program a sorting algorithm that starts from the mast node. If no rule can be applied, we move P one position across the list and try again. After a rule is applied we reset P to be just *Mast*. When we reach the end of the list, the program terminates and the list is correctly ordered. By defining: $swap \triangleq ((\textit{white1})\texttt{orelse}(\textit{red1}))\texttt{orelse}(\textit{red2})$ and $backToMast \triangleq \texttt{Property}(\textit{mast}, \texttt{CrtGraph})$, the strategy is:

```
setPos(backToMast);
while(not(isEmpty(CrtPos)))do
(if(swap)then(swap; setPos(backToMast))else(setPos(NextSuc(CrtPos))))
```

This example illustrates the separation of concerns mentioned in the introduction: to program a flag-sorting algorithm that proceeds from the mast onwards, we do not need to change the rewrite rules (in contrast with other graph rewriting languages where focusing constructs are not available and conditional rewriting is used). More examples can be found in [31].

5   Properties

In this section we discuss termination and completeness of the semantic rules of the strategy language. We refer to [31] for the proofs omitted or just sketched here due to space constraints.

Graph programs are not terminating in general, however we can identify a terminating sublanguage (i.e. a sublanguage for which the semantic rules are terminating) and we can characterise the graph programs in normal form.

Property 1 (Termination). The sublanguage that excludes the while construct is strongly terminating.

Proof. To prove that a graph program in this sublanguage does not generate an infinite derivation with the semantic rules, we interpret configurations as natural numbers and show that this interpretation strictly decreases with each application of a rewrite rule within the semantics. The interpretation of a configuration is defined as follows: $int(\{O_1, \ldots, O_n\}) = \sum_{i=1}^{n} \mathcal{I}(O_i)$ where $\mathcal{I}([S, G]) = size(S)$ and $\mathcal{I}(\langle S, G \rangle) = size(S)$. The size function is defined below (B denotes a Boolean and C a configuration):

$$size(\mathsf{Id}) = 0 \qquad\qquad size(\mathsf{Fail}) = 0 \qquad\qquad size((L \Rightarrow R)_M^N) = 1$$
$$size(\mathsf{one}(T)) = 1$$
$$size(\mathsf{setPos}(F)) = 1 \qquad size(\mathsf{setBan}(F)) = 1 \qquad size(\mathsf{isEmpty}(F)) = 1$$
$$size(S_1; S_2) = 2 + size(S_1) + size(S_2)$$
$$size(C;_2 S) = 1 + int(C) + size(S)$$
$$size(\mathsf{if}(S_1)\mathsf{then}(S_2)\mathsf{else}(S_3)) = 2 + size(S_1) + size(S_2) + size(S_3)$$
$$size(\mathsf{if}_2(B)\mathsf{then}(S_2)\mathsf{else}(S_3)) = 1 + size(S_2) + size(S_3)$$
$$size((S_1)\mathsf{orelse}(S_2)) = 2 + size(S_1) + size(S_2)$$
$$size((C)\mathsf{orelse}_2(S_2)) = 1 + int(C) + size(S_2)$$
$$size(\mathsf{ppick}(S_1, p_1, \ldots, S_i, p_i)) = 1 + size(S_1) + \ldots + size(S_i)$$
$$size(T_1 \| T_2) = 1 + size(T_1) + size(T_2)$$

Property 2 (Characterisation of Normal Forms). Every graph program which is not a value $\left[\mathsf{Id}, G_P^Q\right]$ or $\left[\mathsf{Fail}, G_P^Q\right]$ is reducible using the semantic rules.

Proof. By inspection of the semantic rules, every graph program $[S, G_P^Q]$ different from $[\mathsf{Id}, G_P^Q]$ or $[\mathsf{Fail}, G_P^Q]$ can be matched by a left-hand side of a semantic rule. This is true because S has a top operator which is one of the syntactic constructions and there is a semantic rule which applies to it. Moreover every expression of the form $\langle X, G_P^Q \rangle$ is reducible, because X either contains a graph program $[S, G_P^Q]$ different from $[\mathsf{Id}, G_P^Q]$ or $[\mathsf{Fail}, G_P^Q]$ that so can be matched by a left-hand side of a rule as above, or one of the auxiliary rules can apply.

The language contains non-deterministic operators in each of its syntactic categories: `OneSuc()` for Focusing, `one()` for Applications and `ppick()` for Strategies. For the sublanguage that excludes them, we have the property:

Property 3 (Result Set). Each graph program in the sublanguage that excludes `OneSuc()`, `one()` and `ppick()`, has at most one *result set*.

Intuitively, in a strategy, an application of a transformation with `one()` creates a configuration that is a subset of the configuration computed by the same transformation without `one()`.

Property 4 (Result Set with `one()`). A graph program in the sublanguage that excludes `OneSuc()` and `ppick()` (but contains `one()`) produces a result set that is a (non-strict) subset of the same graph program where all occurrences of `one()` have been removed from its strategy.

With respect to the expressive power of the language, it is easy to state, as in [23], the Turing completeness property.

Property 5 (Completeness). The set of all graph programs $\left[S_{\mathcal{R}}, G_P^Q\right]$ is Turing complete, i.e. can simulate any Turing machine.

It is also interesting to consider which sublanguages of our language are Turing complete.

Property 6 (Complete sublanguage). The sublanguage consisting of graph programs where S is built from Id, Fail, rules $(L \Rightarrow R)_M^N$, sequential composition (;), iteration (while), and orelse is Turing complete.

The same result could be obtained by replacing orelse with the conditional construct if then else. Perhaps more surprising is the fact that Turing machine computations can be simulated by a term rewriting system consisting of just one rule, using a strategy that forces the reduction steps to take place at specific positions, as shown by Dauchet [16]. Given a sequence of transitions in the Turing machine, Dauchet [16] shows how to build a rewrite rule to simulate the transitions, using a strategy to build S-deep-only derivations, selecting the position for rewriting according to the instruction used by the machine. It follows that the sublanguage consisting of focusing operators, sequential composition (;), iteration (while) and rule application $(L \Rightarrow R)_M^N$, together with Id, Fail, setPos() and setBan() is Turing complete. The setPos() construct can be used to simulate Dauchet's strategy, by moving the focus of rewriting to the corresponding subterm after each rewrite step. Building a strategy expression for Dauchet's system is a task left for future work.

6   Related Work and Conclusion

The strategy language defined in this paper is part of the PORGY system [2], an environment for visual modelling of complex systems through port graphs and port graph rewrite rules. PORGY provides tools to build port graphs from scratch, with nodes, ports, edges and associated attributes. It offers also means to visualise traces of rewriting as a derivation tree. The strategy language is used in particular to guide the construction of this derivation tree. The implementation uses the small-step operational semantics given above. Some of these semantic rules require a copy of the graph program; this is done efficiently in PORGY thanks to the cloning functionalities of the underlying TULIP system [5].

Graph rewriting is implemented in a variety of tools. In AGG [19], application of rules can be controlled by defining *layers* and then iterating through and across layers. PROGRES [39] allows users to define the way rules are applied and includes non-deterministic constructs, sequence, conditional and loops. The Fujaba [32] Tool Suite offers a basic strategy language, including conditionals, sequence and method calls, but no parallelism. GROOVE [37] permits to control the application of rules, via a control language with sequence, loop, random choice, try()else() and simple function calls. In GReAT [6] the pattern-matching algorithm always starts from specific nodes called "pivot nodes"; rule execution is sequential and there are conditional and looping structures. Gr-Gen.NET [21] uses the concept of search plans to represent different matching strategies. GP [36] is a rule-based, non-deterministic programming language, where programs are defined by sets of graph rewrite rules and a textual strategy expression. The strategy language has three main control constructs: sequence, repetition and conditional, and uses a Prolog-like backtracking technique.

None of the languages above has focusing constructs. Compared to these systems, the PORGY strategy language clearly separates the issues of selecting positions for rewriting and selecting rules, with primitives for focusing as well as traditional strategy constructs. PORGY also emphasises visualisation and scale, thanks to the TULIP back-end which can handle large graphs with millions of elements and comes with powerful visualisation and interaction features.

The strategy language defined in this paper is strongly inspired by the work on GP and PROGRES, and by strategy languages developed for term rewriting such as ELAN [11] and Stratego [42]. It can be applied to terms as a particular case (since terms are just trees). When applied to trees, the constructs dealing with applications and strategies are similar to those found in ELAN or Stratego. The focusing sublanguage on the other hand can be seen as a lower level version of these languages, because term traversals are not directly available in our language but can be programmed using focusing constructs.

The PORGY environment is still evolving. The strategy language could be enhanced by allowing, for instance, more parallelism (using techniques from the K semantic framework [38]) and focusing constructs to deal with properties of edges. Verification and debugging tools for avoiding conflicting rules or non-termination for instance are planned for future work.

Acknowledgements. We thank the members of the PORGY team for many valuable discussions. This work was supported by Inria's "Associate team" programme: PORGY project.

References

1. Andrei, O.: A Rewriting Calculus for Graphs: Applications to Biology and Autonomous Systems. PhD thesis, Institut National Polytechnique de Lorraine (2008)
2. Andrei, O., Fernández, M., Kirchner, H., Melançon, G., Namet, O., Pinaud, B.: PORGY: Strategy driven interactive transformation of graphs. In: Proceedings of TERMGRAPH 2011, Saarbrucken. EPTCS (April 2011)
3. Andrei, O., Kirchner, H.: A Rewriting Calculus for Multigraphs with Ports. In: Proceedings of RULE 2007. Electronic Notes in Theoretical Computer Science, vol. 219, pp. 67–82 (2008)
4. Andrei, O., Kirchner, H.: A Higher-Order Graph Calculus for Autonomic Computing. In: Lipshteyn, M., Levit, V.E., McConnell, R.M. (eds.) Golumbic Festschrift. LNCS, vol. 5420, pp. 15–26. Springer, Heidelberg (2009)
5. Auber, D.: Tulip – A huge graph visualization framework. In: Mutzel, P., Jünger, M. (eds.) Graph Drawing Software. Mathematics and Visualization, pp. 105–126. Springer (2003)

6. Balasubramanian, D., Narayanan, A., van Buskirk, C.P., Karsai, G.: The Graph Rewriting and Transformation Language: GReAT. ECEASST 1 (2006)
7. Balland, E., Brauner, P., Kopetz, R., Moreau, P.-E., Reilles, A.: Tom: Piggybacking Rewriting on Java. In: Baader, F. (ed.) RTA 2007. LNCS, vol. 4533, pp. 36–47. Springer, Heidelberg (2007)
8. Barendregt, H.: The Lambda Calculus, Its Syntax and Semantics. North-Holland (1981)
9. Barendregt, H., van Eekelen, M., Glauert, J., Kennaway, J.R., Plasmeijer, M., Sleep, M.: Term Graph Rewriting. In: de Bakker, J.W., Nijman, A.J., Treleaven, P.C. (eds.) PARLE 1987. LNCS, vol. 259, pp. 141–158. Springer, Heidelberg (1987)
10. Barthelmann, K.: How to construct a hyperedge replacement system for a context-free set of hypergraphs. Technical report, Universität Mainz, Institut für Informatik (1996)
11. Borovanský, P., Kirchner, C., Kirchner, H., Moreau, P.-E., Ringeissen, C.: An overview of ELAN. ENTCS 15 (1998)
12. Bourdier, T., Cirstea, H., Dougherty, D.J., Kirchner, H.: Extensional and intensional strategies. In: Proceedings Ninth International Workshop on Reduction Strategies in Rewriting and Programming. EPTCS, vol. 15, pp. 1–19 (2009)
13. Bravenboer, M., Kalleberg, K.T., Vermaas, R., Visser, E.: Stratego/XT 0.17. A language and toolset for program transformation. Science of Computer Programming (2008); Special issue on Experimental Systems and Tools
14. Corradini, A., Montanari, U., Rossi, F., Ehrig, H., Heckel, R., Löwe, M.: Algebraic approaches to graph transformation - part i: Basic concepts and double pushout approach. In: Handbook of Graph Grammars and Computing by Graph Transformations, Volume 1: Foundations, pp. 163–246. World Scientific (1997)
15. Courcelle, B.: Graph Rewriting: An Algebraic and Logic Approach. In: van Leeuwen, J. (ed.) Handbook of Theoretical Computer Science, Volume B: Formal Models and Semantics, pp. 193–242. Elsevier, MIT Press (1990)
16. Dauchet, M.: Simulation of Turing Machines by a Left-Linear Rewrite Rule. In: Dershowitz, N. (ed.) RTA 1989. LNCS, vol. 355, pp. 109–120. Springer, Heidelberg (1989)
17. Dijkstra, E.W.: Selected writings on computing - a personal perspective. Texts and monographs in computer science. Springer (1982)
18. Ehrig, H., Engels, G., Kreowski, H.-J., Rozenberg, G.: Handbook of Graph Grammars and Computing by Graph Transformations, vol. 1–3. World Scientific (1997)
19. Ermel, C., Rudolf, M., Taentzer, G.: The AGG approach: Language and environment. In: Ehrig, H., Engels, G., Kreowski, H.-J., Rozenberg, G. (eds.) Handbook of Graph Grammars and Computing by Graph Transformations, Volume 2: Applications, Languages, and Tools, pp. 551–603. World Scientific (1997)
20. Fernández, M., Namet, O.: Strategic programming on graph rewriting systems. In: Proceedings International Workshop on Strategies in Rewriting, Proving, and Programming, IWS 2010. EPTCS, vol. 44, pp. 1–20 (2010)
21. Geiß, R., Batz, G.V., Grund, D., Hack, S., Szalkowski, A.: GrGen: A Fast SPO-Based Graph Rewriting Tool. In: Corradini, A., Ehrig, H., Montanari, U., Ribeiro, L., Rozenberg, G. (eds.) ICGT 2006. LNCS, vol. 4178, pp. 383–397. Springer, Heidelberg (2006)
22. Habel, A., Müller, J., Plump, D.: Double-pushout graph transformation revisited. Mathematical Structures in Computer Science 11(5), 637–688 (2001)
23. Habel, A., Plump, D.: Computational Completeness of Programming Languages Based on Graph Transformation. In: Honsell, F., Miculan, M. (eds.) FOSSACS 2001. LNCS, vol. 2030, pp. 230–245. Springer, Heidelberg (2001)

24. Hanus, M.: Curry: A multi-paradigm declarative language (system description). In: Twelfth Workshop Logic Programming, WLP 1997, Munich (1997)
25. Jones, S.L.P.: Haskell 98 language and libraries: the revised report. Cambridge University Press (2003)
26. Kirchner, C., Kirchner, F., Kirchner, H.: Strategic computations and deductions. In: Reasoning in Simple Type Theory. Studies in Logic and the Foundations of Mathematics, vol. 17, pp. 339–364. College Publications (2008)
27. Lafont, Y.: Interaction nets. In: Proceedings of the 17th ACM Symposium on Principles of Programming Languages (POPL 1990), pp. 95–108. ACM Press (1990)
28. Lucas, S.: Strategies in programming languages today. Electr. Notes Theor. Comput. Sci. 124(2), 113–118 (2005)
29. Martí-Oliet, N., Meseguer, J., Verdejo, A.: Towards a strategy language for Maude. Electr. Notes Theor. Comput. Sci. 117, 417–441 (2005)
30. Martí-Oliet, N., Meseguer, J., Verdejo, A.: A rewriting semantics for Maude strategies. Electr. Notes Theor. Comput. Sci. 238(3), 227–247 (2008)
31. Namet, O.: Strategic Modelling with Graph Rewriting Tools. PhD thesis, King's College London (2011)
32. Nickel, U., Niere, J., Zündorf, A.: The FUJABA environment. In: ICSE, pp. 742–745 (2000)
33. Plasmeijer, M.J., van Eekelen, M.C.J.D.: Functional Programming and Parallel Graph Rewriting. Addison-Wesley (1993)
34. Plotkin, G.D.: A structural approach to operational semantics. J. Log. Algebr. Program. 60-61, 17–139 (2004)
35. Plump, D.: Term graph rewriting. In: Ehrig, H., Engels, G., Kreowski, H.-J., Rozenberg, G. (eds.) Handbook of Graph Grammars and Computing by Graph Transformations, Volume 2: Applications, Languages, and Tools, pp. 3–61. World Scientific (1998)
36. Plump, D.: The Graph Programming Language GP. In: Bozapalidis, S., Rahonis, G. (eds.) CAI 2009. LNCS, vol. 5725, pp. 99–122. Springer, Heidelberg (2009)
37. Rensink, A.: The GROOVE Simulator: A Tool for State Space Generation. In: Pfaltz, J.L., Nagl, M., Böhlen, B. (eds.) AGTIVE 2003. LNCS, vol. 3062, pp. 479–485. Springer, Heidelberg (2004)
38. Rosu, G., Serbanuta, T.-F.: An overview of the K semantic framework. J. Log. Algebr. Program. 79(6), 397–434 (2010)
39. Schürr, A., Winter, A.J., Zündorf, A.: The PROGRES Approach: Language and Environment. In: Ehrig, H., Engels, G., Kreowski, H.-J., Rozenberg, G. (eds.) Handbook of Graph Grammars and Computing by Graph Transformations, Volume 2: Applications, Languages, and Tools, pp. 479–546. World Scientific (1997)
40. Terese. Term Rewriting Systems. Cambridge University Press (2003); Bezem, M., Klop, J.W., de Vrijer, R. (eds)
41. Thiemann, R., Sternagel, C., Giesl, J., Schneider-Kamp, P.: Loops under strategies... continued. In: Proceedings International Workshop on Strategies in Rewriting, Proving, and Programming. EPTCS, vol. 44, pp. 51–65 (2010)
42. Visser, E.: Stratego: A Language for Program Transformation Based on Rewriting Strategies. System Description of Stratego. In: Middeldorp, A. (ed.) RTA 2001. LNCS, vol. 2051, pp. 357–361. Springer, Heidelberg (2001)
43. Visser, E.: A survey of strategies in rule-based program transformation systems. J. Symb. Comput. 40(1), 831–873 (2005)

Improved Termination Analysis of CHR Using Self-sustainability Analysis

Paolo Pilozzi* and Danny De Schreye

Dept. of Computer Science, K.U. Leuven, Belgium
{paolo.pilozzi,danny.deschreye}@cs.kuleuven.be

Abstract. In the past few years, several successful approaches to termination analysis of Constraint Handling Rules (CHR) have been proposed. In parallel to these developments, for termination analysis of Logic Programs (LP), recent work has shown that a stronger focus on the analysis of the cycles in the strongly connected components (SCC) of the program is very beneficial, both for precision and efficiency of the analysis.

In this paper we investigate the benefit of using the cycles of the SCCs of CHR programs for termination analysis. It is a non-trivial task to define the notion of a cycle for a CHR program. We introduce the notion of a self-sustaining set of CHR rules and show that it provides a natural counterpart for the notion of a cycle in LP. We prove that non-self-sustainability of an SCC in a CHR program entails termination for all queries to that SCC. Then, we provide an efficient way to prove that an SCC of a CHR program is non-self-sustainable, providing an additional, new way of proving termination of (part of) the program.

We integrate these ideas into the CHR termination analyser CHRisTA and demonstrate by means of experiments that this extension significantly improves both the efficiency and the performance of the analyser.

1 Introduction

Termination analysis techniques for Logic Programs often make use of dependency analysis. In some approaches (e.g. [5]), this is done in order to detect (mutually) recursive predicates. For non-recursive predicates, no termination proof is needed, so that termination conditions are only expressed and verified for the recursive ones. In other approaches (e.g. [11], [17]) the termination conditions are explicitly expressed in terms of cycles in the strongly connected components (SCCs) of the dependency graph. It was shown in [17] that such an approach is both efficient and precise. On the benchmark of the termination analysis competition (see [22]) it outperformed competing systems.

Recently, in [23] and [12], we adapted termination analysis techniques from Logic Programming (LP) to Constraint Handling Rules (CHR) [1,8] and developed the CHR termination analyser CHRisTA [13]. These techniques and system are based on recursion, rather than on cycles in SCCs.

---

* Supported by I.W.T. and F.W.O. Flanders - Belgium.

G. Vidal (Ed.): LOPSTR 2011, LNCS 7225, pp. 189–204, 2012.

The main motivation that led to the current work is that we wanted to adapt the approach based on SCCs to CHR, with the hope of improving efficiency and precision. It turned out that there were considerable complications to achieve this, caused by the fact that for CHR a useful notion of "cycle" is hard to define.

Consider for example the CHR program, P, consisting of the CHR rule,

$$R @ a, b \Leftrightarrow a.$$

For each application of R, two constraints, a and b, are removed, while a new constraint, a, is added. Since R adds constraints that are required to fire R, the application of R depends on itself. An SCC analysis on the dependency graph of P would therefore consider R to be cyclic, while in fact, for every application of R, a constraint b is removed and never added again. Therefore, at some point in a computation of P, the b constraints will be depleted and thus P will terminate.

A more accurate analysis of cycles in CHR could therefore still rely on an SCC analysis, but could also verify for each SCC whether there exists a dependency for each head of the rules of a component, provided for by the component itself.

Unfortunately, such an approach is still not very accurate. Consider for example the CHR program, P, consisting of the rule,

$$R @ a, a \Leftrightarrow a.$$

For R, which is part of an SCC, both heads are provided for from within the component itself. However, for every application of R, two constraints a are removed, while only one constraint a is added. Repeated application of R will therefore result in depletion of these a constraints, and thus P will terminate.

Therefore, in order for an SCC in CHR to be cyclic, we must guarantee that the component is *self-sustainable*. That is, when the rules of an SCC are applied, they must provide the constraints to fire the same rules again.

In fact, in definite LP, an SCC of the dependency graph of a definite LP program is self-sustainable by default. The reason for this is that definite LP is a single-headed language. Therefore, each cycle in the dependency graph of a definite LP program —as described by the SCCs of the program— corresponds to rule applications that have the potential to be repeated.

In CHR, to characterise the notion of a self-sustainable SCC, we need to identify the multisets of rule applications of an SCC that can be repeated. Consider for example the CHR program, P:

$$R_1 @ a, a \Leftrightarrow b, b. \qquad R_2 @ b \Leftrightarrow a.$$

As can be verified, it is non-terminating. That is, when R_1 is applied once, it adds two b constraints, allowing R_2 to be applied twice. By applying R_2 twice, the input required for application of R_1 is provided for, thus resulting in a loop.

In this paper, we introduce the notion of *self-sustainability* for a set of CHR rules. It provides a rather precise approximation of cyclic behaviour in CHR. We characterise self-sustainability for a set of rules by means of a system of linear inequalities. Each solution to this system identifies a multiset based on the given rules, such that, if it is *traversed* —all rules in the multiset are applied once— it provides all required constraints to potentially traverse the multiset again.

Unfortunately, we cannot use this concept directly as a basis for termination analysis, similar to [11] for LP. The reason is that for any cyclic SCC in a CHR program, the system has infinitely many solutions. In [15], we describe a rather inefficient approach to cope with this problem, resulting in little gain in precision.

Alternatively, we can use the concept of self-sustainability in a "negative" way. If we can show that an SCC in a CHR program is *not* self-sustainable —the system has no solution— then this proves that the rules involved in that SCC cannot be part of an infinite computation. Therefore, we can ignore these rules in the termination analysis. By integrating this idea in the CHRisTA analyser, it turns out that precision and efficiency of the analysis is improved.

Our approach will be restricted to the abstract CHR semantics. It is however widely applicable as most other CHR semantics [18] are instances of the abstract CHR semantics. That is, if an SCC of a CHR program is not self-sustainable for the abstract semantics, it cannot be self-sustainable for any more refined semantics. Note that, under the abstract semantics, the two different kinds of rules —*simpagation* and *propagation* rules— of a CHR program behave as *simplification* rules. Hence, we discuss our approach in terms of simplification only.

Finally, note that a CHR program may contain propagation rules and that these rules do not remove the constraints on which they fire. Under the abstract CHR semantics, propagation rules are infinitely applicable on the same combination of constraints. Represented as a simplification rule, they explicitly replace the constraints on which they fire and thus are considered self-sustainable by default. Our approach can therefore only prove non-self-sustainability of an SCC without propagation. In Section 6, we provide intuitions regarding self-sustainability under the theoretical CHR semantics [8]. This refinement of the abstract CHR semantics additionally considers a *fire-once policy* on propagation rules to overcome the infinite applicability of propagation rules.

Overview of the paper. In Section 2, we discuss the syntax and semantics of abstract CHR. Section 3 defines CHR dependency graphs for an SCC analysis and CHR nets —a more accurate description of the dependencies of a CHR program— for a self-sustainability analysis. In Section 4, we formalise the notion of self-sustainability of an SCC (Section 4.1). Then, we present a test for verifying that an SCC is not self-sustainable (Section 4.2). In Section 5, we evaluate our approach and discuss CHR programs that we can only prove terminating using the test for non-self-sustainability. Finally, in Section 6, we conclude the paper.

2   Preliminaries

We assume familiarity with LP and its main results [2, 9]. By L, we denote the first order language underlying a CHR program P. By $Term_P$ and $Atom_P$, we denote, respectively, the sets of terms and atoms constructible from L.

2.1   The Syntax of Abstract CHR

The rules of a CHR program act on first-order atoms, called *constraints*.

Definition 1 (constraint). Constraints, $c(t_1, \ldots, t_n)$ $(n \geq 0)$, are first-order predicates applied to terms, t_i $(1 \leq i \leq n)$. We distinguish between two kinds of constraints: CHR constraints are user-defined and are solved by the CHR rules of a CHR program; built-in constraints are pre-defined and are solved by a constraint theory (CT) defined in the host language. □

Constraints are kept in a *constraint store*, of which the behaviour corresponds to a *multiset* (or bag) of constraints. Therefore, to formalise the syntax and semantics of CHR, we first recall multiset theory [4, 6].

A *multiset* is a tuple, $M_S = \langle S, m_S \rangle$, where S is a set, called the *underlying set*, and m_S a *multiplicity function*, mapping the elements, e, of S to natural numbers, $m_S(e) \in \mathbb{N}_0$, representing the number of occurrences of e in M_S. Like any function, m_S may be represented as a set, $\{(s, m_S(s)) : s \in S\}$. An example of a multiset is $\langle \{a, b\}, \{(a, 2), (b, 1)\} \rangle$. We introduce the alternative notation to represent this multiset as $[\![a, a, b]\!]$.

If a *universe U* in which the elements of S must live is specified, the definition of a multiset can be simplified to a *multiset indicator function*. Let U be a universe for a multiset $M_S = \langle S, m_S \rangle$. Then, we define the *multiset indicator function*, $\mu_S : U \to \mathbb{N}$, of M_S w.r.t. U as: $\mu_S(u) = m_S(u)$ if $u \in S$ and $\mu_S(u) = 0$ if $u \notin S$. Since it is often more favourable to discuss multisets in terms of some universe, we can —as an abuse of notation— denote a multiset $M_S = \langle S, m_S \rangle$ also by $\langle U, \mu_S \rangle$, or μ_S if its universe is clear from context.

Furthermore, we will need the notion of a *multisubset* in the constraint store since the rules of a CHR program will be applied on multisubsets of constraints. Let U be a universe for the multisets M_{S_1} and M_{S_2}. Then, multiset M_{S_1} is a *multisubset* of multiset M_{S_2}, denoted $M_{S_1} \sqsubseteq M_{S_2}$, iff the multiset indicator functions, μ_{S_1} and μ_{S_2}, w.r.t. U, of M_{S_1} and M_{S_2} respectively, are such that $\mu_{S_1}(u) \leq \mu_{S_2}(u)$, for all u in U. Associated to $\sqsubseteq$, we define *strict multisubsets*: $M_{S_1} \sqsubset M_{S_2} \leftrightarrow M_{S_1} \sqsubseteq M_{S_2} \wedge M_{S_2} \not\sqsubseteq M_{S_1}$.

We are now ready to recall the syntax of *simplification rules*.

Definition 2 (abstract CHR program). *An* abstract CHR program *is a finite set of* simplification rules, R_i.

A simplification rule,

$$R_i @ H_1^i, \ldots, H_{n_i}^i \Leftrightarrow G_1^i, \ldots, G_{k_i}^i \mid B_1^i, \ldots, B_{l_i}^i, C_1^i, \ldots, C_{m_i}^i.$$

is applicable on a multiset of CHR constraints, matching with the head, $[\![H_1^i, \ldots, H_{n_i}^i]\!]$, of the rule, such that the guard, $[\![G_1^i, \ldots, G_{k_i}^i]\!]$, is satisfiable. Upon application, the multiset of constraints matching with the head is replaced by an appropriate instance of the multiset of built-in and CHR constraints, $[\![B_1^i, \ldots, B_{l_i}^i, C_1^i, \ldots, C_{m_i}^i]\!]$, from the body of the rule.

Note that the guard, $[\![G_1^i, \ldots, G_{k_i}^i]\!]$, of a CHR rule may only consist of built-in constraints. Also note that naming a rule by "rulename @" is optional. □

The next example program is an introductory example to CHR(Prolog).

Example 1 (Primes). The program below implements the Sieve of Eratosthenes for finding the prime numbers up to a given number. User-defined $prime/1$ constraints are used to hold the values of the positive integers for which we derive whether they are prime. User-defined $primes/1$ constraints are used to query the program and to generate the $prime/1$ constraints for prime evaluation.

R_1 @ $primes(2) \Leftrightarrow prime(2)$.
R_2 @ $primes(N) \Leftrightarrow N > 2 \mid Np \; is \; N - 1, prime(N), primes(Np)$.
R_3 @ $prime(M), prime(N) \Leftrightarrow div(M, N) \mid prime(M)$.

R_2 generates the numbers for prime evaluation top-down. It replaces a $primes(n)$ constraint, matching with the head $primes(N)$, by the constraints $Np \; is \; n - 1$, $prime(n)$ and $primes(Np)$, if $CT \models n > 2$. R_1 removes a $primes(2)$ constraint, replacing it with the final number for prime evaluation, $prime(2)$. R_3 implements the sieve. For any two matching constraints, $prime(m)$ and $prime(n)$, such that m divides n, we replace both $prime/1$ constraints by $prime(m)$.   □

2.2 The Semantics of Abstract CHR

The abstract CHR semantics is defined as a state transition system.

Definition 3 (abstract CHR state). *An* abstract CHR state *is a multiset, Q, of constraints (with universe $Atom_P$), called a* constraint store.   □

To define the transitions on CHR states, as given by the CHR rules of the program, we introduce the multiset operator *join*, adding multisets together. Let M_A and M_B be multisets with universe U, and let u be an element of U. Then, the multiset indicator function of the *join* $M_C = M_A \uplus M_B$ is given by the sum, $\mu_C(u) = \mu_A(u) + \mu_B(u)$, of the multiset indicator functions of M_A and M_B.

The CHR transition relation represents the consecutive CHR states for a CHR program P and constraint theory CT.

Definition 4 (abstract CHR transition relation). *Let P be a CHR program and CT a constraint theory for the built-in constraints. Let θ represent the substitutions corresponding to the bindings generated when resolving built-in constraints by CT, and let σ represent substitutions for the variables in the head of the rule as a result of matching. Then, the transition relation, $\rightarrow_P$, for P, where Q is a CHR state, is defined by:*

1. *A* solve transition:
   if $Q = [\![b]\!] \uplus Q'$, with b a built-in constraint, and $CT \models b\theta$,
   then, $Q \rightarrow_P Q'\theta$.
2. *Simplification:*
   if $(R_i$ @ $H_1^i, \ldots, H_{n_i}^i \Leftrightarrow G_1^i, \ldots, G_{k_i}^i \mid B_1^i, \ldots, B_{l_i}^i, C_1^i, \ldots, C_{m_i}^i.) \in P$,
   and if $Q = [\![h_1, \ldots, h_{n_i}]\!] \uplus Q'$ and
       $\exists \sigma\theta : CT \models (h_1 = H_1^i\sigma) \wedge \ldots \wedge (h_{n_i} = H_{n_i}^i\sigma) \wedge (G_1^i, \ldots, G_{k_i}^i)\sigma\theta$,
   then, $Q \rightarrow_P ([\![B_1^i, \ldots, B_{l_i}^i, C_1^i, \ldots, C_{m_i}^i]\!] \uplus Q')\sigma\theta$.

Rules in CHR are non-deterministically applied until exhaustion, thus until no more transitions are possible. Rule application is a committed choice. Built-in constraints are assumed to return an answer in finite time and cannot introduce new CHR constraints. If built-ins cannot be solved by CT, the program fails. □

By an *initial CHR state* or *query state* for a CHR program P, we mean any abstract CHR state for P. By a *final CHR state* or *answer state* for P, we mean a CHR state Q, such that no CHR state Q' exists for which $(Q, Q') \in \to_P$.

Example 2 (Primes continued). Executing the program from Example 1 on an initial CHR state $Q_0 = [\![primes(7)]\!]$, we obtain by application of R_2 the next state $Q_1 = [\![Np\ is\ 7-1, prime(7), primes(Np)]\!]$. Solving the built-in constraint, $Np\ is\ 7-1$, binds 6 to Np, yielding the state, $Q_2 = [\![prime(7), primes(6)]\!]$. Proceeding in this way, we obtain the state $Q_{10} = [\![prime(3), primes(2), prime(4),$ $prime(5),\ prime(6),\ prime(7)]\!]$, on which R_2 is no longer applicable. Then, by application of R_1, we obtain $Q_{11} = [\![prime(2),\ prime(3),\ prime(4),\ prime(5),$ $prime(6),\ prime(7)]\!]$.

Now only R_3 is applicable, replacing two *prime*/1 constraints by the constraint dividing the other. For example, we can remove $prime(4)$ since $prime(2)$ divides it. A next state could therefore be $Q_{12} = [\![prime(2),\ prime(3),\ prime(5),$ $prime(6),\ prime(7)]\!]$. For $prime(6)$ there are two prime numbers that divide it. One possibility is that $prime(3)$ is used, and obtain $Q_{13} = [\![prime(3),\ prime(2),$ $prime(5),\ prime(7)]\!]$.

Since on Q_{13} no more rules are applicable, we reach an answer state. This state holds all prime numbers up to 7. □

3   The CHR Dependency Graph and CHR Net

In CHR, the head constraints of a rule represent the constraints *required* for rule application. The body CHR constraints represent the constraints *newly available* after rule application. To represent these sets, we introduce *abstract CHR constraints*.

Definition 5 (abstract CHR constraint). *An* abstract CHR constraint *is a pair* $\mathcal{C} = (C, B)$*, where C is a CHR constraint, and B a conjunction of built-in constraints. The denotation of an abstract CHR constraint is given by a mapping* $\wp : (C, B) \mapsto \{C\sigma \mid \exists \theta : CT \models B\sigma\theta\}$.

In a CHR program P, we distinguish between two types of abstract CHR constraints. An abstract input constraint*, $in_j^i = (H_j^i, G^i)$, represents the CHR constraints that can be used to match the head H_j^i of a CHR rule R_i, without invalidating the guards $G^i = G_1^i \wedge \cdots \wedge G_{k_i}^i$ of R_i. An* abstract output constraint*, $out_j^i = (C_j^i, G^i)$, represents the CHR constraints that become newly available after rule application, and are related to the body CHR constraints C_j^i of R_i.* □

By $\mathcal{I}n_{R_i}$ and $\mathcal{O}ut_{R_i}$, we represent the sets of abstract input and output constraints of a rule R_i. By $\mathcal{I}n_P$ and $\mathcal{O}ut_P$, we represent the abstract input and output constraints of a CHR program P.

Example 3 (abstract CHR constraints of Primes). We revisit the Primes program from Example 1, and derive:

- $\mathcal{I}n_P = \mathcal{I}n_{R_1} \cup \mathcal{I}n_{R_2} \cup \mathcal{I}n_{R_3} = \{in_1^1, in_1^2, in_1^3, in_2^3\}$, where
  $in_1^1 = (primes(2), true)$ $in_1^2 = (primes(N), N > 2)$
  $in_1^3 = (prime(M), div(M, N))$ $in_2^3 = (prime(N), div(M, N))$
- $\mathcal{O}ut_P = \mathcal{O}ut_{R_1} \cup \mathcal{O}ut_{R_2} \cup \mathcal{O}ut_{R_3} = \{out_1^1, out_1^2, out_2^2, out_1^3\}$, where
  $out_1^1 = (prime(2), true)$ $out_1^2 = (prime(N), N > 2)$
  $out_2^2 = (primes(Np), N > 2)$ $out_1^3 = (prime(M), div(M, N))$ □

The rules of a CHR program P relate abstract inputs to abstract outputs. We call this relation the *rule transition relation* of P.

Definition 6 (rule transition relation). *A rule transition of an abstract CHR program, P, is an ordered pair $T_i = (\mathcal{I}n_{R_i}, \mathcal{O}ut_{R_i})$, relating the set of abstract input constraints $\mathcal{I}n_{R_i} = \{in_1^i, \ldots, in_{n_i}^i\}$ of $R_i \in P$ to the set of abstract output constraints $\mathcal{O}ut_{R_i} = \{out_1^i, \ldots, out_{m_i}^i\}$ of R_i. The rule transition relation $\mathcal{T}_P = \{T_i \mid R_i \in P\}$ of P is the set of rule transitions of P.*   □

Example 4 (rule transition relation of Primes). The Primes program P from Example 1 defines three rule transitions:

- $\mathcal{T} = \{T_1, T_2, T_3\}$, where
  $T_1 = (\{in_1^1\}, \{out_1^1\})$ $T_2 = (\{in_1^2\}, \{out_1^2, out_2^2\})$
  $T_3 = (\{in_1^3, in_2^3\}, \{out_1^3\})$ □

Abstract output constraints relate to abstract input constraints by a *match transition relation*. This second kind of relation is the result of a dependency analysis between abstract constraints, relating the constraints newly available after rule application to constraints required for rule application.

Definition 7 (match transition relation). *A match transition of an abstract CHR program P is an ordered pair $M_{(i,j,k,l)} = (out_j^i, in_l^k)$, relating an output $out_j^i = (C_j^i, G^i)$ of $\mathcal{O}ut_P$ to an input $in_l^k = (C_l^k, G^k)$ of $\mathcal{I}n_P$ such that $\exists \theta : CT \models (C_j^i = C_l^k \wedge G^i \wedge G^k)\theta$. The match transition relation $\mathcal{M}_P$ is the set of all match transitions $M_{(i,j,k,l)}$ in P.*   □

Note that a match transition exists for an abstract output and input if the intersection of their denotation is non-empty, i.e. $\wp(out_j^i) \cap \wp(in_l^k) \neq \emptyset$.

Example 5 (match transition relation of Primes). The Primes program, P, from Example 1 defines the match transition relation,
$$\mathcal{M}_P = \{M_{(1,1,3,1)}, M_{(1,1,3,2)}, M_{(2,1,3,1)}, M_{(2,1,3,2)}$$
$$M_{(2,2,1,1)}, M_{(2,2,2,1)}, M_{(3,1,3,1)}, M_{(3,1,3,2)}\}.$$ □

As in LP, the dependency graph of a CHR program is a directed graph where the nodes represent rules, and the directed arcs dependencies between these rules.

Definition 8 (CHR dependency graph). *A CHR dependency graph, $\mathcal{D}_P$, of an abstract CHR program P is an ordered tuple $\langle T, D \rangle$ of nodes T, one for each transition in the rule transition relation, $\mathcal{T}_P$, of P, and directed arcs D, one for each ordered pair of transitions between which a match transition exists in the match transition relation, $\mathcal{M}_P$, of P.* □

For an SCC analysis, a CHR dependency graph is sufficient, however, for self-sustainability we rely in the next sections on a CHR net instead.

Definition 9 (CHR net). *A CHR net, $\mathcal{N}_P$, of an abstract CHR program, P, is a quadruple $\langle \mathcal{I}n_P, \mathcal{O}ut_P, \mathcal{T}_P, \mathcal{M}_P \rangle$. Here, respectively, $\mathcal{I}n_P$ and $\mathcal{O}ut_P$ are the abstract input and output constraints of P, and $\mathcal{T}_P$ and $\mathcal{M}_P$ the rule and match transition relations of P.* □

Note that a CHR net corresponds to a bipartite hypergraph. We illustrate both notions in Figure 1 for the Primes program of Example 1.

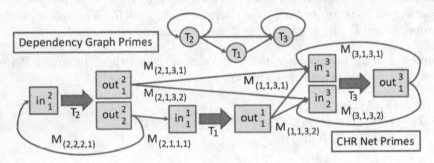

Fig. 1. Dependency graph and CHR net for Primes

4   Self-sustainable SCCs of a CHR Program

To derive the SCCs of a CHR program, we perform an SCC analysis on the dependency graph of the CHR program. This is similar to such an analysis in LP and can be done efficiently using Tarjan's algorithm [21].

Example 6 (SCCs of Primes). Based on $\mathcal{D}_P = (\{T_1, T_2, T_3\}, \{(T_1, T_3), (T_2, T_1), (T_2, T_2), (T_2, T_3), (T_3, T_3)\})$, the dependency graph of Primes of Figure 1, we derive two SCCs, $\{T_2\}$ and $\{T_3\}$. □

4.1   Self-sustainable SCCs

Consider again the non-terminating example program from the Introduction:

$$R_1 @ a, a \Leftrightarrow b, b. \qquad R_2 @ b \Leftrightarrow a.$$

Its dependency graph and CHR net are shown in Figure 2. As can be verified, in Figure 2, $in_1^1 = in_2^1 = out_1^2 = (a, true)$ and $out_1^1 = out_2^1 = in_1^2 = (b, true)$. For the example, there is only one SCC: $\{T_1, T_2\}$.

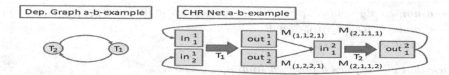

Fig. 2. Dependency graph and CHR net for the a-b-example

What we want to characterise —in the notion of self-sustainability for such an SCC— is whether we can duplicate some of the transitions of this SCC, such that for the resulting multiset of transitions and its *extended CHR net*, it is so that a multisubset of all $M_{(i,j,k,l)}$ match transitions exists, that maps a multisubset of all the out_q^p nodes *onto all of* the in_s^r nodes. Returning to our example, consider the multiset $[\![T_1, T_2, T_2]\!]$ based on $\{T_1, T_2\}$. In Figure 3, we expand the CHR net of Figure 2 to represent all match transitions for this multiset.

It is clear that from the multiset of 8 match transitions in Figure 3, we can select 4 transitions, for example $[\![M_{(1,1,2,1)}, M_{(1,2,2,1)}, M_{(2,1,1,1)}, M_{(2,1,1,2)}]\!]$, such that these transitions define a function from a multisubset of all out_q^p nodes *onto* all in_s^r nodes. This function is represented in Figure 3 in the thick arrows.

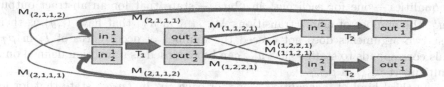

Fig. 3. Expanded CHR net for $[\![T_1, T_2, T_2]\!]$

Note that in this example, the multisubsets of out_q^p nodes on which the function is defined is equal to the entire multiset, $[\![out_1^1, out_2^1, out_1^2, out_1^2]\!]$, of out_q^p nodes. In general, this is not necessary: out_q^p nodes are allowed to be "unused" by in_s^r nodes. It suffices that a CHR constraint is produced for each in_s^r node.

We generalise the construction in the example above in the following definition. In this definition, we start off from a set of CHR rules, C, and its CHR net, $\mathcal{N}_C = \langle \mathcal{I}n_C, \mathcal{O}ut_C, \mathcal{T}_C, \mathcal{M}_C \rangle$. In the definition we will consider a multiset with universe the rule transitions $\mathcal{T}_C$, and will denote it by $\mu_\mathcal{T}$. Note that, given such a multiset $\mu_\mathcal{T}$, there are associated multisets $\mu_{\mathcal{I}n}$ and $\mu_{\mathcal{O}ut}$, with universes $\mathcal{I}n_C$ and $\mathcal{O}ut_C$, respectively. If a transition $T_i \in \mathcal{T}_C$ occurs n_i times in $\mu_\mathcal{T}$, then all its in_j^i and out_l^i abstract constraints occur n_i times in $\mu_{\mathcal{I}n}$ and $\mu_{\mathcal{O}ut}$, respectively.

Given $\mu_\mathcal{T}$, there is also an associated multiset $\mu_\mathcal{M}$, with universe $\mathcal{M}_C$. If transitions T_i and T_k occur respectively n_i and n_k times in $\mu_\mathcal{T}$, then all $M(i,j,k,l) \in \mathcal{M}_C$ occur $n_i \times n_k$ times in $\mu_\mathcal{M}$.

Definition 10 (self-sustainability of a set of CHR rules). *Let C be a set of simplification rules and $\mathcal{N}_C = \langle \mathcal{I}n_C, \mathcal{O}ut_C, \mathcal{T}_C, \mathcal{M}_C \rangle$ the CHR net of C. The set C is self-sustainable iff there exist*

- a non-empty multiset μ_T, with the associated multisets μ_{In}, μ_{Out} and μ_M,
- a multiset $\mu'_{Out} \sqsubseteq \mu_{Out}$, and
- a multiset $\mu'_M \sqsubseteq \mu_M$,

such that μ'_M defines a function from μ'_{Out} onto μ_{In}.   □

Before we illustrate this concept on our running example, Primes, we provide an alternative, numeric characterisation.

Proposition 1 (numeric characterisation self-sustainability). *Let C be a set of simplification rules and $\mathcal{N}_C = \langle \mathcal{I}n_C, \mathcal{O}ut_C, \mathcal{T}_C, \mathcal{M}_C \rangle$ its CHR net. The set C is self-sustainable iff there exist multisets μ_T, with associated multisets μ_{In}, μ_{Out}, μ_M, $\mu'_{Out} \sqsubseteq \mu_{Out}$ and $\mu'_M \sqsubseteq \mu_M$, such that $\sum_{i:T_i \in \mathcal{T}_C} \mu_T(T_i) \geq 1$ and such that*

$$\forall out_j^i \in \mathcal{O}ut_C : \sum_{k,l:M_{(i,j,k,l)} \in \mathcal{M}_C} \mu'_M(M_{(i,j,k,l)}) \leq \mu_T(T_i), \text{ and}$$

$$\forall in_j^i \in \mathcal{I}n_C : \sum_{k,l:M_{(k,l,i,j)} \in \mathcal{M}_C} \mu'_M(M_{(k,l,i,j)}) = \mu_T(T_i).   □$$

Proof. The first inequality expresses that μ_T is non-empty. The second kind of inequalities —one for each out_j^i in $\mathcal{O}ut_C$— state that for an abstract output out_j^i, the number of match transitions $M_{(i,j,k,l)}$ in μ'_M that have (i,j) as their first two arguments does not exceed the number of occurrences of T_i in μ_T. This corresponds to stating that μ'_M can be regarded as a function defined on a multisubset μ'_{Out} of μ_{Out}.

The third kind of inequalities —one for each in_j^i in $\mathcal{I}n_C$— state that for an abstract input constraint in_j^i, the number of match transitions $M_{(k,l,i,j)}$ in μ'_M that have (i,j) as their last two arguments is exactly the number of occurrences of T_i in μ_T. This means that the function defined by μ'_M is onto μ_{In}.   □

Example 7 (self-sustainable SCCs for Primes). Consider the SCCs, $C_1 = \{T_2\}$ and $C_2 = \{T_3\}$, (see Example 6) of the Primes program, P, of Example 1. While the first component of the program is self-sustainable, the second is not. That is, at some point in a computation of C_2, the *prime/1* constraints to fire the rule will be depleted. Consider their CHR nets,

$$\mathcal{N}_{C_1} = \langle \{in_1^2\}, \{out_1^2, out_2^2\}, \{T_2\}, \{M_{(2,2,2,1)}\} \rangle \quad and$$
$$\mathcal{N}_{C_2} = \langle \{in_1^3, in_2^3\}, \{out_1^3\}, \{T_3\}, \{M_{(3,1,3,1)}, M_{(3,1,3,2)}\} \rangle.$$

Let us denote $\mu'_M(M_{(i,j,k,l)})$ as $m_{(i,j,k,l)}$ and $\mu_T(T_i)$ as t_i, where all $m_{(i,j,k,l)}$ and t_i are natural numbers. Then, we can characterise self-sustainability of C_1 and C_2, respectively, by the systems of linear inequalities,

$$t_2 \geq 1 \quad m_{(2,2,2,1)} \leq t_2 \quad m_{(2,2,2,1)} = t_2 \quad and$$
$$t_3 \geq 1 \quad m_{(3,1,3,1)} + m_{(3,1,3,2)} \leq t_3 \quad m_{(3,1,3,1)} = t_3 \quad m_{(3,1,3,2)} = t_3.$$

If a solution to these systems exists, the underlying component is self-sustainable. C_1 is clearly self-sustainable, while C_2 is not.   □

Several comments with respect to Definition 10 and Proposition 1 are in order. First, the statement that $\mu'_{\mathcal{M}}$ defines a function from $\mu'_{\mathcal{O}ut}$ onto $\mu_{\mathcal{I}n}$ is imprecise. In fact, $\mu'_{\mathcal{M}}$ defines a set of functions from $\mu'_{\mathcal{O}ut}$ onto $\mu_{\mathcal{I}n}$. This is because $\mu'_{\mathcal{O}ut}$ and $\mu_{\mathcal{I}n}$ are multisets, that may contain elements more than once.

Therefore, if in^i_j or out^k_l occur multiple times in $\mu_{\mathcal{I}n}$, respectively $\mu'_{\mathcal{O}ut}$, it is unclear which mapping is defined by an element $M_{(k,l,i,j)}$ of $\mu'_{\mathcal{M}}$. Looking back at the a-b-example of Figure 3, there are four different functions in this graph, all corresponding to the multiset $\mu'_{\mathcal{M}} = [\![M_{(1,1,2,1)}, M_{(1,2,2,1)}, M_{(2,1,1,1)}, M_{(2,1,1,2)}]\!]$. The thick lines in the figure represent one of these, but selecting other arcs, with the same labels, produces the other three.

Apart from this, for a fixed multiset $\mu_{\mathcal{T}}$, there can be several mappings $\mu'_{\mathcal{M}}$ that map a multiset $\mu'_{\mathcal{O}ut}$ onto $\mu_{\mathcal{I}n}$. This is because there can be different abstract constraints out^k_l in $\mu_{\mathcal{O}ut}$ that all have a match transition to some abstract constraint in^i_j. This may give a number of alternative candidates for $\mu'_{\mathcal{M}}$.

Finally, by considering different multisets $\mu_{\mathcal{T}}$ based on $\mathcal{T}_C$, we may obtain a very large number of solutions for $\mu'_{\mathcal{M}}$ in Definition 10. In the context of the a-b-example, consider a multiset $\mu_{\mathcal{T}} = [\![T_1, T_1, T_2, T_2, T_2, T_2]\!]$. From the fact that there are twice as many T_2 rules than T_1 rules, it should be intuitively clear that it again allows to find multisets $\mu'_{\mathcal{O}ut}$ and $\mu'_{\mathcal{M}} : \mu'_{\mathcal{O}ut} \to \mu_{\mathcal{I}n}$, with the latter being *onto*. Because of the increased multiplicity of the rule transitions, the number of different functions that $\mu'_{\mathcal{M}}$ represents in this context is much higher than for the previous $\mu_{\mathcal{T}}$. Moreover, it turns out that by increasing the multiplicity of the rule transitions, we can construct concrete functions associated to $\mu'_{\mathcal{M}}$ that cannot be obtained as the union of multiple concrete functions associated to a solution for a $\mu_{\mathcal{T}}$ with lower multiplicity of rule transitions (see [14]).

This observation implies that we are unable to use the notion of a self-sustainable set of rules as a direct basis for a termination analysis. Such an approach would have to identify a finite set of minimal self-sustainable cycles and then prove that all these are terminating. But since in CHR, there are in general an infinite set of minimal cycles, such an approach is not feasible.

In [15], to tackle the problem of an infinite number of minimal cycles, a new concept of minimality is introduced, based on a finite constructive set of solutions —the Hilbert basis— for the inequalities representing self-sustainability. This approach is however slow, with little gain in precision.

Fortunately, there is another way for using the notion of a self-sustaining set of rules. After determining the SCCs of a CHR program, we can verify which SCCs are not self-sustainable, and disregard such SCCs. This observation is based on the following theorems.

Theorem 1. *If a CHR program, P, is not self-sustainable, then P terminates for every query.* □

Theorem 2. *Let P be a CHR program and let C be an SCC of P. If C is not self-sustainable and if $P \setminus C$ terminates for every query, then P terminates for every query.* □

4.2   Non-self-sustainable SCCs of a CHR Program

From here on, we refer to self-sustainable as *'selfs'* and use the matrix form:
$A \times X \leq B$; to represent systems of linear inequalities.

Example 8 (system in matrix form for Primes). We revisit Example 7 and represent for C_1 and C_2, respectively, their systems of inequalities in matrix form:

$$\begin{bmatrix} -1 & 1 \\ 1 & -1 \\ -1 & 1 \\ -1 & 0 \end{bmatrix} \times \begin{bmatrix} t_2 \\ m_{(2,2,2,1)} \end{bmatrix} \leq \begin{bmatrix} 0 \\ 0 \\ 0 \\ -1 \end{bmatrix} \quad and \quad \begin{bmatrix} -1 & 1 & 0 \\ 1 & -1 & 0 \\ -1 & 0 & 1 \\ 1 & 0 & -1 \\ -1 & 1 & 1 \\ -1 & 0 & 0 \end{bmatrix} \times \begin{bmatrix} t_3 \\ m_{(3,1,3,1)} \\ m_{(3,1,3,2)} \end{bmatrix} \leq \begin{bmatrix} 0 \\ 0 \\ 0 \\ 0 \\ 0 \\ -1 \end{bmatrix}.$$

Note that the equalities of the original system are replaced by two inequalities.
E.g. $m_{(2,2,2,1)} = t_2$ is replaced by $m_{(2,2,2,1)} \geq t_2$ and $m_{(2,2,2,1)} \leq t_2$.   □

In order to prove that a component is not *selfs*, we need to prove that no positive
integer solution exists for the variables of the linear inequalities representing that
it is *selfs*. Thus, we need to prove that such a system is *infeasible*.

Definition 11 (feasible). *A system, $S = A \times X \leq B$, is* (in)feasible *iff S has*
(not) *a solution in* $\mathbb{R}^+$.   □

The following lemma is due to Farkas [7].

Lemma 1 (Farkas' lemma). *Let $S = A \times X \leq B$ be a system of linear
inequalities. Then, S is feasible iff $\forall P \geq 0 : A^T \times P \geq 0 \to B^T \times P \geq 0$.
Alternatively, S is infeasible iff $\exists P \geq 0 : A^T \times P \geq 0 \wedge B^T \times P < 0$.*   □

Note that Farkas' lemma is only applicable to the real case: if a real matrix P
exists, such that $P \geq 0 \wedge A^T \times P \geq 0 \wedge B^T \times P < 0$, then, the original system,
S, is infeasible. Therefore, it has no solution in $\mathbb{N} \subset \mathbb{R}^+$ and must be *non-selfs*.

Infeasibility has received much attention in linear programming (see e.g. [3])
and several approaches exist to tackle the problem. It is not in our intention to
improve on these approaches. To evaluate our approach, we do however formulate a simple test, where we first represent infeasibility as a constraint problem
on symbolic coefficients. That is, we introduce a symbolic matrix P', for each
infeasibility problem, of the same dimensions as P, with symbolic coefficients p_i,
one for each position in the matrix. Then, we derive constraints on the symbolic
coefficients, based on $P' \geq 0 \wedge A^T \times P' \geq 0 \wedge B^T \times P' < 0$.

Example 9 (infeasibility for Primes). We revisit Example 8, and formulate infeasibility of the systems of linear inequalities. That is, let P_1 for C_1 be a (4×1)-
matrix of (integer) symbolic coefficients $p_0^1, p_1^1, \ldots, p_3^1$, all greater or equal to 0.
Then, to prove *non-selfs* for C_1, we need to satisfy:

$$\begin{bmatrix} -1 & 1 & -1 & -1 \\ 1 & -1 & 1 & 0 \end{bmatrix} \times P_1 \geq 0 \quad and \quad \begin{bmatrix} 0 & 0 & 0 & -1 \end{bmatrix} \times P_1 < 0.$$

We derive the following problem, where $\forall i \in \{0, 1, \ldots, 3\} : p_i^1 \geq 0$:

$$-p_0^1 + p_1^1 - p_2^1 - p_3^1 \geq 0 \qquad p_0^1 - p_1^1 + p_2^1 \geq 0 \qquad -p_3^1 < 0$$

There is no solution to this problem, thus C_1 is *selfs*. For C_2, we have

$$\begin{bmatrix} -1 & 1 & -1 & 1 & -1 & -1 \\ 1 & -1 & 0 & 0 & 1 & 0 \\ 0 & 0 & 1 & -1 & 1 & 0 \end{bmatrix} \times P_2 \geq 0 \; and \; \begin{bmatrix} 0 & 0 & 0 & 0 & 0 & -1 \end{bmatrix} \times P_2 < 0.$$

We derive the following problem, where $\forall i \in \{0, 1, \ldots, 5\} : p_i^2 \geq 0$:

$$-p_0^2 + p_1^2 - p_2^2 + p_3^2 - p_4^2 - p_5^2 \geq 0 \qquad p_0^2 - p_1^2 + p_4^2 \geq 0$$
$$p_2^2 - p_3^2 + p_4^2 \geq 0 \qquad -p_5^2 < 0$$

One solution is: $p_0^2 = 0$, $p_1^2 = 1$, $p_2^2 = 0$, $p_3^2 = 1$, $p_4^2 = 1$, and $p_5^2 = 1$. Therefore, C_2 is *non-selfs* and thus must be terminating. □

To find a solution to these constraint problems, we can transform them to SAT problems, representing that a component is *non-selfs*, by using a transformation based on signed integer arithmetic. Thus, in this approach, we can only represent integers and not reals. This is still sufficient to represent the infeasibility problem, however, yields an incomplete approach (see Lemma 1): it is not guaranteed that a solution in the integers can be found if the original system is infeasible. Nevertheless, as it turns out, the approach works well in practice.

5   Evaluation

We have implemented the *non-selfs* test using SWI-Prolog [19] and integrated it with CHRisTA, a termination analyser for CHR(Prolog) [13]. This resulted in a new termination analyser, T-CoP [20] (Termination of CHR on top of Prolog).

CHRisTA uses CHR nets as a representation of CHR programs and implements an SCC analysis based on Tarjan's algorithm. To solve the satisfiability problems that CHRisTA is confronted with, we represent the Diophantine constraints as a sufficient SAT problem based on signed integer arithmetic and solve it with MiniSAT2 [10]. Of course, by doing so, we place limits on the domains of variables and integers of the original problem. That is, we limit on the bit-sizes used for representing the variables and the integers of the problem. We reuse this system to prove infeasibility of the linear inequalities that we obtain for *selfs*.

Therefore, integrating the *non-selfs* test into CHRisTA involved the following steps: deriving a system of linear inequalities for *selfs* from the CHR net of an SCC, formulating infeasibility of that system, and transforming the resulting infeasibility problem to a SAT problem. Afterwards, a proof of termination of the remaining SCCs, that are possibly *selfs*, is attempted by the termination proof procedure of CHRisTA.

Table 1 contains a representative set of the results that we obtained. T-CoP, as well as the full benchmark, is available online for reference [20]. Note that

Table 1. Results of *non-selfs* test on terminating SCCs

CHR program	3bit	sec	4bit	sec	CHRisTA (sec)	T-CoP (sec)
factorial	-	0.10	-	0.14	1.50	1.64
mean	+	0.35	+	0.58	1.26	0.58
mergesortscc1	-	0.12	-	0.20	1.44	1.64
mergesortscc2	+	0.19	+	0.34	1.51	0.34
newcase1	+	0.41	+	0.74	timeout	0.74
newcase2	-	1.09	+	2.03	timeout	2.03
primesscc1	-	0.11	-	0.19	1.67	1.86
primesscc2	+	0.18	+	0.33	1.11	0.33

the considered programs of Table 1 contain only a single SCC of a size and complexity similar to the SCCs expected in real-sized CHR programs.

In Table 1, there are SCCs that are *non-selfs*, which are proven terminating (+) using at most a 4 bit representation for the variables and integers of the SAT problems. The SCCs, only proven terminating by CHRisTA, are *selfs*.

Among these programs, there are two programs on which the proof procedure of CHRisTA fails (see [12]). An example, representative for such programs, is $\{(R_1 @ a, a \Leftrightarrow b.), (R_2 @ b \Leftrightarrow a.)\}$. To prove termination, for R_1, the size of a must be greater or equal to b. For R_2, a has to be strictly smaller than b.

Our new approach can handle such programs, and integrated with CHRisTA, improves the precision of the termination analysis as shown in Table 1 under "T-CoP". From these results we may also conclude that the *non-selfs* test improves the efficiency of the analysis: If an SCC cannot be proven *non-selfs*, the overhead is acceptable; if it can be proven *non-selfs*, the gain in speed is relatively high.

Finally, note that the transformational approach of [16], combined with the technique of [17], can be used to handle programs such as newcase 1 and 2. Unfortunately, in the presence of propagation, the transformational approach of [16] cannot be applied.

6   Conclusion and Future Work

In this paper, we developed an approach to detect non-self-sustainability of the SCCs of a CHR program by means of a satisfiability problem. Integrating the approach with CHRisTA [13], resulted in a more efficient analyser, T-CoP. For one, the approach improves the precision of the termination analysis, being able to prove termination of a new class of CHR programs automatically. Furthermore, we improve the efficiency of the termination analyser CHRisTA (cfr. T-CoP).

Future work will consider propagation rules. For self-sustainability of propagation, we will need to assume presence of a fire-once policy. Intuitively, under such a policy, if no new combinations of constraints are ever introduced as a consequence of applying a propagation rule, such that the propagation rule can be applied again on the newly introduced combinations of constraints, then the propagation rule is non-selfs. Informally, this corresponds to verifying whether there does not exist a cycle in the CHR net of a CHR program across an abstract input constraint of the propagation rule and across some abstract output constraint *added* by a rule of the program. If so, the propagation rule can be ignored. Therefore, if we obtain by this approach an SCC that ultimately does not contain propagation, then, we can verify whether the SCC is non-selfs by the approach of this paper. Otherwise, the remaining SCC is still selfs by default.

References

1. Abdennadher, S.: Operational Semantics and Confluence of Constraint Propagation Rules. In: Smolka, G. (ed.) CP 1997. LNCS, vol. 1330, pp. 252–266. Springer, Heidelberg (1997)
2. Apt, K.R.: Logic programming. In: Handbook of Theoretical Computer Science, Volume B: Formal Models and Sematics (B), pp. 493–574 (1990)
3. Beasley, J.E. (ed.): Advances in linear and integer programming. Oxford University Press, Inc. (1996)
4. Blizard, W.D.: Multiset theory. Notre Dame Journal of Formal Logic 30(1), 36–66 (1989)
5. Decorte, S., De Schreye, D., Vandecasteele, H.: Constraint-based termination analysis of logic programs. ACM Trans. Program. Lang. Syst. 21(6), 1137–1195 (1999)
6. Dershowitz, N., Manna, Z.: Proving termination with multiset orderings. Commun. ACM 22(8), 465–476 (1979)
7. Farkas, J.G.: Über die theorie der einfachen ungleichungen. Journal für die Reine und Angewandte Mathematik 124, 1–27 (1902)
8. Frühwirth, T.W.: Theory and practice of constraint handling rules. J. Log. Program. 37(1-3), 95–138 (1998)
9. Lloyd, J.W.: Foundations of Logic Programming, 2nd edn. Springer (1987)
10. MiniSAT (2010), http://minisat.se/
11. Nguyen, M.T., Giesl, J., Schneider-Kamp, P., De Schreye, D.: Termination Analysis of Logic Programs Based on Dependency Graphs. In: King, A. (ed.) LOPSTR 2007. LNCS, vol. 4915, pp. 8–22. Springer, Heidelberg (2008)
12. Pilozzi, P., De Schreye, D.: Termination Analysis of CHR Revisited. In: Garcia de la Banda, M., Pontelli, E. (eds.) ICLP 2008. LNCS, vol. 5366, pp. 501–515. Springer, Heidelberg (2008)
13. Pilozzi, P., De Schreye, D.: Automating Termination Proofs for CHR. In: Hill, P.M., Warren, D.S. (eds.) ICLP 2009. LNCS, vol. 5649, pp. 504–508. Springer, Heidelberg (2009)
14. Pilozzi, P., De Schreye, D.: Scaling termination proofs by a characterization of cycles in CHR. Technical Report CW 541, K.U.Leuven - Dept. of C.S., Leuven, Belgium (2009)
15. Pilozzi, P., De Schreye, D.: Scaling termination proofs by a characterisation of cycles in CHR. In: Termination Analysis, Proceedings of 11th International Workshop on Termination, WST 2010, United Kingdom, July 14-15 (2010)
16. Pilozzi, P., Schrijvers, T., De Schreye, D.: Proving termination of CHR in Prolog: A transformational approach. In: Termination Analysis, Proceedings of 9th International Workshop on Termination, WST 2007, Paris, France (June 2007)
17. Schneider-Kamp, P., Giesl, J., Nguyen, M.T.: The Dependency Triple Framework for Termination of Logic Programs. In: De Schreye, D. (ed.) LOPSTR 2009. LNCS, vol. 6037, pp. 37–51. Springer, Heidelberg (2010)

18. Sneyers, J., Van Weert, P., Schrijvers, T., De Koninck, L.: As time goes by: Constraint Handling Rules – A survey of CHR research between 1998 and 2007. Theory and Practice of Logic Programming 10(1), 1–47 (2010)
19. SWI-Prolog (2010), http://www.swi-prolog.org
20. T-CoP (2010), http://people.cs.kuleuven.be/~paolo.pilozzi?pg=tcop
21. Tarjan, R.E.: Depth-first search and linear graph algorithms. SIAM J. Comput. 1(2), 146–160 (1972)
22. Termination Competition (2010), http://termination-portal.org/
23. Voets, D., De Schreye, D., Pilozzi, P.: A new approach to termination analysis of constraint handling rules. In: Pre-proceedings of Logic Programming, 18th International Symposium on Logic-Based Program Synthesis and Transformation, LOPSTR 2008, Valencia, Spain, pp. 28–42 (2008)

Proving Properties of Co-Logic Programs by Unfold/Fold Transformations

Hirohisa Seki*

Dept. of Computer Science, Nagoya Inst. of Technology,
Showa-ku, Nagoya, 466-8555 Japan
seki@nitech.ac.jp

Abstract. We present a framework for unfold/fold transformation of co-logic programs, where each predicate is annotated as either inductive or coinductive, and the declarative semantics of co-logic programs is defined by an alternating fixpoint model: the least fixpoints for inductive predicates and the greatest fixpoints for coinductive predicates. We show that straightforward applications of conventional program transformation rules are not adequate for co-logic programs, and propose new conditions which ensure the preservation of the intended semantics of co-logic programs through program transformation. We then examine the use of our transformation rules for proving properties of co-logic programs which specify computations over infinite structures. We show by some examples in the literature that our method based on unfold/fold transformation can be used for verifying some properties of Büchi automata and nested automata.

1 Introduction

Co-logic programming (co-LP) is an extension of logic programming recently proposed by Gupta et al. [6] and Simon et al. [20,21], where each predicate in definite programs is annotated as either *inductive* or *coinductive*, and the declarative semantics of co-logic programs is defined by an alternating fixpoint model: the least fixpoints for inductive predicates and the greatest fixpoints for coinductive predicates. Predicates in co-LP are defined over infinite structures such as infinite trees or infinite lists as well as finite ones, and co-logic programs allow us to represent and reason about properties of programs over such infinite structures. Co-LP therefore has interesting applications to reactive systems and verifying properties such as safety and liveness in model checking and so on.

In co-LP, Simon et al. also proposed the operational semantics defined by combining standard SLD-resolution for inductive predicates and *co-SLD* resolution for coinductive predicates. However, to the best of our knowledge, methodologies for reasoning about co-LP such as the familiar unfold/fold transformation rules have not been studied. In fact, as discussed in [16], very few methods which use

---

* This work was partially supported by JSPS Grant-in-Aid for Scientific Research (C) 21500136.

G. Vidal (Ed.): LOPSTR 2011, LNCS 7225, pp. 205–220, 2012.
© Springer-Verlag Berlin Heidelberg 2012

logic programs over infinite structures, have been reported so far for proving properties of infinite computations. The work by Pettorossi, Proietti and Senni [16] is one of few exceptions of such proposals, and they gave a method based on their unfold/fold transformation rules for ω-*programs*, locally stratified programs on infinite lists, for verifying properties of such programs. The semantics of ω-programs is defined based on perfect model semantics [17], where the semantics is defined in terms of iterated least fixpoints only (without using greatest fixpoints).

In this paper we consider a framework for unfold/fold transformation of co-logic programs. We first show that straightforward applications of conventional transformation rules for definite programs such as those by Tamaki and Sato [22] are not adequate for co-logic programs, and then propose new conditions which ensure the preservation of the intended semantics of co-logic programs through program transformation.

One of the motivations of this paper is to further study the applicability of techniques based on unfold/fold transformation not only to program development originally due to Burstall and Darlington [1], but also for proving properties of programs, which goes back to Kott [10] in functional programs. We show by examples that our method based on unfold/fold transformation can be used for verifying some properties of Büchi automata and nested automata.

The organization of this paper is as follows. In Section 2, we summarise some preliminary definitions on co-logic programs. In Section 3, we present our transformation rules, and give some conditions which ensure the correctness of our transformation system. In Section 4, we explain by examples how our proof method via unfold/fold transformations proves properties of co-logic programs. Finally, we discuss about the related work and give a summary of this work in Section 5. [1]

Throughout this paper, we assume that the reader is familiar with the basic concepts of logic programming, which are found in [11].

2   Preliminaries: Co-Logic Programs

In this section, we recall some basic definitions and notations concerning co-logic programs. The details and more examples are found in [6,20,21]. We also explain some preliminaries on constraint logic programming (CLP) (e.g., [9] for a survey).

Since co-logic programming can deal with infinite terms such as infinite lists or trees like $f(f(\dots))$ as well as finite ones, we consider the *complete* (or *infinitary*) Herbrand base [11,8], denoted by HB_P^*, where P is a program. Therefore, an equation $X = f(X)$, for example, has a solution $X = f(f(\dots))$. The equality theory in co-logic programming is the same as that of the standard one [11], except that we have no axiom corresponding to "occur-check", i.e., $\forall (t \neq X)$ for each term t containing X and different from X. The theory of

---

[1]   Due to space constraints, we omit most proofs and some details, which will appear in the full paper.

equality in this case is studied by Colmerauer [3]. Throughout this paper, we assume that there exists at least one constant and one function symbol of arity ≥ 1, thus HB_P^* is non-empty.

It is well-known that pure logic programming (LP for short) can be seen as an instance of the CLP scheme obtained by considering the following simple translation: let $\gamma : p(\tilde{t}_0) \leftarrow p_1(\tilde{t}_1), \ldots, p_n(\tilde{t}_n)$ be a clause in LP, where $\tilde{t}_0, \ldots, \tilde{t}_n$ are tuples of terms. Then, γ is mapped into the following *pure* CLP clause:

$$p(\tilde{x}_0) \leftarrow \tilde{x}_0 = \tilde{t}_0 \wedge \tilde{x}_1 = \tilde{t}_1 \wedge \cdots \wedge \tilde{x}_n = \tilde{t}_n \, [\!] \, p_1(\tilde{x}_1), \ldots, p_n(\tilde{x}_n),$$

where $\tilde{x}_0, \ldots, \tilde{x}_n$ are tuples of new and distinct variables, and $[\!]$ means conjunction ("$\wedge$"). Therefore, in examples we will use the conventional representation of an LP clause as a shorthand for a pure CLP clause, for the sake of readability.

In the following, for a CLP clause γ of the form: $H \leftarrow c \, [\!] \, B_1, \ldots, B_n$, the head H and the body $B_1, \ldots, B_n$ are denoted by $hd(\gamma)$ and $bd(\gamma)$, respectively. We call c the *constraint* of γ. A conjunction $c \, [\!] \, B_1, \ldots, B_n$ is said to be a *goal* (or a *query*). The predicate symbol of the head of a clause is called the *head predicate* of the clause.

The set of all clauses in a program P with the same predicate symbol p in the head is called the definition of p and denoted by $Def(p, P)$. We say that a predicate p *depends on* a predicate q in P iff either (i) $p = q$, (ii) there exists in P a clause of the form: $p(\ldots) \leftarrow c \, [\!] \, B$ such that predicate q occurs in B or (iii) there exists a predicate r such that p depends on r in P and r depends on q in P. The *extended definition* [14] of p in P, denoted by $Def^*(p, P)$, is the conjunction of the definitions of all the predicates on which p depends in P.

2.1   Syntax

In co-logic programming, predicate symbols are annotated as either inductive or coinductive.[2] We denote by $\mathcal{P}^{in}$ ($\mathcal{P}^{co}$) the set of inductive (coinductive) predicates in P, respectively. However, there is one restriction, referred to as the *stratification restriction*: Inductive and coinductive predicates are not allowed to be mutually recursive. This notion is defined formally in Def. 1. The following is an example of co-logic programs.

Example 1. [21]. Suppose that predicates *member* and *drop* are annotated as inductive, while predicate *comember* is annotated as coinductive.

$member(H, [H|\_]) \leftarrow$ $\qquad\qquad\qquad$ $drop(H, [H|T], T) \leftarrow$
$member(H, [\_|T]) \leftarrow member(H, T)$ $\qquad$ $drop(H, [\_|T], T_1) \leftarrow drop(H, T, T_1)$
$comember(X, L) \leftarrow drop(X, L, L_1), comember(X, L_1)$

The definition of *member* is a conventional one, and, since it is an inductive predicate, its meaning is defined in terms of the least fixpoint (Sect. 2.2). Therefore, the prefix ending in the desired element H must be finite. The similar thing also holds for predicate *drop*.

---

[2] We call an atom, A, an *inductive* (a *coinductive*) atom when the predicate of A is an inductive (a coinductive) predicate, respectively.

On the other hand, predicate *comember* is coinductive, whose meaning is defined in terms of the greatest fixpoint (Sect. 2.2). Therefore, it is true if and only if the desired element X occurs an infinite number of times in the list L. Hence it is false when the element does not occur in the list or when the element only occurs a finite number of times in the list.   □

Definition 1. Co-logic Program

A *co-logic program* P is a definite program such that it satisfies the *stratification restriction*, namely, it is possible to decompose the set $\mathcal{P}$ of all predicates in P into a collection (called a *stratification*) of mutually disjoint sets $\mathcal{P}_1, \ldots, \mathcal{P}_r$ $(1 \leq r)$, called *strata*, so that for every clause

$$p(\tilde{x}_0) \leftarrow c \, [\!] \, p_1(\tilde{x}_1), \ldots, p_n(\tilde{x}_n),$$

in P, we have that $\sigma(p) \geq \sigma(p_i)$ if p and p_i have the same inductive/coinductive annotations, and $\sigma(p) > \sigma(p_i)$ otherwise, where $\sigma(q) = i$, if the predicate symbol q belongs to $\mathcal{P}_i$.   □

2.2 Semantics of Co-Logic Programs

The declarative semantics of a co-logic program is a stratified interleaving of the least fixpoint semantics and the greatest fixpoint semantics.

In this paper, we consider the complete Herbrand base HB_P^* as the set of elements in the domain of a *structure* $\mathcal{D}$ (i.e., a complete Herbrand interpretation [11]), and as a constraint a set of equations of the form $s = t$, where s and t are finite terms.[3]

Given a structure $\mathcal{D}$ and a constraint c, $\mathcal{D} \models c$ denotes that c is true under the interpretation for constraints provided by $\mathcal{D}$. Moreover, if θ is a ground *substitution* (i.e., a mapping of variables on the domain $\mathcal{D}$, HB_P^* in this case) and $\mathcal{D} \models c\theta$ holds, then we say that c is *satisfiable*, and θ is called a *solution* (or ground *satisfier*) of c, where $c\theta$ denotes the application of θ to the variables in c. We refer to [3] for an algorithm for checking constraint satisfiability.

Suppose that P is a program and I is an interpretation (i.e., a subset of HB_P^*). The operator $\mathcal{T}_{P,I}$ assigns to every set T of ground atoms a new set $\mathcal{T}_{P,I}(T)$ of ground atoms. Intuitively, I represents facts currently known to be true and $\mathcal{T}_{P,I}(T)$ contains *new* facts (i.e., facts not contained in I), whose truth can be immediately derived from the program P assuming that all facts in I hold and assuming that all facts in T are true.

Definition 2. For set T of ground atoms we define:

$\mathcal{T}_{P,I}(T) = \{ A : A \notin I,$ and there is a ground substitution θ and a clause in P

$\qquad H \leftarrow c \, [\!] \, B_1, \cdots, B_n, \ n \geq 0,$ such that

$\qquad$ (i) $A = H\theta$, (ii) θ is a solution of c, and

$\qquad$ (iii) for every $1 \leq i \leq n$, either $B_i\theta \in I$ or $B_i\theta \in T \}$.   □

---

[3] In Example 5 in Sect. 4, we use *unequation* [8] (or *inequation* [4]) ($\neq$) only in a limited form.

We consider $\mathcal{T}_{P,I}$ to be the operator defined on the set of all subsets of HB_P^*, ordered by standard inclusion. Next, two subsets $T_{P,I}^{\uparrow\omega}$ and $T_{P,I}^{\downarrow\omega}$ of complete Herbrand base are defined by iterating the operators $\mathcal{T}_{P,I}$.

Definition 3. Let I be an interpretation. Define:

$$T_{P,I}^{\uparrow 0} = \emptyset \text{ and } T_{P,I}^{\downarrow 0} = HB_P^* ;$$

$$T_{P,I}^{\uparrow n+1} = \mathcal{T}_{P,I}(T_{P,I}^{\uparrow n}) \text{ and } T_{P,I}^{\downarrow n+1} = \mathcal{T}_{P,I}(T_{P,I}^{\downarrow n}) ;$$

$$T_{P,I}^{\uparrow\omega} = \cup_{n<\omega} T_{P,I}^{\uparrow n} \text{ and } T_{P,I}^{\downarrow\omega} = \cap_{n<\omega} T_{P,I}^{\downarrow n}.$$

□

Finally, the model $M(P)$ of a co-logic program P is defined as follows:

Definition 4. Suppose that a co-logic program P has a stratification $\mathcal{P}_i$ ($1 \leq i \leq m$), where $\mathcal{P}_i$ is a set of predicates with stratum i. We denote by P_i the union of the definitions of the predicates in $\mathcal{P}_i$, and let $P_{<i}$ be the union of P_j such that $\mathcal{P}_j$ is a strictly lower stratum than $\mathcal{P}_i$.
Let:

$$M(P_i) = \begin{cases} T_{P_i, M(P_{<i})}^{\uparrow\omega}, & \text{if } \mathcal{P}_i \text{ is inductive,} \\ T_{P_i, M(P_{<i})}^{\downarrow\omega}, & \text{if } \mathcal{P}_i \text{ is coinductive.} \end{cases}$$

Then, the model of P, denoted by $M(P)$, is the union of all models $M(P_i)$, for $i = 1, \ldots, m$. □

We note that Simon et al. [20,21] defined the semantics $M(P)$ in terms of the greatest fixpoint of $\mathcal{T}_{P,I}$. On the other hand, we use $T_{P,I}^{\downarrow\omega}$ instead. This is justified by the notable fact that $gfp(\mathcal{T}_{P,I}) = T_{P,I}^{\downarrow\omega}$ in the complete Herbrand base HB_P^* [11].

3   Unfold/fold Transformation of Co-Logic Programs

We first explain our transformation rules, and then give some conditions imposed on the transformation rules which are necessary for correctness of transformation. Our transformation rules are formulated in the framework of CLP, following that by Etalle and Gabbrielli [5]. The set of variables in an expression E is denoted by $var(E)$. Given atoms A, H, we write $A = H$ as a shorthand for: (i) $a_1 = t_1 \wedge \cdots \wedge a_n = t_n$, if, for some predicate symbol p and $n \geq 0$, $A \equiv p(a_1, \ldots, a_n)$ and $H \equiv p(t_1, \ldots, t_n)$ (where $\equiv$ denotes syntactic equality), (ii) *false*, otherwise. This notation readily extends to conjunctions of atoms.

Let c_i be constraints, G_i be conjunctions of atoms, and $\gamma_i : A_i \leftarrow c_i \,[\!]\, G_i$ be clauses ($i = 1, 2$). We say that a goal $c_1 \,[\!]\, G_1$ is an *instance* of the goal $c_2 \,[\!]\, G_2$ iff, for any solution θ of c_1, there exists a solution τ of c_2 such that $G_1\theta \equiv G_2\tau$. We write $\gamma_1 \simeq \gamma_2$ iff, for any $i, j \in [1,2]$ and for any solution θ of c_i, there exists a solution τ of c_j such that $A_i\theta \equiv A_j\tau$ and $G_i\theta$ and $G_j\tau$ are equal as multisets.

3.1 Transformation Rules

A sequence of programs $P_0, \ldots, P_n$ is said to be a *transformation sequence* with a given initial program P_0, if each P_i $(1 \leq i \leq n)$ is obtained from P_{i-1} by applying a transformation rule as described below. It is well-known that an unconditional application of transformation rules will not preserve the semantics of the initial program P_0. We thus impose some conditions on our transformation rules as well as the definition of an initial program P_0.

Our motivation of this paper is to use program transformation rules for proving properties of a given system represented in a co-logic program P_0. We thus assume that there exist two kinds of predicate symbols in P_0: *base* predicates and *defined* predicates. A base predicate depends on base predicates only in P_0, and it is intended to represent the structure and its behaviours of the given system. Its definition is therefore assumed to remain unchanged during program transformation. On the other hand, a defined predicate, which is also defined in terms of base predicates only in P_0, is intended to represent a property of the given system such as safety and liveness properties. Its definition will be changed during program transformation so that its truth value in $M(P_n)$ will be easily known. The following is a formal definition of an initial program P_0:

Definition 5. Initial Program

We call a co-logic program P_0 an *initial program* if it satisfies the following conditions:

1. P_0 is divided into two disjoint sets of clauses, P_b and *Defs*. The predicates defined in P_b are called *base* predicates, while those defined in *Defs* are called *defined* predicates. The set of base (defined) predicates is denoted by $\mathcal{P}_b$ ($\mathcal{P}_{def}$), respectively.
2. The defined predicates appear neither in P_b nor in the bodies of the clauses in *Defs*.
3. Let γ be a clause in *Defs* and h be its head predicate. Then, h is annotated as *inductive* iff there exists at least one inductive atom in each body of the definition of h, while h is annotated as *coinductive* iff every predicate symbol occurring in each body of the definition of h is annotated as *coinductive*. □

From the above definition, an initial program consists of two layers: P_b and *Defs*. The conditions 1 and 2 are the same as those in the original framework by Tamaki and Sato [22], where a base (defined) predicate is called an *old* (*new*) predicate, respectively.

As mentioned above, the role of the predicate annotations such as "base" and "defined" is to specify the conditions for applying our transformation rules; they are orthogonal to the predicate annotations of "inductive" and "coinductive", which are used to represent the intended semantics of each predicate in co-logic programs. In particular, from the above condition 2, the meaning of a *defined* predicate is determined only by the base predicates on which it depends, irrelevant of whether its annotation is inductive or coinductive, as far as an initial program P_0 is concerned. However, the definition of a defined predicate will be changed during program transformation, which requires an extra condition 3 for enforcing the annotation of either "inductive" or "coinductive" on a *defined* predicate (see Example 2). On the other hand, since we assume that the definition of every *base* predicate in P_0 remains *untransformed* at any step in a transformation sequence as noted above, the following transformation rules are supposed to be applied to clauses whose head predicates are *defined*.

Our transformation rules consist of standard ones such as unfolding, folding and replacement rule. In particular, unfolding and folding are the same as those in Etalle and Gabbrielli [5], which are CLP counterparts of those by Tamaki and Sato [22] for definite programs. The definitions of the transformation rules are given modulo reordering of the bodies of clauses, and we assume that the clauses of a program have been renamed so that they do not share variables pairwise.

R1. Unfolding. Let γ be a clause in a program P of the form: $A \leftarrow c \, [\!] \, H, G$, where c is a constraint, H is an atom, and G is a (possibly empty) conjunction of atoms. Let $\gamma_1, \ldots, \gamma_k$ with $k \geq 0$, be all clauses of P such that $c \wedge c_i \wedge (H = hd(\gamma_i))$ is satisfiable for $i = 1, \ldots, k$, where c_i is the constraint of γ_i.
By *unfolding* γ w.r.t. H, we derive from P the new program P' by replacing γ by $\eta_1, \ldots, \eta_k$, where η_i is the clause $A \leftarrow c \wedge c_i \wedge (H = hd(\gamma_i)) \, [\!] \, bd(\gamma_i), G$, for $i = 1, \ldots, k$. We say that clauses $\gamma_1, \ldots, \gamma_k$ are *unfolding clauses* and clauses $\eta_1, \ldots, \eta_k$ are *derived from* γ. In particular, if $k = 0$, i.e., there exists no clause in P such that $c \wedge c_i \wedge (H = hd(\gamma_i))$ is satisfiable, then we derive from P the new program P' by deleting clause γ.

R2. Folding. Let γ be a clause in P of the form: $A \leftarrow c_A \, [\!] \, K, G$, where K and G are conjunctions of atoms. Let δ be a clause in *Defs* of the form: $D \leftarrow c_D \, [\!] \, M$, where M is a non-empty conjunction of atoms, and $c_A \, [\!] \, K$ is an instance of $c_D \, [\!] \, M$.
Suppose that there exists a constraint e such that $var(e) \subseteq var(D) \cup var(\gamma)$, and the following conditions hold:

   (i) $\mathcal{D} \models \exists_{-var(A,G,M)} \, c_A \wedge e \wedge c_D \leftrightarrow \exists_{-var(A,G,M)} \, c_A \wedge (M = K)$, where $\exists_{-var(E)} \, \phi$ is the notation, due to [9], to denote the existential closure of the formula ϕ except for the variables in E which remain unquantified, and
   (ii) there is no clause $\delta' : B \leftarrow c_B \, [\!] \, M'$ in *Defs* such that $\delta' \neq \delta$ and $c_A \wedge e \wedge (D = B) \wedge c_B$ is satisfiable.

By applying *folding* to γ w.r.t. K using clause δ, we get the clause $\eta: A \leftarrow c_A \wedge e \, [\!] \, D, G$, and we say that η is *derived from* γ. We derive from P the new program $P' = (P \setminus \{\gamma\}) \cup \{\eta\}$.
The clause γ is called the *folded clause* and δ the *folding clause* (or *folder clause*).

R3. Replacement Rule. We consider the following two rules depending on the annotation of a predicate.

  –The set of the *useless* inductive predicates[4] of a program P is the maximal set U of inductive predicates of P such that a predicate p is in U if, for the body of each clause of $Def(p, P)$, it has an inductive atom whose predicate is in U. By applying the *replacement rule* to P w.r.t. the useless inductive

---

[4] This notion is originally due to Pettorossi and Proietti [14], where the rule is called *clause deletion rule*.

predicates in P, we derive the new program P' from P by removing the definitions of the useless inductive predicates.

–Let $p(\tilde{t})$ be a coinductive atom, and γ be a clause (modulo $\simeq$) in P of the form: $p(\tilde{t}) \leftarrow p(\tilde{t})$. By applying the *replacement rule* to P w.r.t. $p(\tilde{t})$, we derive from P the new program $P' = P \setminus \{\gamma\} \cup \{p(\tilde{t}) \leftarrow\}$.

We note that we do not explicitly consider *definition introduction* [14] as a transformation rule, which allows us to introduce the definition of a new defined predicate during program transformation. The reason is that defined predicates introduced in the course of transformation can be assumed to be present in an initial program from scratch as in [22].

Etalle and Gabbrielli [5] show that we can always replace any clause γ in a program P by a clause γ' without affecting the results of the transformations, provided that $\gamma \simeq \gamma'$. Since this operation is useful to *clean up* the constraints, thereby presenting a clause in a more readable form, we will use it in examples.

Example 2. Consider the following program P_0, where we assume that p (a) is a base (defined) predicate, respectively. Consider the transformation sequence from program P_0 to P_1, which is obtained by applying unfolding to the second clause in P_0 and then applying folding to the resultant clause:

$$P_0 : p \leftarrow p \qquad\qquad P_1 : p \leftarrow p$$
$$a \leftarrow p \qquad\qquad\qquad a \leftarrow a$$

The above transformation is a legitimate transformation in the sense of Tamaki-Sato [22] as well as Seki [19], since the application of folding is *admissible* (formally defined in Sect. 3.2), which informally means that the application is not "immediate self-folding". We thus have that $lfp(P_0) = lfp(P_1) = \emptyset$ and $gfp(P_0) = gfp(P_1) = \{p, a\}$, i.e., the above transformation preserves the least fixpoint and the greatest fixpoint of the initial program.

Table 1 shows the alternating fixpoint semantics of P_0 and P_1, depending on the predicate annotations. Since the definition of *defined* predicate a in P_0 is non-recursive, the truth value of a is determined by those base predicates with lower strata (i.e., p), regardless of whether a is inductive or not. After the unfold/fold transformation, however, the annotation of a does matter to its truth value in P_1, where a is recursively defined. We note that condition 3 in Def. 5 enforces such a restriction on the predicate annotation of a *defined* predicate.   □

Table 1. The Alternating Fixpoint Semantics in Example 2: The rows in bold letters mean the cases where the semantics is preserved

annotations		$M(P_0)$	$M(P_1)$
p: inductive	a: inductive	$\emptyset$	$\emptyset$
	a: coinductive	$\emptyset$	$\{a\}$
p: coinductive	a: inductive	$\{p, a\}$	$\{p\}$
	a: coinductive	**$\{p, a\}$**	**$\{p, a\}$**

3.2 Conditions on Transformation Rules

In the following, we will explain the conditions imposed on unfolding and folding rules to preserve the alternating fixpoint semantics of a co-logic program. Our application conditions of folding are given depending on whether the annotation of the head predicate of a folded clause is inductive or not.

Let f be an *inductive* defined predicate. We call an inductive predicate p *primitive*, if, for some coinductive predicate q on which f depends in P_0, q depends on p in P_0. We denote by $\mathcal{P}_{pr}$ the set of all the primitive predicates, i.e., $\mathcal{P}_{pr} = \{p \in \mathcal{P}^{in} \mid f$ depends on q in P_0, q depends on p in $P_0, f \in \mathcal{P}^{in} \cap \mathcal{P}_{def}, q \in \mathcal{P}^{co}\}$. We call an inductive predicate p *non-primitive*, if it is not primitive. We call an atom with non-primitive (primitive) predicate symbol a non-primive (primitive) atom, respectively.

Example 3. Consider the following program P_0, where the annotations of predicates p, q and f are given in the table below. From the above definition, p is a *primitive* atom.

			$\mathcal{P}^{in}$	$\mathcal{P}^{co}$
$P_0 : p \leftarrow$	$P_1 : p \leftarrow$			
$q \leftarrow p, q$	$q \leftarrow p, q$	$\mathcal{P}_b$	p	q
$f \leftarrow p, q$	$f \leftarrow f$	$\mathcal{P}_{def}$	f	$-$

We consider the transformation sequence from program P_0 to P_1, which is obtained by applying unfolding to the third clause in P_0 w.r.t. p, followed by unfolding the derived clause w.r.t. q, and then applying folding to the resultant clause. Then, we have that $M(P_0) = \{p, q, f\}$, while $M(P_1) \not\models f$. Thus, the above transformation does not preserve the alternating fixpoint semantics. We note that, in the transformation sequence, unfolding is applied to $f \leftarrow p, q$ w.r.t. the *primitive* atom p.   □

The above example leads us to the following notion of marking, which will be used to specify our application condition when the head predicate of a folded clause is annotated as inductive. Let $P_0, \ldots, P_n$ be a transformation sequence. We first mark each clause in $Defs \subseteq P_0$ "*not TS-foldable*". Let η be a clause in P_{k+1} $(0 \le k < n)$.

(i) Suppose that η is derived from $\gamma \in P_k$ by unfolding $\gamma : A \leftarrow c \, [\!] \, H, G$ w.r.t. H. If H is a *non-primitive* inductive atom and A is an inductive atom, then we mark η "*TS-foldable*". Otherwise, η inherits the mark of γ.

(ii) Suppose that η is derived from $\gamma \in P_k$ by folding. Then, η inherits the mark of γ.

(iii) Suppose that η is not involved in the derivation from P_k to P_{k+1}. Then, η inherits its mark in P_k.

Next, we consider an application condition when the head predicate of a folded clause is coinductive. The next notion is due to [19], which is originally introduced to give a condition on folding to preserve the finite failure set.

Definition 6. Inherited Atom [19], Fair Folding

Let $P_0, \ldots, P_n$ be a transformation sequence starting from P_0, and η a clause in P_i $(0 \le i \le n)$ whose head predicate is a defined predicate. Then, an atom in the body of η is called an atom *inherited from* P_0 if one of the following conditions is satisfied:

1. $\eta \in Defs$. Then, each atom in $bd(\eta)$ is *inherited from* P_0.
2. Suppose that η is derived by unfolding $\gamma \in P_{i-1}$ w.r.t. H. Thus, γ and η are of the form: $A \leftarrow c_A [\![H, G$ and $A \leftarrow c \wedge c_i \wedge (H = hd(\gamma_i)) [\![bd(\gamma_i), G$, respectively, where γ_i is an unfolding clause. Then, each atom B occurring in G of η is *inherited from* P_0, if B in G of γ is inherited from P_0.
3. Suppose that η is derived by folding γ in P_{i-1}. Thus, γ and η are of the form: $A \leftarrow c_A [\![K, G$ and $A \leftarrow c_A \wedge e [\![D, G$, respectively, for a folding clause δ with $hd(\delta) = D$. Then, each atom B in G of η is *inherited from* P_0, if B in G of γ is inherited from P_0.

Moreover, the application of folding is said to be *fair*, if, in condition 3, there is no atom in K which is inherited from P_0.   □

Intuitively, an inherited atom is an atom such that it was in the body of some clause in *Defs* and no unfolding has been applied to it.

Conditions on Folding. Let $P_0, \ldots, P_n$ be a transformation sequence with an initial program P_0. Suppose that P_k $(0 < k \le n)$ is derived from P_{k-1} by folding $\gamma \in P_{k-1}$. The application of folding is said to be *admissible* if the following conditions are satisfied:

(1) P_k satisfies the stratification restriction,
(2) if $hd(\gamma)$ is an inductive atom, then γ is marked "TS-foldable" in P_{k-1}, and
(3) if $hd(\gamma)$ is a coinductive atom, then the application of folding is fair.   □

We call condition (2) in the above as *TS-folding condition*.

The following shows the correctness of our transformation system.

Proposition 1. Correctness of Transformation
Let P_0 be an initial co-logic program, and $P_0, \ldots, P_n$ $(0 \le n)$ a transformation sequence, where every application of folding is admissible. Then, the alternating fixpoint semantics $M(P_0)$ of P_0 is preserved, i.e., $M(P_n) = M(P_0)$.   □

Remark 1. Table 2 summarizes the conditions imposed on the transformation rules. The results on the least fixpoint semantics $T_P^{\uparrow \omega}$ for definite programs (column $T_P^{\uparrow \omega}$ in the table) are due to Tamaki-Sato [22], while those on the finite failure set (FF) (column $T_P^{\downarrow \omega}$) are found in [19]. It is known (e. g., [11]) that $FF(P) = \overline{T_P^{\downarrow \omega}}$, the complement of $T_P^{\downarrow \omega}$ for definite program P, and that $gfp(P) = T_P^{\downarrow \omega}$ in complete (infinitary) Herbrand base HB_P^*. Therefore, one might think that unfold/fold transformation simply using fair folding would be appropriate for co-logic programs. However, Example 2 and Example 3 show that it is not the case, since a co-logic program consists of the definitions of predicates

Table 2. Conditions Imposed on the Transformations Rules: The application conditions in column $M(P)$ are newly given in this paper. We denote by h a head predicate of an unfolded clause to which folding is applied.

trans. rule	$T_P^{\uparrow\omega}$	$T_P^{\downarrow\omega}$	$M(P)$
definition	conditions 1&2 in Def. 5	the same	Def. 5
folding	TS-folding [22]	fair folding [19]	strat. restriction & TS-folding, if $h \in \mathcal{P}^{in}$ fair folding, if $h \in \mathcal{P}^{co}$

defined either inductively or coinductively, and its semantics is defined in terms of alternating the least fixpoints and the greatest fixpoints. Proposition 1 gives some additional conditions (shown in column $M(P)$) which are necessary to preserve the alternating fixpoint semantics of co-logic programs.   □

4   Proving Properties of Co-Logic Programs

In this section, we explain how our transformation rules given in Sect. 3 will be utilized to prove properties of co-logic programs. We use two examples studied in the literature: one is the non-emptiness of the language accepted by a Büchi automaton [16]. The other is an example of proving a liveness property of a self-correcting system, a nested finite and infinite automaton [20].

Let P be a co-inductive program and *prop* be a predicate specifying a property of interest which is defined in terms of the predicates in P. Then, in order to check whether or not $M(P) \models \exists X\, prop(X)$, our approach is very simple: we first introduce a defined predicate f defined by clause C_f of the form: $f \leftarrow prop(X)$, where the annotation of predicate f is determined according to Def. 5. We then apply the transformation rules for co-logic programs given in Sect. 3 to $P \cup \{C_f\}$ as (possibly a subset of) an initial program P_0, constructing a transformation sequence $P_0, \ldots, P_n$ so that the truth value of f in $M(P_n)$ will be easily known. For example, if the definition of f in P_{n-1} consists of a single self-recursive clause $f \leftarrow f$, we will apply the replacement rule to it, obtaining P_n from P_{n-1}, where $Def(f, P_n) = \emptyset$ (i.e., $M(P_n) \models \neg f$) if f is inductive, $Def(f, P_n) = \{f \leftarrow\}$ (i.e., $M(P_n) \models f$) otherwise.

Example 4. Adapted from [16]. A Büchi automaton $\mathcal{A}$ is a nondeterministic finite automaton $\langle \Sigma, Q, q_0, \delta, F \rangle$, where Σ is the input alphabet, Q is the set of states, q_0 is the initial state, $\delta \subseteq Q \times \Sigma \times Q$ is the transition relation, and F is the set of final states. A *run* of the automaton $\mathcal{A}$ on an infinite input word $w = a_0 a_1 \cdots \in \Sigma^\omega$ is an infinite sequence $\rho = \rho_0 \rho_1 \cdots \in Q^\omega$ of states such that ρ_0 is the initial state q_0 and, for all $n \geq 0$, $\langle \rho_n, a_n, \rho_{n+1} \rangle \in \delta$. Let $Inf(\rho)$ denote the set of states that occur infinitely often in the infinite sequence ρ of states. An infinite word $w \in \Sigma^\omega$ is *accepted* by $\mathcal{A}$ if there exists a run ρ of $\mathcal{A}$ on w such that $Inf(\rho) \cap F \neq \emptyset$. The language accepted by $\mathcal{A}$ is the subset of Σ^ω, denoted by $\mathcal{L}(\mathcal{A})$, of the infinite words accepted by $\mathcal{A}$.

In order to check whether or not the language $\mathcal{L}(\mathcal{A})$ is empty, we consider the following co-logic program which includes the definition of a unary predicate $accepting\_run$ such that: $\mathcal{L}(\mathcal{A}) \neq \emptyset$ iff $\exists X \, accepting\_run(X)$.

1. $automata([X|Xs], [Q, Q'|Qs]) \leftarrow trans(Q, X, Q'), automata(Xs, [Q'|Qs])$
2. $run([Q_0|Qs]) \leftarrow initial(Q_0), automata(Xs, [Q_0|Qs])$
3. $accepting\_run(Qs) \leftarrow run(Qs), final(Q), comember(Q, Qs)$

where, for all $q, q_1, q_2 \in Q$, for all $a \in \Sigma$,

(i) $initial(q)$ iff $q = q_0$,   (ii) $trans(q_1, a, q_2)$ iff $\langle q_1, a, q_2 \rangle \in \delta$,
(iii) $final(q)$ iff $q \in F$.

We note that predicates $automata$ and $comember$ (defined in Example 1) are annotated as coinductive, while the other predicates are inductive. Now, let us consider a Büchi automaton $\mathcal{A}$ such that: $\Sigma = \{a, b\}, Q = \{1, 2\}, q_0 = 1, \delta = \{\langle 1, a, 1 \rangle, \langle 1, b, 1 \rangle, \langle 1, a, 2 \rangle, \langle 2, a, 2 \rangle\}, F = \{2\}$ which can be represented in Fig. 1. For this automaton $\mathcal{A}$, program $P_\mathcal{A}$ consists of clauses 1-3 and the following clauses 4-9 that encode the initial state (clause 4), the final state (clause 5) and the transition relation (clauses 6-9):

4. $initial(1) \leftarrow$ 6. $trans(1, a, 1) \leftarrow$ 8. $trans(1, a, 2) \leftarrow$
5. $final(2) \leftarrow$ 7. $trans(1, b, 1) \leftarrow$ 9. $trans(2, a, 2) \leftarrow$

To check the non-emptiness of $\mathcal{L}(\mathcal{A})$, we introduce the following clause:
$C_f : \quad f \leftarrow accepting\_run(X)$,
where f is a inductive defined predicate. By some unfolding steps, from C_f we obtain:

10. $f \leftarrow automata(Xs, [1|Qs]), comember(2, [1|Qs])$

By some unfolding steps, from clause 10 we get as one of the derived clauses:

11. $f \leftarrow automata(Xs, [2|Qs]), comember(2, [1, 2|Qs])$

By applying unfolding to clause 11 w.r.t. $comember(2, [1, 2|Qs])$, we get:

12. $f \leftarrow automata(Xs, [2|Qs]), comember(2, Qs)$

Now, we introduce:
$C_g : \quad g \leftarrow automata(Xs, [2|Qs]), comember(2, Qs)$
where g is a coinductive defined predicate. Since clause 12 satisfies the conditions of TS-folding, we apply folding to that clause with folder clause C_g and get:

13. $f \leftarrow g$

By some unfolding steps, from C_g we obtain as one of the derived clauses:

14. $g \leftarrow automata(Xs, [2|Qs]), comember(2, Qs)$

Then, we can apply fair folding to clause 14 and obtain:

15. $g \leftarrow g$

Applying the replacement rule to the above w.r.t. a coinductive atom g, we obtain:

16. $g \leftarrow$

Then, we apply unfolding to clause 13 with unfolding clause 16, obtaining:

17. $f \leftarrow$

This means that $M(P_\mathcal{A}) \models \exists X \, accepting\_run(X)$, namely $\mathcal{L}(\mathcal{A}) \neq \emptyset$.

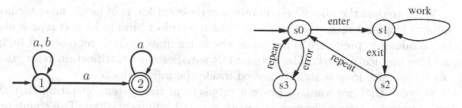

Fig. 1. Examples: a Büchi automaton (left) [16]; a self-correcting system (right) [20]

It will be interesting to compare our approach with the original method by Pettorossi, Proietti and Senni [16]. In their method, (i) the given problem is encoded in an ω-*program* P, a locally stratified program on infinite lists, then (ii) their transformation rules in [16] are applied to P, deriving a *monadic* ω-program T, and finally (iii) the decision procedure in [15] is applied to T to check whether or not $PERF(T) \models \exists prop(X)$, where $PERF(T)$ is the perfect model of T.

To represent $Inf(\rho) \cap F \neq \emptyset$, their ω-program first introduces a defined predicate $rejecting(Qs)$, meaning that run Qs is rejecting (not accepting). Then, the non-emptyness of the language is represented by using nested negations. This process of definition introduction will be inevitable, because their semantics is based on perfect model semantics which is defined by a least model characterization. On the other hand, co-logic programming allows us to use predicate *comember* in this case, which makes the representation succinct and easy to understand. □

The next example due to [20] is on proving a liveness property of a self-correcting system in Fig. 1 (right).

Example 5. Adapted from [20]. The following co-logic program P encodes the self correcting system in Fig. 1:

1. $state(s0, [s0, is1|T]) \leftarrow enter, work, state(s1, T)$
2. $state(s1, [s1|T]) \leftarrow exit, state(s2, T)$
3. $state(s2, [s2|T]) \leftarrow repeat, state(s0, T)$
4. $state(s0, [s0|T]) \leftarrow error, state(s3, T)$
5. $state(s3, [s3|T]) \leftarrow repeat, state(s0, T)$

6. $work \leftarrow work$ 8. $enter \leftarrow$ 10. $repeat \leftarrow$
7. $work \leftarrow$ 9. $exit \leftarrow$ 11. $error \leftarrow$

where $state$ is a coinductive predicate, while the other predicates are inductive. The above program P encodes the system consisting of four states s0, s1, s2 and s3. The system starts in state s0, enters state s1, performs a finite amount of work in state s1. This inner loop state in s1 is denote by is1 (clause 1). The system then exits to state s2. From state s2 the system transitions back to state s0, and repeats the entire loop again, an infinite number of times. However, the system might encounter an error, causing a transition to state s3; corrective action is taken, returning back to s0 (this can also happen infinitely often).

We note that the above system uses two different kinds of loops: an outermost infinite loop and an inner finite loop. The outermost infinite loop is represented by coinductive predicate *state*, while the inner finite loop is represented by inductive predicate *work*. The program P satisfies the stratification restriction, since the finite loop is strictly nested inside the infinite loop.

Suppose that we want to prove a property of the system, ensuring that the system must traverse through the work state s2 infinitely often. The counterexamples to the property can be specified as: $\exists T\ state(s0, T),\ absent(s2, T)$, where *absent* is an inductive predicate which checks that the state s2 is not present in the infinite list T infinitely often. Predicate *absent* is the negation of predicate *comember*, which is defined as follows:

12. $absent(s2, [\_|T]) \leftarrow absent(s2, T)$
13. $absent(s2, [H|T]) \leftarrow H \neq s2, no\_occ(s2, T)$
14. $no\_occ(s2, [H|T]) \leftarrow H \neq s2, no\_occ(s2, T)$

where $no\_occ(s2, T)$ is a coinductive predicate which checks that the state s2 never occurs in the infinite list T.

Let $P_b = P \cup \{12, 13, 14\}$. We first introduce the following clause:
$C_f:\ f \leftarrow state(s0, T),\ absent(s2, T),$
where f is an inductive defined predicate. By some unfolding steps, from C_f we obtain:
15. $f \leftarrow state(s3, T'),\ no\_occ(s2, T')$

We then introduce:
$C_g:\ g \leftarrow state(s3, T),\ no\_occ(s2, T)$
where g is a coinductive defined predicate. Since clause 15 satisfies the TS-folding condition, we apply folding to that clause with folder clause C_g and get:
16. $f \leftarrow g$

By some unfolding steps, from C_g we obtain:
17. $g \leftarrow state(s3, T'),\ no\_occ(s2, T')$

Then, we can apply fair folding to clause 17 and obtain:
18. $g \leftarrow g$

Applying the replacement rule to clause 18 w.r.t. g, we obtain:
19. $g \leftarrow$

Then, we apply unfolding to clause 16 w.r.t. g, obtaining the following:
20. $f \leftarrow$

This means that $M(P_b \cup \{C_f, C_g\}) \models f$, which implies that there exists a counterexample to the original property; in fact, $T = [s0, s3|T]$ satisfies the body of the definition clause C_f.

It will be interesting to compare our approach with the operational semantics of co-logic programs by Simon [20,21], which is a combination of standard SLD-resolution for inductive predicates, and *co-SLD* resolution for coinductive predicates. A new inference rule called *coinductive hypothesis rule* is introduced in co-SLD resolution. Stated informally, if a current subgoal A in a derivation is a coinductive atom, and it appears as an ancestor of the derivation, A is supposed to be true.

It is easy to see that folding together with replacement rule w.r.t. coinductive predicates plays the same role as coinductive hypothesis rule in the above example. In general, however, the pair of folding and the replacement rule is more powerful than coinductive hypothesis rule, in that folding allows several atoms in a folded clause to be folded, while only a single atom is concerned in coinductive hypothesis rule.

Moreover, the pair of folding together with replacement rule for *inductive* predicates works as a loop-check mechanism for inductive predicates, while such a mechanism is not employed in the operational semantics by Simon et al. Although it is possible to incorporate such a mechanism into their operational semantics, it would inevitably make the description of the semantics more complicated.   □

5   Related Work and Concluding Remarks

We have proposed a new framework for unfold/fold transformation of co-logic programs. We have explained that straightforward applications of conventional unfold/fold rules by Tamaki-Sato [22] and Seki [19] are not adequate for co-logic programs, and proposed new conditions which ensure the preservation of the intended semantics of co-logic programs through transformation.

For reasoning about infinite structures, Simon et al. have proposed a new operational semantics using a new inference rule called *coinductive hypothesis rule* [20], as explained at the end of Example 5 in Sect. 4. On the other hand, our approach for reasoning about such structures is based on a framework for familiar unfold/fold transformation rules.

As discussed in [16], very few methods have been reported so far for proving properties of infinite computations by using logic programs over infinite structures, and as a notable exception, we have discussed the work by Pettorossi, Proietti and Senni [16] in Example 4. Although the example is taken from their paper, the proof illustrated here has shown that, as far as this particular example is concerned, our method using co-logic programs can prove the given property in a simpler and more intuitive manner.

On the other hand, the framework by Pettorossi, Proietti and Senni is very general in that arbitrary first order formulas are allowed to specify properties of a given program, while our framework is restricted in that only definite co-logic programs are allowed, although we use *unequation* ($\neq$) only in a limited form in Example 5. One direction for future work is therefore to extend the current framework to allow negation. Min and Gupta [12], for example, have already proposed such an extension, where they extend co-LP with negation. We hope that our results reported in this paper will be a contribution to promote further cross-fertilization between program transformation and model checking.

Acknowledgement. The author would like to thank anonymous reviewers for their constructive and useful comments on the previous version of the paper. This work has been partially supported by the Kayamori Foundation of Information Science Advancement.

References

1. Burstall, R.M., Darlington, J.: A Transformation System for Developing Recursive Programs. J. ACM 24(1), 44–67 (1977)
2. Clarke, E.M., Grumberg, O., Peled, D.A.: Model Checking. MIT Press (1999)
3. Colmerauer, A.: Prolog and Infinite Trees. In: Logic Programming, pp. 231–251. Academic Press (1982)
4. Colmerauer, A.: Equations and Inequations in Finite and Infinite Trees. In: Proc. FGCS 1984, Tokyo, pp. 85–99 (1984)
5. Etalle, S., Gabbrielli, M.: Transformations of CLP Modules. Theor. Comput. Sci., 101–146 (1996)
6. Gupta, G., Bansal, A., Min, R., Simon, L., Mallya, A.: Coinductive Logic Programming and Its Applications. In: Dahl, V., Niemelä, I. (eds.) ICLP 2007. LNCS, vol. 4670, pp. 27–44. Springer, Heidelberg (2007)
7. Jaffar, J., Lassez, J.-L., Maher, M. J.: A theory of complete logic programs with equality. The Journal of Logic Programming 1(3), 211–223 (1984)
8. Jaffar, J., Stuckey, P.: Semantics of infinite tree logic programming. Theoretical Computer Science 46, 141–158 (1986)
9. Jaffar, J., Maher, M.J.: Constraint Logic Programming: A Survey. J. Log. Program. 19/20, 503–581 (1994)
10. Kott, L.: Unfold/fold program transformations. In: Nivat, M., Reynolds, J.C. (eds.) Algebraic Methods in Semantics, ch. 12, pp. 411–434. Cambridge University Press (1985)
11. Lloyd, J.W.: Foundations of Logic Programming, 2nd edn. Springer (1987)
12. Min, R., Gupta, G.: Coinductive Logic Programming with Negation. In: De Schreye, D. (ed.) LOPSTR 2009. LNCS, vol. 6037, pp. 97–112. Springer, Heidelberg (2010)
13. Pettorossi, A., Proietti, M.: Transformation of Logic Programs: Foundations and Techniques. J. Logic Programming 19/20, 261–320 (1994)
14. Pettorossi, A., Proietti, M.: Perfect Model Checking via Unfold/Fold Transformations. In: Palamidessi, C., Moniz Pereira, L., Lloyd, J.W., Dahl, V., Furbach, U., Kerber, M., Lau, K.-K., Sagiv, Y., Stuckey, P.J. (eds.) CL 2000. LNCS (LNAI), vol. 1861, pp. 613–628. Springer, Heidelberg (2000)
15. Pettorossi, A., Proietti, M., Senni, V.: Deciding Full Branching Time Logic by Program Transformation. In: De Schreye, D. (ed.) LOPSTR 2009. LNCS, vol. 6037, pp. 5–21. Springer, Heidelberg (2010)
16. Pettorossi, A., Proietti, M., Senni, V.: Transformations of logic programs on infinite lists. Theory and Practice of Logic Programming 10, 383–399 (2010)
17. Przymusinski, T.C.: On the Declarative and Procedural Semantics of Logic Programs. J. Automated Reasoning 5(2), 167–205 (1989)
18. Seki, H.: On Inductive and Coinductive Proofs via Unfold/Fold Transformations. In: De Schreye, D. (ed.) LOPSTR 2009. LNCS, vol. 6037, pp. 82–96. Springer, Heidelberg (2010)
19. Seki, H.: Unfold/Fold Transformation of Stratified Programs. Theoretical Computer Science 86, 107–139 (1991)
20. Simon, L., Mallya, A., Bansal, A., Gupta, G.: Coinductive Logic Programming. In: Etalle, S., Truszczyński, M. (eds.) ICLP 2006. LNCS, vol. 4079, pp. 330–345. Springer, Heidelberg (2006)
21. Simon, L.E.: Extending Logic Programming with Coinduction, Ph.D. Dissertation, University of Texas at Dallas (2006)
22. Tamaki, H., Sato, T.: Unfold/Fold Transformation of Logic Programs. In: Proc. 2nd Int. Conf. on Logic Programming, pp. 127–138 (1984)
23. Tamaki, H., Sato, T.: A Generalized Correctness Proof of the Unfold/Fold Logic Program Transformation, Technical Report, No. 86-4, Ibaraki Univ. (1986)

Probabilistic Termination of CHRiSM Programs

Jon Sneyers* and Danny De Schreye

Dept. of Computer Science, K.U. Leuven, Belgium
{jon.sneyers,danny.deschreye}@cs.kuleuven.be

Abstract. Termination analysis has received considerable attention in Logic Programming for several decades. In recent years, probabilistic extensions of Logic Programming languages have become increasingly important. Languages like PRISM, CP-Logic, ProbLog, and CHRiSM have been introduced and proved very useful for addressing problems in which a combination of logical and probabilistic reasoning is required. As far as we know, the termination of probabilistic logical programs has not received any attention in the community so far.

Termination of a probabilistic program is not a crisp notion. Given a query, such a program does not simply either terminate or not terminate, but it terminates with a certain probability.

In this paper, we explore this problem in the context of CHRiSM, a probabilistic extension of CHR. We formally introduce the notion of probabilistic termination. We study this concept on the basis of a number of case studies. We provide some initial sufficient conditions to characterize probabilistically terminating programs and queries. We also discuss some challenging examples that reveal the complexity and interest of more general settings. The paper is intended as a first step in a challenging and important new area in the analysis of Logic Programs.

Keywords: Termination Analysis, Probabilistic LP, Constraint Handling Rules.

1 Introduction

Termination analysis has received considerable attention from the Logic Programming research community. Over several decades, many key concepts have been defined and formally studied (e.g. acceptability [3] and various variants of it), several powerful techniques for automatic verification of termination have been designed (e.g. the query-mapping pairs [15], constraint-based termination [7], the dependency-pairs framework [20]) and systems, implementing these techniques, currently provide very refined analysis tools for the termination property (e.g. cTI [17], Aprove [12], Polytool [21]). The main goals have been to support the study of total correctness of programs, to facilitate debugging tasks and to provide termination information for program optimization techniques such as partial evaluation and other transformation systems [16,14,29].

---

* This research is supported by F.W.O. Flanders.

G. Vidal (Ed.): LOPSTR 2011, LNCS 7225, pp. 221–236, 2012.
© Springer-Verlag Berlin Heidelberg 2012

Several Logic Programming related languages have been considered in all this work. Most research has addressed pure Prolog, but other languages, such as full Prolog [25,28], CLP [18], and CHR [8,30,22] have also been considered.

In the past decade, probabilistic extensions of Logic Programming have become increasingly more important. Languages like PRISM [24] and ProbLog [6] extend Logic Programming with probabilistic reasoning and allow to tackle applications that require combinations of logical and probabilistic inference. Probabilistic termination analysis for imperative languages [19,13] and for rewrite systems [5,4] has already been studied in the past. However, as far as we know, termination analysis in the context of (rule-based) probabilistic logic programming has not yet received any attention, with the exception of some comments on probabilistic termination in [10]. The work of Bournez and Garnier (a.o. [5,4]) is closest to what is presented in this paper, although it focusses on *positive* almost sure termination (i.e., the mean derivation length is finite, which is a stronger notion of termination than the one we consider) for probabilistic abstract reduction systems and for probabilistic rewrite systems. They provide general soundness and completeness results. These results are of theoretical interest, since many programming languages have a formal operational semantics that can be seen as an abstract reduction system. However, in practice, theorems like the guaranteed existence (for probabilistically terminating programs) of a "level mapping" function (under certain conditions) do not necessarily provide a way to actually find such functions and prove probabilistic termination. In particular, it is not straightforward to get an intuitive understanding and to see which kind of programs terminate probabilistically, and why they do so.

It is the aim of the current paper to provide an initial investigation of the problem, mostly based on case studies, for the special instance of one specific programming language. Our study will be performed in the context of the probabilistic-logical language CHRiSM [26], a language based on CHR [9,27,11] and PRISM [24].

CHR – Constraint Handling Rules – is a high-level language extension based on multi-headed rules. Originally, CHR was designed as a special-purpose language to implement constraint solvers (see e.g. [2]), but in recent years it has matured into a general purpose programming language. Being a language *extension*, CHR is implemented on top of an existing programming language, which is called the *host language*. An implementation of CHR in host language X is called CHR(X). For instance, several CHR(Prolog) systems are available [27].

PRISM – PRogramming In Statistical Modeling – is a probabilistic extension of Prolog. It supports several probabilistic inference tasks, including sampling, probability computation, and expectation-maximization (EM) learning.

In [26], a new formalism was introduced, called CHRiSM, short for CHance Rules Induce Statistical Models. It is based on CHR(PRISM) and it combines the advantages of CHR and those of PRISM.

By way of motivation, let us consider a simple example of a CHRiSM program and briefly look at its termination properties.

Example 1 (Repeated Coin Flipping) *Suppose we flip a fair coin, and if the result is* tail, *we stop, but if it is* head, *we flip the coin again. We can model this process in* CHRiSM *as follows:*

```
flip <=> head:0.5 ; tail:0.5.
head <=> flip.
```

In Section 2 we formally introduce the syntax and semantics of CHRiSM, *but intuitively, the first rule states that the outcome of a* flip *event has an equal probability of 0.5 to result in* head *or in* tail. *The second rule is not probabilistic (or has probability 1): if there is* head, *we always* flip *again.*

The program would typically be activated by the query flip. *In a computation for the query* flip, *the first rule will either select* head *or* tail, *with equal probability. The choice is not backtrackable.*

It is unclear whether we should consider this program as terminating or non-terminating. The program has an infinite derivation in which the coin always lands on head. *However, out of all the possible derivations, there is only one infinite one and its probability of being executed is* $\lim_{n\to\infty}(0.5)^n = 0$. *So, if we execute this program, the probability of it terminating is equal to 1. Therefore, it is debatable whether we should call the program non-terminating and we will need the more refined notion of probabilistic termination.* □

In this paper we will study the notion of probabilistic termination. Most of this study is based on a number of case studies, where we compute the probabilities of termination for specific programs. We also compute the expected number of rule applications for some of these programs. For some classes of programs, we are able to generalize the results of our examples and formulate and prove probabilistic termination theorems. However, for the more complex programs, our study reveals that the mathematical equations modelling the probability of termination sometimes become too complex to solve. This is not surprising since even universal termination, let alone probabilistic termination, is undecidable in general.

The paper is organized as follows. In Section 2 we recall the syntax and semantics of CHRiSM. In Section 3 we define probabilistic termination of a CHRiSM program. We relate it to universal termination and we study some simple examples. Section 4 introduces a class of programs that behave like Markov chains, for which we provide a termination criterion. In Section 5 we discuss more complex examples and show that solving such examples provides a challenge for currently available mathematical techniques. We conclude in Section 6.

2   CHRiSM

In this section we briefly recall the CHRiSM programming language, in order to make this paper as self-contained as possible given the limited space. However we encourage the reader to refer to [26] for a more detailed description.

A CHRiSM program $\mathcal{P}$ consists of a sequence of *chance rules*. Chance rules rewrite a multiset $\mathbb{S}$ of data elements, which are called (CHRiSM) *constraints* (mostly for historical reasons). Syntactically, a constraint $c(t_1, .., t_n)$ looks like a Prolog predicate: it has a functor c of some arity n and arguments $t_1, .., t_n$ which are Prolog terms. The multiset $\mathbb{S}$ of constraints is called the *constraint store* or just *store*. The initial store is called the *query* or *goal*, the final store (obtained by exhaustive rule application) is called the *answer* or *result*.

We use $\{\!\{\ \}\!\}$ to denote multisets, $\uplus$ for multiset union, $\subseteq$ for multiset subset, and $\bar{\exists}_A B$ to denote $\exists x_1, \ldots, x_n : B$, with $\{x_1, \ldots, x_n\} = vars(B) \setminus vars(A)$, where A and B are arbitrary formulae and $vars(A)$ are the (free) variables in A; if A is omitted it is empty.

Chance rules. A chance rule has the following form: P ?? Hk \ Hr <=> G | B. where P is a probability expression (as defined below), Hk is a conjunction of (kept head) constraints, Hr is a conjunction of (removed head) constraints, G is a guard condition, and B is the body of the rule. If Hk is empty, the rule is called a *simplification* rule and the backslash is omitted; if Hr is empty, the rule is called a *propagation* rule, written as "P ?? Hk ==> G | B". If both Hk and Hr are non-empty, the rule is called a *simpagation* rule. The guard G is optional; if it is omitted, the "|" is also omitted. The body B is a conjunction of CHRiSM constraints, Prolog goals, and probabilistic disjunctions (as defined below).

Intuitively, the meaning of a chance rule is as follows: If the constraint store $\mathbb{S}$ contains elements that match the head of the rule (i.e. if there is a (satisfiable) matching substitution θ such that $(\theta(\text{Hk}) \uplus \theta(\text{Hr})) \subseteq \mathbb{S}$), and furthermore, the guard $\theta(\text{G})$ is satisfied, then we can consider rule application. The subset of $\mathbb{S}$ that corresponds to the head of the rule is called a rule *instance*. Depending on the probability expression P, the rule instance is either ignored or it actually leads to a rule application. Every rule instance may only be considered once.

Rule application has the following effects: the constraints matching Hr are removed from the constraint store and then the body B is executed, that is, Prolog goals are called and CHRiSM constraints are added into the store.

Probability expressions. In this paper, we assume probabilities to be fixed numbers. The CHRiSM system also supports learnable probabilities and other types of probability expressions. We refer to [26] for an overview.

Probabilistic disjunction. The body B of a CHRiSM rule may contain probabilistic disjunctions: "D1:P1 ; ... ; Dn:Pn" indicates that a disjunct Di is chosen with probability Pi. The probabilities should sum to 1 (otherwise a compile-time error occurs). Unlike CHR$^\vee$ disjunctions [1], which create a choice point, probabilistic disjunctions are *committed-choice*: once a disjunct is chosen, the choice is not undone later. However, when later on in a derivation, the same disjunction is reached again, the choice can of course be different.

Operational Semantics. The operational semantics of a CHRiSM program $\mathcal{P}$ is given by a state-transition system that resembles the abstract operational semantics ω_t of CHR [27]. The execution states are defined analogously, except that we additionally define a unique failed execution state, which is denoted by "*fail*" (because we don't want to distinguish between different failed states). We use the symbol $\omega_t^{??}$ to refer to the (abstract) operational semantics of CHRiSM.

Definition 1 (identifiers). *An* identified constraint *$c\#i$ is a CHRiSM constraint c associated with some unique integer i. This number serves to differentiate between copies of the same constraint. We introduce the functions $chr(c\#i) = c$ and $id(c\#i) = i$, and extend them to sequences and sets, e.g.:*

$$chr(S) = \{c \mid c\#i \in S\}$$

Definition 2 (execution state). *An* execution state *σ is a tuple $\langle \mathbb{G}, \mathbb{S}, \mathbb{B}, \mathbb{T}\rangle_n$. The* goal *$\mathbb{G}$ is a multiset of constraints to be rewritten to solved form. The* store *$\mathbb{S}$ is a set of identified constraints that can be matched with rules in the program $\mathcal{P}$. Note that $chr(\mathbb{S})$ is a multiset although $\mathbb{S}$ is a set. The* built-in store *$\mathbb{B}$ is the conjunction of all Prolog goals that have been called so far. The* history *$\mathbb{T}$ is a set of tuples, each recording the identifiers of the CHRiSM constraints that fired a rule and the rule number. The history is used to prevent trivial non-termination: a rule instance is allowed to be considered only once. Finally, the* counter *$n \in \mathbb{N}$ represents the next free identifier.*

We use $\sigma, \sigma_0, \sigma_1, \dots$ to denote execution states and Σ to denote the set of all execution states. We use $\mathcal{D}_{\mathcal{H}}$ to denote the theory defining the host language (Prolog) built-ins and predicates used in the CHRiSM program. For a given program $\mathcal{P}$, the transitions are defined by the binary relation $\rightarrowtail_{\mathcal{P}} \subseteq \Sigma \times \Sigma$ shown in Figure 1. Every transition is annotated with a probability.

Execution of a query Q proceeds by exhaustively applying the transition rules, starting from an initial state (root) of the form $\sigma_Q = \langle Q, \emptyset, true, \emptyset\rangle_0$ and performing a random walk in the directed acyclic graph defined by the transition relation $\rightarrowtail_{\mathcal{P}}$, until a leaf node is reached, which is called a final state. We use Σ_f to denote the set of final states. The probability of a path from an initial state to the state σ is simply the product of the probabilities along the path.

We assume a given execution strategy ξ that fixes the non-probabilistic choices in case multiple transitions are applicable (see section 4.1 of [26]).

We use $\sigma_0 \xrightarrow{p}{}^*_{\mathcal{P}} \sigma_f$ to denote all k different derivations from σ_0 to σ_f:

$$\sigma_0 \xrightarrow{p_{1,1}}_{\mathcal{P}} \sigma_{1,1} \xrightarrow{p_{1,2}}_{\mathcal{P}} \sigma_{1,2} \xrightarrow{p_{1,3}}_{\mathcal{P}} \dots \xrightarrow{p_{1,l_1}}_{\mathcal{P}} \sigma_f$$

$$\vdots$$

$$\sigma_0 \xrightarrow{p_{k,1}}_{\mathcal{P}} \sigma_{k,1} \xrightarrow{p_{k,2}}_{\mathcal{P}} \sigma_{k,2} \xrightarrow{p_{k,3}}_{\mathcal{P}} \dots \xrightarrow{p_{k,l_k}}_{\mathcal{P}} \sigma_f$$

where

$$p = \sum_{i=1}^{k} \prod_{j=1}^{l_i} p_{i,j}.$$

1. **Fail.** $\langle \{\!\!\{ b \}\!\!\} \uplus \mathbb{G}, \mathbb{S}, \mathbb{B}, \mathbb{T} \rangle_n \xrightarrow[\mathcal{P}]{1} fail$
   where b is a built-in (Prolog) constraint and $\mathcal{D}_\mathcal{H} \models \neg \bar{\exists}(\mathbb{B} \wedge b)$.

2. **Solve.** $\langle \{\!\!\{ b \}\!\!\} \uplus \mathbb{G}, \mathbb{S}, \mathbb{B}, \mathbb{T} \rangle_n \xrightarrow[\mathcal{P}]{1} \langle \mathbb{G}, \mathbb{S}, b \wedge \mathbb{B}, \mathbb{T} \rangle_n$
   where b is a built-in (Prolog) constraint and $\mathcal{D}_\mathcal{H} \models \bar{\exists}(\mathbb{B} \wedge b)$.

3. **Introduce.** $\langle \{\!\!\{ c \}\!\!\} \uplus \mathbb{G}, \mathbb{S}, \mathbb{B}, \mathbb{T} \rangle_n \xrightarrow[\mathcal{P}]{1} \langle \mathbb{G}, \{ c \# n \} \cup \mathbb{S}, \mathbb{B}, \mathbb{T} \rangle_{n+1}$
   where c is a CHRiSM constraint.

4. **Probabilistic-Choice.** $\langle \{\!\!\{ d \}\!\!\} \uplus \mathbb{G}, \mathbb{S}, \mathbb{B}, \mathbb{T} \rangle_n \xrightarrow[\mathcal{P}]{p_i} \langle \{\!\!\{ d_i \}\!\!\} \uplus \mathbb{G}, \mathbb{S}, \mathbb{B}, \mathbb{T} \rangle_n$
   where d is a probabilistic disjunction of the form $d_1 : p_1 \; ; \; \ldots \; ; \; d_k : p_k$ or of the
   form $\mathtt{P} \; \mathtt{??} \; d_1 \; ; \; \ldots \; ; \; d_k$, where the probability distribution given by $\mathtt{P}$ assigns
   the probability p_i to the disjunct d_i.

5. **Maybe-Apply.** $\langle \mathbb{G}, H_1 \uplus H_2 \uplus \mathbb{S}, \mathbb{B}, \mathbb{T} \rangle_n \xrightarrow[\mathcal{P}]{1-p} \langle \mathbb{G}, H_1 \uplus H_2 \uplus \mathbb{S}, \mathbb{B}, \mathbb{T} \cup \{h\} \rangle_n$
       $\langle \mathbb{G}, H_1 \uplus H_2 \uplus \mathbb{S}, \mathbb{B}, \mathbb{T} \rangle_n \xrightarrow[\mathcal{P}]{p} \langle B \uplus \mathbb{G}, H_1 \uplus \mathbb{S}, \theta \wedge \mathbb{B}, \mathbb{T} \cup \{h\} \rangle_n$
   where the r-th rule of $\mathcal{P}$ is of the form $\mathtt{P} \; \mathtt{??} \; H_1' \; \backslash \; H_2' \; \mathtt{<=>} \; \mathtt{G} \; | \; \mathtt{B}$,
   θ is a matching substitution such that $chr(H_1) = \theta(H_1')$ and $chr(H_2) = \theta(H_2')$,
   $h = (r, id(H_1), id(H_2)) \notin \mathbb{T}$, and $\mathcal{D}_\mathcal{H} \models \mathbb{B} \rightarrow \bar{\exists}_\mathbb{B}(\theta \wedge G)$. If $\mathtt{P}$ is a number, then
   $p = \mathtt{P}$. Otherwise p is the probability assigned to the success branch of $\mathtt{P}$.

Fig. 1. Transition relation $\xrightarrow[\mathcal{P}]{}$ of the abstract operational semantics $\omega_t^{??}$ of CHRiSM

If σ_0 is an initial state and σ_k is a final state, then we call these derivations an
explanation set with total probability p for the query σ_0 and the result σ_k. Note
that if $k = 0$, i.e. there is no derivation from σ_0 to σ_f, then $p = 0$. We define a
function *prob* to give the probability of an explanation set: $prob(\sigma_0 \xrightarrow[\mathcal{P}]{p}{}^* \sigma_k) = p$.

If σ_0 is an initial state and there exist infinite sequences s_i of transitions

$$\sigma_0 \xrightarrow[\mathcal{P}]{p_{i,1}} \sigma_{i,1} \xrightarrow[\mathcal{P}]{p_{i,2}} \sigma_{i,2} \xrightarrow[\mathcal{P}]{p_{i,3}} \cdots$$

then we call these sequences infinite derivations from σ_0. We use $\sigma_0 \xrightarrow[\mathcal{P}]{p}{}^* \infty$ to
denote the (possibly infinite) set D of infinite derivations from σ_0, where

$$p = \lim_{l \to \infty} \sum_{i=1}^{|D|} \prod_{j=1}^{l} p_{i,j}.$$

Note that if all rule probabilities are 1 and the program contains no probabilistic disjunctions — i.e. if the CHRiSM program is actually just a regular CHR
program — then the $\omega_t^{??}$ semantics boils down to the ω_t semantics of CHR.

3   Probabilistic Termination

In the contexts of Prolog and CHR, the usual notion of termination is *universal
termination*: a program $\mathcal{P}$ terminates for a query Q if Q does not have an infinite
derivation for $\mathcal{P}$.

For some CHRiSM programs, universal termination is too strong. In order to
be able to execute (sample) a program, *probabilistic* termination is sufficient.

Definition 3 (Probabilistic termination). *A program $\mathcal{P}$ probabilistically terminates for a query Q with probability p if the probability of the event that the computation for Q in $\mathcal{P}$ halts is equal to p, i.e.*

$$p = \sum_{\sigma \in \Sigma_f} prob(\sigma_Q \rightarrowtail_{\mathcal{P}}^* \sigma) = 1 - prob(\sigma_Q \rightarrowtail_{\mathcal{P}}^* \infty).$$

A program $\mathcal{P}$ probabilistically terminates for a query Q if the program probabilistically terminates for Q with probability 1. A program $\mathcal{P}$ probabilistically terminates if it probabilistically terminates for all finite queries.

Note that the above definition does not give a general practical method to compute the termination probability for a given query, and like universal termination, probabilistic termination is undecidable. Note also that in general the number of finite derivations may be infinite.

Although many CHRiSM programs do not universally terminate (so they have infinite derivations) universal termination is of course sufficient for probabilistic termination. This already provides us with a practical way of proving probabilistic termination of one class of CHRiSM programs. In general, we can associate to any CHRiSM program $\mathcal{P}$ a corresponding $CHR^\vee$ program [1], $CHR^\vee(\mathcal{P})$, by removing all the probability information from $\mathcal{P}$. As already mentioned, apart from removing the probability factors, this transformation changes committed choice disjunctions into backtrackable disjunctions. However, from a perspective of proving universal termination, this is not a problem, because we need to prove that all derivations are finite anyway.

Proposition 1. *For any CHRiSM program $\mathcal{P}$ and query Q, $\mathcal{P}$ is probabilistically terminating for Q if $CHR^\vee(\mathcal{P})$ is universally terminating for Q.*

Proof. If $CHR^\vee(\mathcal{P})$ has no infinite derivation for Q, then $\mathcal{P}$ has no infinite derivation for Q. Thus the probability of halting is one.

Proving that $CHR^\vee(\mathcal{P})$ universally terminates for Q can be done using the techniques presented in [8], [30], or [22]. The latter technique has been automated and implemented [23]. Of course, these techniques were developed for CHR, rather than for $CHR^\vee$, but again, a $CHR^\vee$ program can easily transformed into a CHR program with the same universal termination behavior.

Example 2 (Low-power Countdown). *Consider the CHRiSM program:*

```
0.9 ?? countdown(N) <=> N > 1 | countdown(N-1).
0.9 ?? countdown(0) <=> writeln('Happy New Year!').
```

representing a New Year countdown device with low battery power, which may or may not display its New Year wishes starting from a query countdown(10). *At every tick, there is a 10% chance that the battery dies and the countdown stops.*

Consider the CHR program obtained by omitting the probabilities. The ranking techniques of all three of the approaches presented in [8,30,22], prove universal termination for that program and the same query. Thus, Low-power New Years Countdown will terminate (universally and probabilistically) as well. □

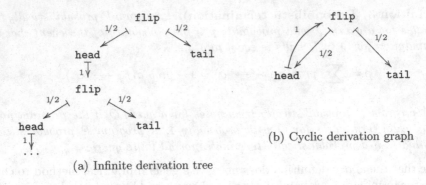

(b) Cyclic derivation graph

(a) Infinite derivation tree

Fig. 2. Repeated coin flipping example

However, the main attention in this paper will be addressed to the case in which the CHRiSM program does not universally terminate.

Example 3 (Repeated Coin Flipping cont'd) *Consider again Example 1.*

```
flip <=> head:0.5 ; tail:0.5.
head <=> flip.
```

Recall that the only infinite derivation has probability $\lim_{n \to \infty} (0.5)^n = 0$, *so that we can conclude that the program probabilistically terminates.*

Figure 2 shows the derivation tree for this program, assuming the query flip. *This derivation tree corresponds to a finite cyclic derivation graph. There is only one cycle in this graph and its probability is less than one.* ☐

Example 4 (Basic a-to-a) *In terms of termination behavior, the coin flipping program is equivalent to the following program, given the query* a:

```
0.5 ?? a <=> a.
```

To analyze the probabilistic termination of programs we have to compute the termination probability as the sum of the probabilities of all terminating derivations. It is not difficult to see that for the above program, just as in the original coin flipping program, the termination probability is $\sum_{i=1}^{\infty} (0.5)^i = 1$. ☐

The above example can be generalized to the case where the probability is some p between 0 and 1, instead of 0.5. Consider the rule "p ?? a <=> a." where p is a fixed probability. All terminating derivations consist of n rule applications followed by one non-applied rule instance. Thus, the probability of such a derivation is $p^n(1-p)$. Hence, the total termination probability s for the above program is $s = \sum_{n=0}^{\infty} p^n(1-p)$. To solve this infinite sum it suffices to note that $s - ps = 1 - p$, so if $p < 1$ we have $s = (1-p)/(1-p) = 1$. If $p = 1$ we get $s = 0$ since every term is zero.

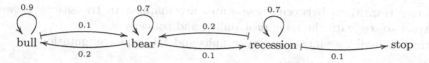

Fig. 3. Markov chain representing the economy

We can compute the expected number of rule applications (i.e., the average run time) as follows, assuming $p < 1$ so the probability of non-termination is zero: $\sum_{n=0}^{\infty} p^n (1-p)n = p/(1-p)$. So e.g. if $p = 3/4$, then the expected number of rule applications is 3.

4 Markov Chains and MC-Type Computations

Example 5 (Bull-bear) *Consider the following* CHRiSM *program, which implements a Markov chain modeling the evolution of some market economy:*

```
s(T,bull) ==> (s(T+1,bull):0.9 ; s(T+1,bear):0.1).
s(T,bear) ==> (s(T+1,bear):0.7 ; s(T+1,bull):0.2 ; s(T+1,recession):0.1).
s(T,recession) ==> (s(T+1,recession):0.7 ; s(T+1,bear):0.2 ; stop:0.1).
```

Figure 3 illustrates the transitions. In every state the most likely option is to stay in that state. Also, a bull market can become a bear market, a bear market can recover to a bull market or worsen into a recession, and in a recession we can recover to a bear market or the market economy transition system may come to an end (e.g. a socialist revolution happened).

In this case, the program terminates probabilistically. This can be shown as follows. From every execution state, the probability of termination has a nonzero lower bound, as can be verified from Fig.3. Indeed, it is easy to see that in every state, the probability of terminating "immediately" (i.e. by taking the shortest path to the final state "stop") is at least 0.001. Now, every infinite derivation has to visit one of the states in Fig.3 infinitely often. Let p_x be the total probability of all derivations that visit state x infinitely often, then the probablity of non-termination is at most $p_{bull} + p_{bear} + p_{recession}$. We now show that $p_x = 0$. Consider all subderivations of the form $x \rightarrowtail y_1 \rightarrowtail y_2 \rightarrowtail ... \rightarrowtail x$ where all intermediate $y_i \neq x$. The total probability for all such subderivations has to be less than 0.999, since there is a subderivation $x \xrightarrow{p} {}^* stop$ with $p \geq 0.001$ (where all intermediate steps are different from x). This means that the probability p_x is bounded from above by

$$\lim_{n \to \infty} (0.999)^n = 0$$

so the probability of termination is one. □

In the previous two examples, the transition graph is essentially finite, in the following sense: there exists a finite set of abstractions of states and probabilistic transitions between these abstract states and an abstraction function from execution states to abstract states, such that for all reachable executions states, concrete transitions between these states are mapped to transitions between abstract states, with the same probability, and conversely.

More formally, we introduce MC-graphs and MC-type computations.

Definition 4 (MC-graph). *An MC-graph is an annotated, directed graph, (V, A, L), consisting of a finite set of vertices $V = \{v_1, \ldots, v_n\}$, a set of arcs, $A \subseteq V \times V$, and a function $L : A \to]0, 1]$. We refer to L as the probability labeling for A.*

It should be clear that an MC-graph represents a Markov Chain.

Definition 5 (MC-type computation). *Let $\mathcal{P}$ be a CHRiSM program and Q a query to $\mathcal{P}$. The computation for $\mathcal{P}$ and Q consists of all possible transitions $\sigma \overset{p}{\underset{\mathcal{P}}{\rightarrowtail}} \sigma'$, reachable from the initial state $\langle Q, \emptyset, true, \emptyset \rangle_0$.*

The computation for $\mathcal{P}$ and Q is an MC-type computation, if there exists an MC-graph (V, A, L) and a function $\alpha : \Sigma \to V$, such that:

- *If there is a transition $\sigma \overset{p}{\underset{\mathcal{P}}{\rightarrowtail}} \sigma'$ of type **Probabilistic-Choice** or **Maybe-Apply** (see Fig. 1) in the computation for $\mathcal{P}$ and Q, where $p > 0$ (we omit impossible transitions), then $(\alpha(\sigma), \alpha(\sigma')) \in A$ and $L((\alpha(\sigma), \alpha(\sigma'))) = p$.*
- *If $\alpha(\sigma) \in V$, $(\alpha(\sigma), v) \in A$ and $L((\alpha(\sigma), v)) = p$, then there exists a reachable execution state $\sigma' \in \Sigma$, such that $\alpha(\sigma') = v$ and $\sigma \overset{p}{\underset{\mathcal{P}}{\rightarrowtail}} \sigma'$ is in the computation for $\mathcal{P}$ and Q.*

Example 6 (a-to-a, bull-bear cont'd) *In the a-to-a example, $V = \{a, stop\}$, $A = \{(a, a), (a, stop)\}$, $L((a, a)) = p$, and $L((a, stop)) = 1 - p$. All reachable non-final execution states are mapped to a, all final states are mapped to stop.*

In the bull-bear example, the MC-graph is essentially as represented in Figure 3, with the addition of the loop-arcs on the vertices bull, bear and recession. The mapping α maps an execution state of the form $\langle \mathbb{G}, \mathbb{S}, \mathbb{B}, \mathbb{T} \rangle_n$ to the node x iff $s(\_,x) \in \mathbb{G}$ or $s(\_,x)\#k \in \mathbb{S}$ and $(\_, k) \notin \mathbb{T}$. □

Note that a leaf node (a node without outgoing edges) in the MC-graph corresponds to final states. We have the following criterion for probabilistic termination of MC-type computations.

Theorem 1. *Let $\mathcal{P}$ be a CHRiSM program and Q a query, such that the computation for $\mathcal{P}$ and Q is MC-type and let (V, A, L) be the associated MC-graph. The program $\mathcal{P}$ probabilistically terminates for Q if for every node $v_i \in V$, there is a path in A from v_i to a leaf node.*

Proof. The argument is identical to the one for the bull-bear example above.

It is tempting (but wrong) to think that a CHRiSM program $\mathcal{P}$ is probabilistically terminating if every cycle in its derivation graph has a probability $p < 1$. This is a tempting idea, because, for such programs, every infinite derivation has probability zero. However, even for MC-type computations, this is wrong.

Example 7 (Infinite Coin Flipping) *The following program terminates with probability zero, although every infinite derivation has probability zero:*

```
flip <=> head:0.5 ; tail:0.5.
head <=> flip.
tail <=> flip.
```

□

5   The Drunk Guy on a Cliff and More

Now let us analyze a more challenging toy example in which the transition graph is essentially infinite. Consider the following rule:

```
0.5 ?? a <=> a, a.
```

We are trying to find the termination probability s for the query a. With probability 0.5, the program terminates immediately (the rule instance is not applied), and with probability 0.5, it terminates if and only if it (independently) terminates two times in a row for the query a. If terminating once has probability s, then terminating twice has probability s^2, so we get the following equation:

$$s = 0.5 + 0.5s^2$$

The only solution to this equation is $s = 1$. Somewhat counter-intuitively, the above program does terminate probabilistically.

In general, consider the above rule with an arbitrary fixed probability p:

```
p ?? a <=> a, a.
```

We can compute the termination probability as follows. Again, either the program terminates immediately or it has to terminate twice from the query a, so $s = (1 - p) + ps^2$. Solving for s, we get $s = 1$ or $s = \frac{1-p}{p}$, so taking into account that $0 \leq s \leq 1$ (since s is a probability), we have $s = 1$ if $p \leq 1/2$ and $s = (1 - p)/p$ if $p \geq 1/2$ (see Fig. 4).

The "drunk guy on a cliff" puzzle is defined as follows. There is a sheer cliff, and a drunk guy is facing the cliff. He is staggering drunkenly back and forth. One single step forward from his current location will send him hurtling into the abyss, a step backward will bring him closer to safety. The chance of him staggering backwards (at any time) is p, the chance of him staggering forwards is $1 - p$. What is the chance that he will eventually fall into the abyss?

We can model the drunk guy on a cliff as follows:

```
dist(0) <=> true.
dist(N) <=> N > 0 |  dist(N+1):p ; dist(N-1):(1-p).
```

with the initial query dist(1). The only way to terminate is by reaching dist(0), i.e. by falling into the abyss. So the drunk guy on a cliff puzzle boils down to computing the termination probability of the above program.

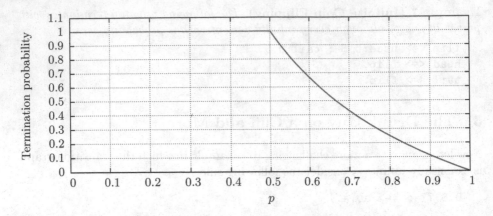

Fig. 4. Termination probability for the rule "p ?? a <=> a,a"

Fig. 5. Derivation graph for the drunk guy on a cliff program

For example, consider the case $p = 1/2$ and the query dist(1). The probability of termination $s = s(\text{dist}(1))$ can be computed as follows:

$$s(\text{dist}(i)) = \frac{1}{2}s(\text{dist}(i-1)) + \frac{1}{2}s(\text{dist}(i+1))$$

Given that $s(\text{dist}(0)) = 1$, it is easy to verify that

$$s(\text{dist}(1)) = \frac{1}{2} + \frac{1}{2}s(\text{dist}(2)) = \frac{2}{3} + \frac{1}{3}s(\text{dist}(3)) = \frac{3}{4} + \frac{1}{4}s(\text{dist}(4)) = \ldots = 1$$

It turns out that we have already solved this problem. Consider again the program consisting of the single rule "p ?? a <=> a,a", and consider an $\omega_t^{??}$ execution state $\sigma = \langle \mathbb{G}, \mathbb{S}, \mathbb{B}, \mathbb{T} \rangle_n$ in a derivation starting from the query "a". The probability of termination from state σ only depends on the number of a/0 constraints for which the rule could still be applied, which is the following number: $|\mathbb{G} \uplus \{a\#n \in \mathbb{S} \mid (1,n) \notin \mathbb{T}\}|$. Let us call this number the *distance to termination* and denote it with $d(\sigma)$. When considering a rule instance, there are two options: with probability p, the distance increases by one (the rule is applied), and with probability $1 - p$, the distance decreases by one (the rule is not applied). The program terminates as soon as $d(\sigma) = 0$.

An alternative way to compute the termination probability of the above programs is as follows. All terminating derivations consist of n rule instances that are applied (or distances that are increased) and $n + 1$ rule instances that are not applied (or distances that are decreased), for some number n. For example, for $n = 3$ we have the following five derivations: $+++----, ++-+---,$ $++--+--, +-++---,$ and $+-+-+--$ where "+" means "applied" (or incremented) and "$-$" means "not applied" (or decremented). For a given number n of applied rule instances, the number of possible derivations is given by the n-th Catalan number[1] $C_n = (2n)!/(n!(n+1)!)$. This gives us the following formula to compute the termination probability s:

$$ s = \sum_{n=0}^{\infty} p^n (1-p)^{n+1} \frac{(2n)!}{n!(n+1)!} $$

The above infinite sum converges to the same values for s we already calculated above. However, this alternative formulation allows us to compute the expected number of rule applications (of the rule "p ?? a <=> a,a"). The expected number of rule applications is given by:

$$ E_p = \sum_{n=0}^{\infty} p^n (1-p)^{n+1} \frac{(2n)!}{n!(n+1)!} n $$

Figure 6 shows the expected number of rule applications E_p as a function of p. As p gets closer to $1/2$, the value for E_p grows quickly: for $p = \frac{2^k-1}{2^{k+1}}$, we have $E_p = \frac{2^k-1}{2}$. The border case $p = 1/2$ is interesting: the termination probability is 1 but the expected number of rule applications is $+\infty$. As p gets larger than $1/2$, the termination probability drops and so does the expected number of rule applications of the (increasingly rare) terminating derivations.

Multi-headed Rules. So far, we have only considered single-headed rules. One of the simplest "interesting" cases is the following rule: (with less than three a's in the body, the program terminates universally)

```
p ?? a, a <=> a, a, a.
```

Given the query "a,a", there are two rule instances to be considered: (a#1, a#2) and (a#2, a#1). If both instances are not applied (probability $(1-p)^2$), the program terminates; otherwise we are further away from termination.

We define the termination distance $d(\sigma)$ of an execution state $\sigma = \langle \mathbb{G}, \mathbb{S}, \mathbb{B}, \mathbb{T} \rangle_n$ as a pair $d(\sigma) = (n, m)$ where $n = |\mathbb{G}|$ is the number of "not yet considered" a's and $m = |\mathbb{S}|$ is the number of "considered" a's. Given the query "a,a", the initial state has distance $(2, 0)$.

The termination probability $s(n, m)$ of a state with distance (n, m) can be computed as follows. If $n = 0$, the rule can no longer be applied so we have

---

[1] The Catalan numbers are named after the Belgian mathematician Eugène Charles Catalan (1814-1894). They are sequence A000108 in The On-Line Encyclopedia of Integer Sequences (http://oeis.org/A000108).

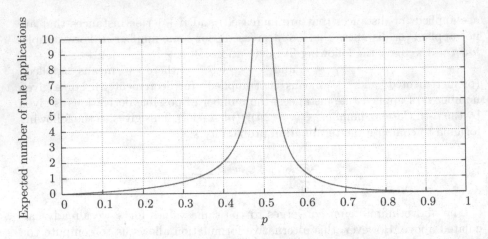

Fig. 6. Expected number of rule applications (if terminating) for "p ?? a <=> a,a"

termination. If $n > 0$, we can take one a/0 and consider all matching rule instances. Since there are m possible partner constraints and two (symmetric) occurrences of the active constraint, there are $2m$ rule instances to consider. If none of these rule instances are applied, we just add the a/0 constraint to the store, so the new distance will be $(n-1, m+1)$. However, if one of the rule instances is applied, we get a new distance $(n+2, m-1)$. So the termination probability $s(n, m)$ is given by the following equations:

$$\begin{cases} s(0,m) = 1 \\ s(n,m) = (1-p)^{2m}s(n-1, m+1) + (1 - (1-p)^{2m})s(n+2, m-1) \end{cases}$$

Note that if $m = 0$, there are no partner constraints so there are no rule instances to consider, i.e. $s(n, 0) = s(n-1, 1)$.

Unfortunately, we have not found a closed-form solution for the above equations. The example shows that, while the complexity of the programs increases, the mathematical models representing the probability of termination become too complex to be solved by standard techniques.

6   Conclusion and Future Work

In this paper we presented the results of an initial investigation of the concept of probabilistic termination of CHRiSM programs. This research has mostly taken the form of a number of small case studies, in which we attempt to reveal the intuitions concerning the concept of probabilistic termination and present some ways of (manually) proving probabilistic termination. In the process, for some of the cases, we also study the expected number of rule applications.

For some classes of programs, we have generalised the observations in our case studies and we formulated and proved termination conditions. In particular, for universally terminating programs we obviously also get probabilistic termination. Therefore, techniques developed to prove universal termination of CHR are sufficient to prove probabilistic termination of corresponding CHRiSM programs. We also identified the class of MC-type programs and formulated and proved a sufficient probabilistic termination condition for it.

For more general classes of programs, termination proofs may become quite complex. We elaborated on a few more complex (but still toy) cases, where sometimes we are able to solve the problem, but in other cases, we observe that the equations expressing probabilistic termination are too complex to be solved with standard techniques. In this exploratory paper we have focused on intuition and examples. It would be interesting to investigate to what extent the theoretical work of [5,4] can be transferred to our setting.

Finally, note that although this work has been in the context of CHRiSM, most of the ideas are also applicable to other probabilistic logic programming languages like PRISM and ProbLog. In most of the probabilistic inference algorithms used in the implementations of these languages (e.g. probability computation, learning), it is assumed that the program universally terminates. An interesting direction for future work is to generalize these algorithms such that they can handle (certain classes of) probabilistically terminating programs.

References

1. Abdennadher, S.: A language for experimenting with declarative paradigms. In: Frühwirth, T., et al. (eds.) RCoRP 2000, bis (2000)
2. Abdennadher, S., Rigotti, C.: Automatic generation of CHR constraint solvers. TPLP 5(4-5), 403–418 (2005)
3. Apt, K.R., Pedreschi, D.: Reasoning about termination of pure Prolog programs. Inf. Comput. 106(1), 109–157 (1993)
4. Bournez, O., Garnier, F.: Proving Positive Almost Sure Termination Under Strategies. In: Pfenning, F. (ed.) RTA 2006. LNCS, vol. 4098, pp. 357–371. Springer, Heidelberg (2006)
5. Bournez, O., Kirchner, C.: Probabilistic Rewrite Strategies. Applications to ELAN. In: Tison, S. (ed.) RTA 2002. LNCS, vol. 2378, pp. 252–266. Springer, Heidelberg (2002)
6. De Raedt, L., Kimmig, A., Toivonen, H.: ProbLog: A probabilistic Prolog and its application in link discovery. In: IJCAI, pp. 2462–2467 (2007)
7. Decorte, S., De Schreye, D., Vandecasteele, H.: Constraint-based automatic termination analysis of logic programs. ACM TOPLAS 21(6), 1137–1195 (1999)
8. Frühwirth, T.: Proving Termination of Constraint Solver Programs. In: Apt, K.R., Kakas, A.C., Monfroy, E., Rossi, F. (eds.) Compulog Net WS 1999. LNCS (LNAI), vol. 1865, pp. 298–317. Springer, Heidelberg (2000)
9. Frühwirth, T.: Constraint Handling Rules. Cambridge University Press (2009)
10. Frühwirth, T., Di Pierro, A., Wiklicky, H.: Probabilistic Constraint Handling Rules. In: Comini, M., Falaschi, M. (eds.) WFLP 2002. ENTCS, vol. 76. Elsevier (2002)
11. Frühwirth, T., Raiser, F. (eds.): Constraint Handling Rules: Compilation, Execution, and Analysis. Books on Demand GmbH, Norderstedt (2011)
12. Giesl, J., Thiemann, R., Schneider-Kamp, P., Falke, S.: Mechanizing and improving dependency pairs. J. Autom. Reasoning 37(3), 155–203 (2006)

13. Hurd, J.: A Formal Approach to Probabilistic Termination. In: Carreño, V.A., Muñoz, C.A., Tahar, S. (eds.) TPHOLs 2002. LNCS, vol. 2410, pp. 230–245. Springer, Heidelberg (2002)
14. Leuschel, M., Martens, B., De Schreye, D.: Controlling generalization amd polyvariance in partial deduction of normal logic programs. ACM TOPLAS 20(1), 208–258 (1998)
15. Lindenstrauss, N., Sagiv, Y., Serebrenik, A.: Proving Termination for Logic Programs by the Query-Mapping Pairs Approach. In: Bruynooghe, M., Lau, K.-K. (eds.) Program Development in CL. LNCS, vol. 3049, pp. 453–498. Springer, Heidelberg (2004)
16. Martens, B., De Schreye, D., Bruynooghe, M.: Sound and complete partial deduction with unfolding based on well-founded measures. In: FGCS, pp. 473–480 (1992)
17. Mesnard, F., Bagnara, R.: cTI: A constraint-based termination inference tool for ISO-Prolog. TPLP 5(1-2), 243–257 (2005)
18. Mesnard, F., Ruggieri, S.: On proving left termination of constraint logic programs. ACM Trans. Comput. Log. 4(2), 207–259 (2003)
19. Monniaux, D.: An Abstract Analysis of the Probabilistic Termination of Programs. In: Cousot, P. (ed.) SAS 2001. LNCS, vol. 2126, pp. 111–126. Springer, Heidelberg (2001)
20. Nguyen, M.T., Giesl, J., Schneider-Kamp, P., De Schreye, D.: Termination Analysis of Logic Programs Based on Dependency Graphs. In: King, A. (ed.) LOPSTR 2007. LNCS, vol. 4915, pp. 8–22. Springer, Heidelberg (2008)
21. Nguyen, M.T., De Schreye, D., Giesl, J., Schneider-Kamp, P.: Polytool: polynomial interpretations as a basis for termination analysis of logic programs. TPLP 11, 33–63 (2011)
22. Pilozzi, P., De Schreye, D.: Termination Analysis of CHR Revisited. In: Garcia de la Banda, M., Pontelli, E. (eds.) ICLP 2008. LNCS, vol. 5366, pp. 501–515. Springer, Heidelberg (2008)
23. Pilozzi, P., De Schreye, D.: Automating Termination Proofs for CHR. In: Hill, P.M., Warren, D.S. (eds.) ICLP 2009. LNCS, vol. 5649, pp. 504–508. Springer, Heidelberg (2009)
24. Sato, T.: A glimpse of symbolic-statistical modeling by PRISM. Journal of Intelligent Information Systems 31, 161–176 (2008)
25. Schneider-Kamp, P., Giesl, J., Ströder, T., et al.: Automated termination analysis for logic programs with cut. TPLP 10(4-6), 365–381 (2010)
26. Sneyers, J., Meert, W., Vennekens, J., Kameya, Y., Sato, T.: CHR(PRISM)-based probabilistic logic learning. TPLP 10(4-6) (2010)
27. Sneyers, J., Van Weert, P., Schrijvers, T., De Koninck, L.: As time goes by: Constraint Handling Rules — a survey of CHR research between 1998 and 2007. TPLP 10(1), 1–47 (2010)
28. Ströder, T., Emmes, F., Schneider-Kamp, P., Giesl, J., Fuhs, C.: A Linear Operational Semantics for Termination and Complexity Analysis of ISO Prolog. In: Vidal, G. (ed.) LOPSTR 2011. LNCS, vol. 7225, pp. 235–250. Springer, Heidelberg (2012)
29. Vidal, G.: A Hybrid Approach to Conjunctive Partial Evaluation of Logic Programs. In: Alpuente, M. (ed.) LOPSTR 2010. LNCS, vol. 6564, pp. 200–214. Springer, Heidelberg (2011)
30. Voets, D., De Schreye, D., Pilozzi, P.: A new approach to termination analysis of constraint handling rules. In: LOPSTR, pp. 28–42 (2008)

A Linear Operational Semantics for Termination and Complexity Analysis of ISO Prolog*

Thomas Ströder[1], Fabian Emmes[1], Peter Schneider-Kamp[2], Jürgen Giesl[1], and Carsten Fuhs[1]

[1] LuFG Informatik 2, RWTH Aachen University, Germany
{stroeder,emmes,giesl,fuhs}@informatik.rwth-aachen.de
[2] IMADA, University of Southern Denmark, Denmark
petersk@imada.sdu.dk

Abstract. We present a new operational semantics for Prolog which covers all constructs in the corresponding ISO standard (including "non-logical" concepts like cuts, meta-programming, "all solution" predicates, dynamic predicates, and exception handling). In contrast to the classical operational semantics for logic programming, our semantics is *linear* and not based on search trees. This has the advantage that it is particularly suitable for automated program analyses such as termination and complexity analysis. We prove that our new semantics is equivalent to the ISO Prolog semantics, i.e., it computes the same answer substitutions and the derivations in both semantics have essentially the same length.

1 Introduction

We introduce a new *state*-based semantics for Prolog. Any query Q corresponds to an initial state s_Q and we define a set of *inference rules* which transform a state s into another state s' (denoted $s \rightsquigarrow s'$). The evaluation of Q is modeled by repeatedly applying inference rules to s_Q (i.e., by the derivation $s_Q \rightsquigarrow s_1 \rightsquigarrow s_2 \rightsquigarrow \ldots$). Essentially, our states s represent the list of those goals that still have to be proved. But in contrast to most other semantics for Prolog, our semantics is *linear* (or *local*), since each state contains all information needed for the next evaluation step. So to extend a derivation $s_0 \rightsquigarrow \ldots \rightsquigarrow s_i$, one only has to consider the last state s_i. Thus, even the effect of cuts and other built-in predicates becomes local.

This is in contrast to the standard semantics of Prolog (as specified in the ISO standard [11,13]), which is defined using a *search tree* built by SLD resolution with a depth-first left-to-right strategy. To construct the next node of the tree, it is not sufficient to regard the node that was constructed last, but due to backtracking, one may have to continue with ancestor goals that occurred much "earlier" in the tree. Advanced features like cuts or exceptions require even more sophisticated analyses of the current search tree. Even worse, "all solution" predicates like findall result in several search trees and the coordination of these trees is highly non-trivial, in particular in the presence of exceptions.

* Supported by DFG grant GI 274/5-3, DFG Research Training Group 1298 (*Algo-Syn*), G.I.F. grant 966-116.6, and the Danish Natural Science Research Council.

We show that our linear semantics is *equivalent* to the standard ISO semantics of Prolog. It does not only yield the same answer substitutions, but we also obtain the same *termination* behavior and even the same *complexity* (i.e., the length of the derivations in our semantics corresponds to the number of unifications performed in the standard semantics). Hence, instead of analyzing the termination or complexity of a Prolog program w.r.t. the standard semantics, one can also analyze it w.r.t. our semantics.

Compared to the ISO semantics, our semantics is much more suitable for such (possibly automated) analyses. In particular, our semantics can also be used for symbolic evaluation of *abstract* states (where the goals contain *abstract variables* representing arbitrary terms). Such abstract states can be generalized ("widened") and instantiated, and under certain conditions one may even *split up* the lists of goals in states [19,20]. In this way, one can represent all possible evaluations of a program by a finite graph, which can then be used as the basis for e.g. termination analysis. In the standard Prolog semantics, such an abstraction of a query in a search tree would be problematic, since the remaining computation does not only depend on this query, but on the whole search tree.

In [19,20] we already used a preliminary version of our semantics for termination analysis of a subset of Prolog containing definite logic programming and cuts. Most previous approaches for termination (or complexity [9]) analysis were restricted to definite programs. Our semantics was a key contribution to extend termination analysis to programs with cuts. The corresponding implementation in the prover AProVE resulted in the most powerful tool for automated termination analysis of logic programming so far, as shown at the *International Termination Competition*.[1] These experimental results are the main motivation for our work, since they indicate that such a semantics is indeed suitable for automated termination analysis. However, it was unclear how to extend the semantics of [19,20] to full Prolog and how to prove that this semantics is really equivalent to the ISO semantics. These are the contributions of the current paper.

Hence, this paper forms the basis which will allow the extension of automated termination techniques to *full* Prolog. Moreover, many termination techniques can be adapted to infer upper bounds on the complexity [12,18,22]. Thus, the current paper is also the basis in order to adapt termination techniques such that they can be used for automated complexity analysis of full Prolog.

There exist several other alternative semantics for Prolog. However, most of them (e.g., [2,4,5,6,7,8,14,15,17]) only handle subsets of Prolog and it is not clear how to extend these semantics in a straightforward way to full Prolog.

Alternative semantics for *full* Prolog were proposed in [3,10,16]. However, these semantics seem less suitable for automated termination and complexity analysis than ours: The states used in [3] are considerably more complex than ours and it is unclear how to abstract the states of [3] for automated termination analysis as in [19,20]. Moreover, [3] does not investigate whether their semantics also yields the same complexity as the ISO standard. The approach in [10] is close to the ISO standard and thus, it has similar drawbacks as the ISO semantics,

---

[1] See http://www.termination-portal.org/wiki/Termination_Competition

since it also works on search trees. Finally, [16] specifies standard Prolog in rewriting logic. Similar to us, [16] uses a list representation for states. However, their approach cannot be used for complexity analysis, since their derivations can be substantially longer than the number of unifications needed to evaluate the query. Since [16] does not use explicit markers for the scope of constructs like the cut, it is also unclear how to use their approach for automated termination analysis, where one would have to abstract and to split states.

The full set of all inference rules of our semantics (for all 112 built-in predicates of ISO Prolog) can be found in [21]. Due to lack of space, in the paper we restrict ourselves to the inference rules for the most representative predicates. Sect. 2 shows the rules needed for definite logic programs. Sect. 3 extends them for predicates like the cut, negation-as-failure, and call. In Sect. 4 we handle "all solution" predicates and Sect. 5 shows how to deal with dynamic predicates like assertz and retract. Sect. 6 extends our semantics to handle exceptions (using catch and throw). Finally, Sect. 7 contains our theorems on the equivalence of our semantics to the ISO semantics. All proofs can be found in [21].

2   Definite Logic Programming

See e.g. [1] for the basics of logic programming. As in ISO Prolog, we do not distinguish between predicate and function symbols. For a term $t = f(t_1, \ldots, t_n)$, let $root(t) = f$. A *query* is a sequence of terms, where $\Box$ denotes the empty query. A *clause* is a pair $h :- B$ where the *head* h is a term and the *body* B is a query. If B is empty, then one writes just "h" instead of "$h :- \Box$".[2] A *Prolog program* $\mathcal{P}$ is a finite sequence of clauses.[3]

We often denote the application of a *substitution* σ by $t\sigma$ instead of $\sigma(t)$. A substitution σ is the *most general unifier* (*mgu*) of s and t iff $s\sigma = t\sigma$ and, whenever $s\gamma = t\gamma$ for some other unifier γ, there is a δ with $X\gamma = X\sigma\delta$ for all $X \in \mathcal{V}(s) \cup \mathcal{V}(t)$.[4] As usual, "$\sigma\delta$" is the composition of σ and δ, where $X\sigma\delta = (X\sigma)\delta$. If s and t have no *mgu* σ, we write $mgu(s, t) = fail$.

A Prolog program without built-in predicates is called a *definite* logic program. Our aim is to define a *linear* operational semantics where each state contains all information needed for backtracking steps. In addition, a state also contains a list of all *answer substitutions* that were found up to now. So a state has the form $\langle G_1 \mid \ldots \mid G_n \; ; \; \delta_1 \mid \ldots \mid \delta_m \rangle$ where $G_1 \mid \ldots \mid G_n$ is a sequence of goals and $\delta_1 \mid \ldots \mid \delta_m$ is a sequence of answer substitutions. We do not include the clauses from $\mathcal{P}$ in the state since they remain static during the evaluation.

---

[2] In ISO Prolog, whenever an empty query $\Box$ is reached, this is replaced by the built-in predicate true. However, we also allow empty queries to ease the presentation.

[3] More precisely, $\mathcal{P}$ are just the program clauses for *static* predicates. In addition to $\mathcal{P}$, a Prolog program may also contain clauses for *dynamic* predicates and *directives* to specify which predicates are dynamic. As explained in Sect. 5, these directives and the clauses for dynamic predicates are treated separately by our semantics.

[4] While the ISO standard uses unification with occurs check, our semantics could also be defined in an analogous way when using unification without occurs check.

$$\frac{\square_\delta \mid S \; ; \; A}{S \; ; \; A \mid \delta} \; (\text{Success}) \qquad \frac{(t,Q)_\delta \mid S \; ; \; A}{(t,Q)_\delta^{c_1} \mid \cdots \mid (t,Q)_\delta^{c_a} \mid S \; ; \; A} \; (\text{Case}) \quad \begin{array}{l} \text{if } \mathit{defined}_\mathcal{P}(t) \text{ and} \\ \mathit{Slice}_\mathcal{P}(t) = \\ (c_1, \ldots, c_a) \end{array}$$

$$\frac{(t,Q)_\delta^{h \, :- \, B} \mid S \; ; \; A}{(B\sigma, Q\sigma)_{\delta\sigma} \mid S \; ; \; A} \; (\text{Eval}) \quad \begin{array}{l} \text{if} \\ \sigma = \\ mgu(t,h) \end{array} \qquad \frac{(t,Q)_\delta^{h \, :- \, B} \mid S \; ; \; A}{S \; ; \; A} \; (\text{Backtrack}) \quad \begin{array}{l} \text{if} \\ mgu(t,h) = \\ \mathit{fail}. \end{array}$$

Fig. 1. Inference Rules for Definite Logic Programs

Essentially, a *goal* is just a *query*, i.e., a sequence of terms. However, to compute answer substitutions, a goal G is labeled by a substitution which collects the effects of the unifiers that were used during the evaluation up to now. So if $(t_1, \ldots, t_k)$ is a query, then a goal has the form $(t_1, \ldots, t_k)_\delta$ for a substitution δ. In addition, a goal can also be labeled by a clause c, where the goal $(t_1, \ldots, t_k)_\delta^c$ means that the next resolution step has to be performed using the clause c.

The *initial state* for a query $(t_1, \ldots, t_k)$ is $\langle (t_1, \ldots, t_k)_\varnothing \; ; \; \varepsilon \rangle$, i.e., the query is labeled by the identity substitution $\varnothing$ and the current list of answer substitutions is ε (i.e., it is empty). This initial state can be transformed by *inference rules* repeatedly. The inference rules needed for definite logic programs are given in Fig. 1. Here, Q is a query, S stands for a sequence of goals, A is a list of answer substitutions, and we omitted the delimiters "$\langle$" and "$\rangle$" for readability.

To illustrate these rules, we use the following program where $member(t_1, t_2)$ holds whenever t_1 unifies with any member of the list t_2. Consider the query $member(U, [1])$.[5] Then the corresponding initial state is $\langle member(U, [1])_\varnothing \; ; \; \varepsilon \rangle$.

$member(X, [X | \_])$. (1) $\qquad\qquad$ $member(X, [\_ | XS]) :- member(X, XS)$. (2)

When evaluating a goal $(t, Q)_\delta$ where $root(t) = p$, one tries all clauses $h :- B$ with $root(h) = p$ in the order they are given in the program. Let $\mathit{defined}_\mathcal{P}(t)$ indicate that $root(t)$ is a user-defined predicate and let $\mathit{Slice}_\mathcal{P}(t)$ be the list of all clauses from $\mathcal{P}$ whose head has the same root symbol as t. However, in the clauses returned by $\mathit{Slice}_\mathcal{P}(t)$, all occurring variables are renamed to fresh ones. Thus, if $\mathit{defined}_\mathcal{P}(t)$ and $\mathit{Slice}_\mathcal{P}(t) = (c_1, \ldots, c_a)$, then we use a (Case) rule which replaces the current goal $(t, Q)_\delta$ by the new list of goals $(t, Q)_\delta^{c_1} \mid \ldots \mid (t, Q)_\delta^{c_a}$. As mentioned, the label c_i in such a goal means that the next resolution step has to be performed using the clause c_i. So in our example, $member(U, [1])_\varnothing$ is replaced by the list $member(U, [1])_\varnothing^{(1)'} \mid member(U, [1])_\varnothing^{(2)'}$, where $(1)'$ and $(2)'$ are freshly renamed variants of the clauses (1) and (2).

To evaluate a goal $(t, Q)_\delta^{h \, :- \, B}$, one has to check whether there is a $\sigma = mgu(t, h)$. In this case, the (Eval) rule replaces t by B and σ is applied to the whole goal. Moreover, σ will contribute to the answer substitution, i.e., we replace δ by $\delta\sigma$. Otherwise, if t and h are not unifiable, then the goal $(t, Q)_\delta^{h \, :- \, B}$ is removed from the state and the next goal is tried (Backtrack). An empty goal $\square_\delta$ corresponds to a successful leaf in the SLD tree. Thus, the (Success) rule removes such an empty goal and adds the substitution δ to the list A of answer

---

[5] As usual, $[t_1, \ldots, t_n]$ abbreviates $.(t_1, .(\ldots, .(t_n, []) \ldots))$ and $[t \mid ts]$ stands for $.(t, ts)$

$$\text{member}(U,[1])_\varnothing \; ; \; \varepsilon$$

$$\text{CASE} \quad \frac{}{\text{member}(U,[1])_\varnothing^{(1)'} \mid \text{member}(U,[1])_\varnothing^{(2)'} \; ; \; \varepsilon}$$

$$\text{EVAL} \quad \frac{}{\square_{\{U/1,\,X'/1\}} \mid \text{member}(U,[1])_\varnothing^{(2)'} \; ; \; \varepsilon}$$

$$\text{SUCCESS} \quad \frac{}{\text{member}(U,[1])_\varnothing^{(2)'} \; ; \; \{U/1,\,X'/1\}}$$

$$\text{EVAL} \quad \frac{}{\text{member}(U,[])_{\{X'/U,\,XS'/[]\}} \; ; \; \{U/1,\,X'/1\}}$$

$$\text{CASE} \quad \frac{}{\text{member}(U,[])_{\{X'/U,\,XS'/[]\}}^{(1)''} \mid \text{member}(U,[])_{\{X'/U,\,XS'/[]\}}^{(2)''} \; ; \; \{U/1,\,X'/1\}}$$

$$\text{BACKTRACK} \quad \frac{}{\text{member}(U,[])_{\{X'/U,\,XS'/[]\}}^{(2)''} \; ; \; \{U/1,\,X'/1\}}$$

$$\text{BACKTRACK} \quad \frac{}{\varepsilon \; ; \; \{U/1,\,X'/1\}}$$

Fig. 2. Evaluation for the Query member$(U,[1])$

substitutions (we denote this by "$A \mid \delta$"). Fig. 2 shows the full evaluation of the initial state $\langle \text{member}(U,[1])_\varnothing \; ; \; \varepsilon \rangle$. Here, $(1)'$ and $(1)''$ (resp. $(2)'$ and $(2)''$) are fresh variants of (1) (resp. (2)) that are pairwise variable disjoint. So for example, X and XS were renamed to X' and XS' in $(2)'$.

3   Logic and Control

In Fig. 3, we present inference rules to handle some of the most commonly used pre-defined predicates of Prolog: the cut (!), negation-as-failure ($\backslash+$), the predicates call, true, and fail, and the Boolean connectives $Conn$ for conjunction $(',')$, disjunction $(';')$, and implication $('->')$.[6] As in the ISO standard, we require that in any *clause* $h :- B$, the term h and the terms in B may not contain variables at *predication positions*. A position is a *predication position* iff the only function symbols that may occur above it are the Boolean connectives from $Conn$. So instead of a clause q$(X) :- X$ one would have to use q$(X) :- $call$(X)$.

The effect of the cut is to remove certain backtracking possibilities. When a cut in a clause $h :- B_1, !, B_2$ with $root(h) = p$ is reached, then one does not backtrack to the remaining clauses of the predicate p. Moreover, the remaining backtracking possibilities for the terms in B_1 are also disregarded. As an example, we consider a modified member program.

member$(X,[X|\_]) :- \; !.$ (3)   member$(X,[_|XS]) :- $ member$(X,XS).$ (4)

In our semantics, the elimination of backtracking steps due to a cut is accomplished by removing goals from the state. Thus, we re-define the (CASE) rule in Fig. 3. To evaluate $p(\ldots)$, one again considers all program clauses $h :- B$ where $root(h) = p$. However, every cut in B is labeled by a fresh natural number m. For any clause c, let $c[!/!_m]$ result from c by replacing all (possibly labeled) cuts ! on *predication positions* by $!_m$. Moreover, we add a *scope delimiter* $?_m$ to make the end of their scope explicit. As the initial query Q might also

---

[6] The inference rules for true and the connectives from $Conn$ are straightforward and thus, we only present the rule for $','$ in Fig. 3. See [21] for the set of all rules.

$$\frac{(t, Q)_\delta \mid S \; ; \; A}{(t, Q)_\delta^{c_1[!/!m]} \mid \cdots \mid (t, Q)_\delta^{c_a[!/!m]} \mid ?_m \mid S \; ; \; A} \; \text{(CASE)} \quad \begin{array}{l} \text{if } \textit{defined}_P(t), \; \textit{Slice}_P(t) = \\ (c_1, \ldots, c_a), \text{ and } m \text{ is fresh} \end{array}$$

$$\frac{(!_m, Q)_\delta \mid S' \mid ?_m \mid S \; ; \; A}{Q_\delta \mid ?_m \mid S \; ; \; A} \; \text{(CUT)} \qquad\qquad \frac{(',' (t_1, t_2), Q)_\delta \mid S \; ; \; A}{(t_1, t_2, Q)_\delta \mid S \; ; \; A} \; \text{(CONJ)}$$

$$\frac{?_m \mid S \; ; \; A}{S \; ; \; A} \; \text{(FAILURE)} \qquad \frac{(\text{call}(t), Q)_\delta \mid S \; ; \; A}{(t[\mathcal{V}/\text{call}(\mathcal{V}), \, !/!_m], Q)_\delta \mid ?_m \mid S \; ; \; A} \; \text{(CALL)} \quad \begin{array}{l} \text{if } t \notin \mathcal{V} \\ \text{and } m \text{ is} \\ \text{fresh.} \end{array}$$

$$\frac{(\text{fail}, Q)_\delta \mid S \; ; \; A}{S \; ; \; A} \; \text{(FAIL)} \qquad \frac{(\backslash + (t), Q)_\delta \mid S \; ; \; A}{(\text{call}(t), !_m, \text{fail})_\delta \mid Q_\delta \mid ?_m \mid S \; ; \; A} \; \text{(NOT)} \quad \begin{array}{l} \text{where } m \text{ is} \\ \text{fresh.} \end{array}$$

Fig. 3. Inference Rules for Programs with Pre-defined Predicates for Logic and Control

contain cuts, we also label them and construct the corresponding initial state $\langle (Q \; [!/!_0])_\varnothing \mid ?_0 \; ; \; \varepsilon \rangle$.

In our example, consider the query $\text{member}(U, [1, 1])$. Its corresponding initial state is $\langle \text{member}(U, [1, 1])_\varnothing \mid ?_0 \; ; \; \varepsilon \rangle$. Now the (CASE) rule replaces the goal $\text{member}(U, [1, 1])_\varnothing$ by $\text{member}(U, [1, 1])_\varnothing^{(3)'[!/!_1]} \mid \text{member}(U, [1, 1])_\varnothing^{(4)'[!/!_1]} \mid ?_1$. Here, $(3)'$ is a fresh variant of the rule (3) and $(3)'[!/!_1]$ results from $(3)'$ by labeling all cuts with 1, i.e., $(3)'[!/!_1]$ is the rule $\text{member}(X', [X'|\_]) :- !_1$.

Whenever a cut $!_m$ is evaluated in the current goal, the (CUT) rule removes all backtracking goals up to the delimiter $?_m$ from the state. The delimiter itself must not be removed, since the current goal might still contain more occurrences of $!_m$. So after evaluating the goal $\text{member}(U, [1, 1])_\varnothing^{(3)'[!/!_1]}$ to $(!_1)_{\{U/1, \, X'/1\}}$, the (CUT) rule removes all remaining goals in the list up to $?_1$.

When a predicate has been evaluated completely (i.e., when $?_m$ becomes the current goal), then this delimiter is removed. This corresponds to a failure in the evaluation, since it only occurs when all solutions have been computed. Fig. 4 shows the full evaluation of the initial state $\langle \text{member}(U, [1, 1])_\varnothing \mid ?_0 \; ; \; \varepsilon \rangle$.

The built-in predicate call allows meta-programming. To evaluate a term $\text{call}(t)$ (where $t \notin \mathcal{V}$, but t may contain connectives from $\textit{Conn}$), the (CALL)

	$\text{member}(U, [1, 1])_\varnothing \mid ?_0 \; ; \; \varepsilon$
CASE	$\text{member}(U, [1, 1])_\varnothing^{(3)'[!/!_1]} \mid \text{member}(U, [1, 1])_\varnothing^{(4)'[!/!_1]} \mid ?_1 \mid ?_0 \; ; \; \varepsilon$
EVAL	$(!_1)_{\{U/1, \, X'/1\}} \mid \text{member}(U, [1, 1])_\varnothing^{(4)'[!/!_1]} \mid ?_1 \mid ?_0 \; ; \; \varepsilon$
CUT	$\square_{\{U/1, \, X'/1\}} \mid ?_1 \mid ?_0 \; ; \; \varepsilon$
SUCCESS	$?_1 \mid ?_0 \; ; \; \{U/1, \, X'/1\}$
FAILURE	$?_0 \; ; \; \{U/1, \, X'/1\}$
FAILURE	$\varepsilon \; ; \; \{U/1, \, X'/1\}$

Fig. 4. Evaluation for the Query $\text{member}(U, [1, 1])$

rule replaces $\mathsf{call}(t)$ by $t[\mathcal{V}/\mathsf{call}(\mathcal{V}), !/!_m]$. Here, $t[\mathcal{V}/\mathsf{call}(\mathcal{V}), !/!_m]$ results from t by replacing all variables X on predication positions by $\mathsf{call}(X)$ and all (possibly labeled) cuts on predication positions by $!_m$. Moreover, a delimiter $?_m$ is added to mark the scope of the cuts in t.

Another simple built-in predicate is fail, whose effect is to remove the current goal. By the cut, call, and fail, we can now also handle the "negation-as-failure" operator $\backslash+$: the (NOT) rule replaces the goal $(\backslash+(t), Q)_\delta$ by the list $(\mathsf{call}(t), !_m, \mathsf{fail})_\delta \mid Q_\delta \mid$ $?_m$. Thus, Q_δ is only executed if $\mathsf{call}(t)$ fails.

As an example, consider a program with the fact a and the rule $\mathsf{a} : - \, \mathsf{a}$. We regard the query $\backslash+(',' (\mathsf{a},!))$. The evaluation in Fig. 5 shows that the query terminates and fails (since we do not obtain any answer substitution).

	$\backslash+(',' (\mathsf{a},!))_\varnothing \mid ?_0 \; ; \; \varepsilon$
NOT	$(\mathsf{call}(',' (\mathsf{a},!)), !_1, \mathsf{fail})_\varnothing \mid ?_1 \mid ?_0 \; ; \; \varepsilon$
CALL	$(',' (\mathsf{a},!_2), !_1, \mathsf{fail})_\varnothing \mid ?_2 \mid ?_1 \mid ?_0 \; ; \; \varepsilon$
CONJ	$(\mathsf{a}, !_2, !_1, \mathsf{fail})_\varnothing \mid ?_2 \mid ?_1 \mid ?_0 \; ; \; \varepsilon$
CASE	$(\mathsf{a}, !_2, !_1, \mathsf{fail})^{\mathsf{a}}_\varnothing \mid (\mathsf{a}, !_2, !_1, \mathsf{fail})^{\mathsf{a}:-\mathsf{a}}_\varnothing \mid ?_2 \mid ?_1 \mid ?_0 \; ; \; \varepsilon$
EVAL	$(!_2, !_1, \mathsf{fail})_\varnothing \mid (\mathsf{a}, !_2, !_1, \mathsf{fail})^{\mathsf{a}:-\mathsf{a}}_\varnothing \mid ?_2 \mid ?_1 \mid ?_0 \; ; \; \varepsilon$
CUT	$(!_1, \mathsf{fail})_\varnothing \mid ?_2 \mid ?_1 \mid ?_0 \; ; \; \varepsilon$
CUT	$\mathsf{fail}_\varnothing \mid ?_1 \mid ?_0 \; ; \; \varepsilon$
FAIL	$?_1 \mid ?_0 \; ; \; \varepsilon$
FAILURE	$?_0 \; ; \; \varepsilon$
FAILURE	$\varepsilon \; ; \; \varepsilon$

Fig. 5. Evaluation for the Query $\backslash+(',' (\mathsf{a},!))$

4 "All Solution" Predicates

We now consider the unification predicate $=$ and the predicates $\mathsf{findall}$, bagof, and setof, which enumerate all solutions to a query. Fig. 6 gives the inference rules for $=$ and $\mathsf{findall}$ (bagof and setof can be modeled in a similar way, cf. [21]).

We extend our semantics in such a way that the collection process of such "all solution" predicates is performed just like ordinary evaluation steps of a program. Moreover, we modify our concept of *states* as little as possible.

A call of $\mathsf{findall}(r, t, s)$ executes the query $\mathsf{call}(t)$. If $\sigma_1, \ldots, \sigma_n$ are the resulting answer substitutions, then finally the list $[r\sigma_1, \ldots, r\sigma_n]$ is unified with s.

We model this behavior by replacing a goal $(\mathsf{findall}(r, t, s), Q)_\delta$ with the list $\mathsf{call}(t) \mid \%^{r, [], s}_{Q, \delta}$. Here, $\%^{r, \ell, s}_{Q, \delta}$ is a $\mathsf{findall}$-*suspension* which marks the "scope" of $\mathsf{findall}$-statements, similar to the markers $?_m$ for cuts in Sect. 3. The $\mathsf{findall}$-suspension fulfills two tasks: it collects all answer terms (r instantiated with an answer substitution of t) in its list ℓ and it contains all information needed to continue the execution of the program after all solutions have been found.

If a goal is evaluated to $\square_\theta$, its substitution θ would usually be added to the list of answer substitutions of the state. However, if the goals contain a $\mathsf{findall}$-suspension $\%^{r, \ell, s}_{Q, \delta}$, we instead insert $r\theta$ at the end of the list of answers ℓ using the

$$\frac{(\mathsf{findall}(r,t,s),Q)_\delta \mid S \; ; \; A}{\mathsf{call}(t)_\varnothing \mid \%_{Q,\delta}^{r,[],s} \mid S \; ; \; A} \; \text{(Findall)} \qquad \frac{\%_{Q,\delta}^{r,\ell,s} \mid S \; ; \; A}{(\ell{=}s,Q)_\delta \mid S \; ; \; A} \; \text{(FoundAll)}$$

$$\frac{\Box_\theta \mid S' \mid \%_{Q,\delta}^{r,\ell,s} \mid S \; ; \; A}{S' \mid \%_{Q,\delta}^{r,\ell\mid r\theta,s} \mid S \; ; \; A} \; \text{(FindNext)} \quad \begin{array}{l} \text{if } S' \text{ contains no} \\ \text{findall-suspensions} \end{array}$$

$$\frac{(t_1 = t_2, Q)_\delta \mid S \; ; \; A}{(Q\sigma)_{\delta\sigma} \mid S \; ; \; A} \; \text{(UnifySuccess)} \quad \text{if } \sigma = mgu(t_1, t_2)$$

$$\frac{(t_1 = t_2, Q)_\delta \mid S \; ; \; A}{S \; ; \; A} \; \text{(UnifyFail)} \; \begin{array}{l}\text{if}\\ mgu(t_1,t_2)=\\ fail\end{array} \qquad \frac{\Box_\delta \mid S \; ; \; A}{S \; ; \; A \mid \delta} \; \text{(Success)} \; \begin{array}{l}\text{if } S \text{ con-}\\ \text{tains no}\\ \text{findall-}\\ \text{suspensions}\end{array}$$

Fig. 6. Additional Inference Rules for Prolog Programs with findall

(FindNext) rule (denoted by "$\ell \mid r\theta$").[7] To avoid overlapping inference rules, we modify the (Success) rule such that it is only applicable if (FindNext) is not.

When $\mathsf{call}(t)$ has been fully evaluated, the first element of the list of goals is a findall-suspension $\%_{Q,\delta}^{r,\ell,s}$. Before continuing the evaluation of Q, we unify the list of collected solutions ℓ with the expected list s (using the built-in predicate $=$).

As an example, for the Prolog program defined by the clauses (3) and (4), an evaluation of the query $\mathsf{findall}(U, \mathsf{member}(U, [1]), L)$ is given in Fig. 7.

	$\mathsf{findall}(U, \mathsf{member}(U,[1]), L)_\varnothing \mid ?_0 \; ; \; \varepsilon$
Findall	$\mathsf{call}(\mathsf{member}(U,[1]))_\varnothing \mid \%_{\Box,\varnothing}^{U,[],L} \mid ?_0 \; ; \; \varepsilon$
Call	$\mathsf{member}(U,[1])_\varnothing \mid ?_1 \mid \%_{\Box,\varnothing}^{U,[],L} \mid ?_0 \; ; \; \varepsilon$
Case	$\mathsf{member}(U,[1])_\varnothing^{(3)'[!/!_2]} \mid \mathsf{member}(U,[1])_\varnothing^{(4)'[!/!_2]} \mid ?_2 \mid ?_1 \mid \%_{\Box,\varnothing}^{U,[],L} \mid ?_0 \; ; \; \varepsilon$
Eval	$(!_2)_{\{U/1,\, X'/1\}} \mid \mathsf{member}(U,[1])_\varnothing^{(4)'[!/!_2]} \mid ?_2 \mid ?_1 \mid \%_{\Box,\varnothing}^{U,[],L} \mid ?_0 \; ; \; \varepsilon$
Cut	$\Box_{\{U/1,\, X'/1\}} \mid ?_2 \mid ?_1 \mid \%_{\Box,\varnothing}^{U,[],L} \mid ?_0 \; ; \; \varepsilon$
FindNext	$?_2 \mid ?_1 \mid \%_{\Box,\varnothing}^{U,[1],L} \mid ?_0 \; ; \; \varepsilon$
Failure	$?_1 \mid \%_{\Box,\varnothing}^{U,[1],L} \mid ?_0 \; ; \; \varepsilon$
Failure	$\%_{\Box,\varnothing}^{U,[1],L} \mid ?_0 \; ; \; \varepsilon$
FoundAll	$([1]{=}L)_\varnothing \mid ?_0 \; ; \; \varepsilon$
UnifySuccess	$\Box_{\{L/[1]\}} \mid ?_0 \; ; \; \varepsilon$
Success	$?_0 \; ; \; \{L/[1]\}$
Failure	$\varepsilon \; ; \; \{L/[1]\}$

Fig. 7. Evaluation for the Query $\mathsf{findall}(U, \mathsf{member}(U, [1]), L)$

---

[7] As there may be nested findall calls, we use the first findall-suspension in the list.

5   Dynamic Predicates

Now we also consider built-in predicates which modify the program clauses for some predicate p at runtime. This is only possible for "new" predicates which were not defined in the program and for predicates where the program contains a dynamic directive before their first clause (e.g., ":− dynamic p/1"). Thus, we consider a program to consist of two parts: a static part $\mathcal{P}$ containing all program clauses for static predicates and a dynamic part, which can be modified at runtime and initially contains all program clauses for dynamic predicates.

Therefore, we extend our states by a list $\mathcal{D}$ which stores all clauses of dynamic predicates, where each of these clauses is labeled by a natural number. We now denote a state as $\langle S \; ; \; \mathcal{D} \; ; \; A \rangle$ where S is a list of goals and A is a list of answer substitutions. The inference rules for the built-in predicates asserta, assertz, and retract in Fig. 8 modify the list $\mathcal{D}$.[8] Of course, the (CASE) rule also needs to be adapted to take the clauses from $\mathcal{D}$ into account (here, "$\mathcal{P} \mid \overline{\mathcal{D}}$" stands for appending the lists $\mathcal{P}$ and $\overline{\mathcal{D}}$). All other previous inference rules do not depend on the new component $\mathcal{D}$ of the states.

$$\frac{(t,Q)_\delta \mid S \; ; \; \mathcal{D} \; ; \; A}{(t,Q)_\delta^{c_1[!/!m]} \mid \cdots \mid (t,Q)_\delta^{c_a[!/!m]} \mid ?_m \mid S \; ; \; \mathcal{D} \; ; \; A} \; (\text{CASE}) \quad \begin{array}{l} \text{if } \mathit{defined}_\mathcal{P}(t), \\ \mathit{Slice}_{(\mathcal{P}\mid\overline{\mathcal{D}})}(t) = (c_1,\dots,c_a), \\ \overline{\mathcal{D}} \text{ is } \mathcal{D} \text{ without clause labels}, \\ \text{and } m \text{ is fresh} \end{array}$$

$$\frac{(\text{asserta}(c),Q)_\delta \mid S \; ; \; \mathcal{D} \; ; \; A}{Q_\delta \mid S \; ; \; (c,m) \mid \mathcal{D} \; ; \; A} \; (\text{ASSA}) \; \begin{array}{l} \text{if } m \in \mathbb{N} \\ \text{is fresh} \end{array} \qquad \frac{(\text{assertz}(c),Q)_\delta \mid S \; ; \; \mathcal{D} \; ; \; A}{Q_\delta \mid S \; ; \; \mathcal{D} \mid (c,m) \; ; \; A} \; (\text{ASSZ}) \; \begin{array}{l} \text{if } m \in \mathbb{N} \\ \text{is fresh} \end{array}$$

$$\frac{(\text{retract}(c),Q)_\delta \mid S \; ; \; \mathcal{D} \; ; \; A}{\not\vdash_{Q,\delta}^{c,(c_1,m_1)} \mid \cdots \mid \not\vdash_{Q,\delta}^{c,(c_a,m_a)} \mid S \; ; \; \mathcal{D} \; ; \; A} \; (\text{RETRACT}) \quad \begin{array}{l} \text{if } \mathit{Slice}_\mathcal{D}(c) = \\ ((c_1,m_1),\dots,(c_a,m_a)) \end{array}$$

$$\frac{\not\vdash_{Q,\delta}^{c,(c',m)} \mid S \; ; \; \mathcal{D} \; ; \; A}{(Q\sigma)_{\delta\sigma} \mid S \; ; \; \mathcal{D} \setminus (c',m) \; ; \; A} \; (\text{RETSUC}) \; \begin{array}{l} \text{if } \sigma = \\ \mathit{mgu}(c,c') \end{array} \qquad \frac{\not\vdash_{Q,\delta}^{c,(c',m)} \mid S \; ; \; \mathcal{D} \; ; \; A}{S \; ; \; \mathcal{D} \; ; \; A} \; (\text{RETFAIL}) \; \begin{array}{l} \text{if} \\ \mathit{mgu}(c,c') \\ = \mathit{fail} \end{array}$$

Fig. 8. Additional Inference Rules for Prolog Programs with Dynamic Predicates

For a clause[9] c, the effect of asserta(c) resp. assertz(c) is modeled by inserting (c,m) at the beginning resp. the end of the list $\mathcal{D}$, where m is a fresh number, cf. the rules (ASSA) and (ASSZ). The labels in $\mathcal{D}$ are needed to uniquely identify each clause as demonstrated by the following query for a dynamic predicate p.

---

[8] The inference rules for the related predicate abolish are analogous, cf. [21].

[9] For asserta(c), assertz(c), and retract(c), we require that the body of the clause c may not be empty (i.e., instead of a fact p(X) one would have to use p(X) :− true). Moreover, c may not have variables on predication positions.

$$\underbrace{\mathsf{assertz(p(a))},\ \mathsf{assertz(p(b))},\ \mathsf{retract(p}(X)),\ X=\mathsf{a},\ \mathsf{retract(p(b))},\ \mathsf{assertz(p(b))},\ \mathsf{fail}}_{Q}$$

So first the two clauses $\mathsf{p(a)}$ and $\mathsf{p(b)}$ are asserted, i.e., $\mathcal{D}$ contains $(\mathsf{p(a)},1)$ and $(\mathsf{p(b)},2)$. When $\mathsf{retract(p}(X))$ is executed, one collects all p-clauses from $\mathcal{D}$, since these are the only clauses which might be removed by this $\mathsf{retract}$-statement.

To this end, we extend the function $Slice$ such that $Slice_{\mathcal{D}}(c)$ returns fresh variants of all labeled clauses c' from $\mathcal{D}$ where $root(head(c)) = root(head(c'))$. An execution of $(\mathsf{retract}(c),Q)_\delta$ then creates a new *retract marker* for every clause in $Slice_{\mathcal{D}}(c) = ((c_1,m_1),\ldots,(c_a,m_a))$, cf. the (RETRACT) inference rule in Fig. 8. Such a retract marker $:\!\!\nleftarrow_{Q,\delta}^{c,(c_i,m_i)}$ denotes that the clause with label m_i should be removed from $\mathcal{D}$ if c unifies with c_i by some mgu σ. Moreover, then the computation continues with the goal $(Q\sigma)_{\delta\sigma}$, cf. (RETSUC). If c does not unify with c_i, then the retract marker is simply dropped by the rule (RETFAIL).

So in our example, we create the two retract markers $:\!\!\nleftarrow_{Q,\varnothing}^{\mathsf{p}(X),(\mathsf{p(a)},1)}$ and $:\!\!\nleftarrow_{Q,\varnothing}^{\mathsf{p}(X),(\mathsf{p(b)},2)}$, where Q are the last four terms of the query. Since $\mathsf{p}(X)$ unifies with $\mathsf{p(a)}$, the first clause $(\mathsf{p(a)},1)$ is retracted from $\mathcal{D}$. Due to the unifier $\{X/\mathsf{a}\}$, the term $(X=\mathsf{a})[X/\mathsf{a}]$ is satisfied. Hence, $\mathsf{retract(p(b))}$ and $\mathsf{assertz(p(b))}$ are executed, i.e., the clause $(\mathsf{p(b)},2)$ is removed from $\mathcal{D}$ and a new clause $(\mathsf{p(b)},3)$ is added to $\mathcal{D}$. When backtracking due to the term fail at the end of the query, the execution of $\mathsf{retract(p}(X))$ is again successful, i.e., the retraction described by the marker $:\!\!\nleftarrow_{Q,\varnothing}^{\mathsf{p}(X),(\mathsf{p(b)},2)}$ succeeds since $\mathsf{p}(X)$ also unifies with the clause $(\mathsf{p(b)},2)$.

	$(\mathsf{assertz(p(a))},\mathsf{assertz(p(b))},\mathsf{retract(p}(X)),Q)_\varnothing \mid ?_0 ; \varepsilon ; \varepsilon$	
AssZ	$(\mathsf{assertz(p(b))},\mathsf{retract(p}(X)),Q)_\varnothing \mid ?_0 ; (\mathsf{p(a)},1) ; \varepsilon$	
AssZ	$(\mathsf{retract(p}(X)),Q)_\varnothing \mid ?_0 ; (\mathsf{p(a)},1) \mid (\mathsf{p(b)},2) ; \varepsilon$	
RETRACT	$:\!\!\nleftarrow_{Q,\varnothing}^{\mathsf{p}(X),(\mathsf{p(a)},1)} \mid :\!\!\nleftarrow_{Q,\varnothing}^{\mathsf{p}(X),(\mathsf{p(b)},2)} \mid ?_0 ; (\mathsf{p(a)},1) \mid (\mathsf{p(b)},2) ; \varepsilon$	
RETSUC	$(Q[X/\mathsf{a}])_{\{X/\mathsf{a}\}} \mid :\!\!\nleftarrow_{Q,\varnothing}^{\mathsf{p}(X),(\mathsf{p(b)},2)} \mid ?_0 ; (\mathsf{p(b)},2) ; \varepsilon$	
$\vdots$		
RETSUC	$(\mathsf{assertz(p(b))},\mathsf{fail})_{\{X/\mathsf{a}\}} \mid :\!\!\nleftarrow_{Q,\varnothing}^{\mathsf{p}(X),(\mathsf{p(b)},2)} \mid ?_0 ; \varepsilon ; \varepsilon$	
AssZ	$\mathsf{fail}_{\{X/\mathsf{a}\}} \mid :\!\!\nleftarrow_{Q,\varnothing}^{\mathsf{p}(X),(\mathsf{p(b)},2)} \mid ?_0 ; (\mathsf{p(b)},3) ; \varepsilon$	
FAIL	$:\!\!\nleftarrow_{Q,\varnothing}^{\mathsf{p}(X),(\mathsf{p(b)},2)} \mid ?_0 ; (\mathsf{p(b)},3) ; \varepsilon$	
RETSUC	$(Q[X/\mathsf{b}])_{\{X/\mathsf{b}\}} \mid ?_0 ; (\mathsf{p(b)},3) ; \varepsilon$	
$\vdots$		
FAILURE	$\varepsilon ; (\mathsf{p(b)},3) ; \varepsilon$	

Fig. 9. Evaluation for a Query using $\mathsf{assertz}$ and $\mathsf{retract}$

However, this retract-statement does not modify $\mathcal{D}$ anymore, since $(\mathsf{p(b)}, 2)$ is no longer contained in $\mathcal{D}$. Due to the unifier $\{X/\mathsf{b}\}$, the next term $(X = \mathsf{a})[X/\mathsf{b}]$ is not satisfiable and the whole query fails. However, then $\mathcal{D}$ still contains $(\mathsf{p(b)}, 3)$. Hence, afterwards a query like $\mathsf{p}(X)$ would yield the answer substitution $\{X/\mathsf{b}\}$. See Fig. 9 for the evaluation of this example using our inference rules.

6   Exception Handling

Prolog provides an *exception handling mechanism* by means of two built-in predicates throw and catch. The unary predicate throw is used to "throw" exception terms and the predicate catch can react on thrown exceptions.

When reaching a term $\mathsf{catch}(t, c, r)$, the term t is called. During this call, an exception term e might be thrown. If e and c unify with the mgu σ, the recover term r is instantiated by σ and called. Otherwise, the effect of the catch-call is the same as a call to $\mathsf{throw}(e)$. If no exception is thrown during the execution of $\mathsf{call}(t)$, the catch has no other effect than this call.

To model the behavior of catch and throw, we augment each goal in our states by context information for every catch-term that led to this goal. Such a catch-*context* is a 5-tuple (m, c, r, Q, δ), consisting of a natural number m which marks the scope of the corresponding catch-term, a catcher term c describing which exception terms to catch, a recover term r which is evaluated in case of a caught exception, as well as a query Q and a substitution δ describing the remainder of the goal after the catch-term. In general, we denote a list of catch-contexts by C and write $Q_{\delta, C}$ for a goal with the query Q and the annotations δ and C.

To evaluate $(\mathsf{catch}(t, c, r), Q)_{\delta, C}$, we append the catch-context (m, c, r, Q, δ) (where m is a fresh number) to C (denoted by "$C \mid (m, c, r, Q, \delta)$") and replace the catch-term by $\mathsf{call}(t)$, cf. (CATCH) in Fig. 10. To identify the part of the list of goals that is caused by the evaluation of this call, we add a scope marker $?_m$.

When a goal $(\mathsf{throw}(e), Q)_{\theta, C \mid (m, c, r, Q', \delta)}$ is reached, we drop all goals up to the marker $?_m$. If c unifies with a fresh variant e' of e using an mgu σ, we replace the current goal by the instantiated recover goal $(\mathsf{call}(r\sigma), Q'\sigma)_{\delta\sigma, C}$ using the rule (THROWSUCCESS). Otherwise, in the rule (THROWNEXT), we just drop the last catch-context and continue with the goal $(\mathsf{throw}(e), Q)_{\theta, C}$. If an exception is thrown without a catch-context, then this corresponds to a program error. To this end, we extend the set of states by an additional element ERROR.

Since we extended goals by a list of catch-contexts, we also need to adapt all previous inference rules slightly. Except for (SUCCESS) and (FINDNEXT), this is straightforward[10] since the previous rules neither use nor modify the catch-contexts. As catch-contexts can be converted into goals, findall-suspensions % and retract-markers :/ have to be annotated with lists of catch-contexts, too.

An interesting aspect is the interplay of nested catch- and findall-calls. When reaching a goal $\square_{\theta, C \mid (m, c, r, Q, \delta)}$ which results from the evaluation of a catch-term,

---

[10] However, several built-in predicates (e.g., call and findall) impose "error conditions". If their arguments do not have the required form, an exception is thrown. Thus, the rules for these predicates must also be extended appropriately, cf. [21].

$$\frac{(\text{catch}(t,c,r),Q)_{\delta,C} \mid S \; ; \; \mathcal{D} \; ; \; A}{\text{call}(t)_{\varnothing,\,C|(m,c,r,Q,\delta)} \mid ?_m \mid S \; ; \; \mathcal{D} \; ; \; A} \; (\text{CATCH}) \; \text{where } m \text{ is fresh}$$

$$\frac{(\text{throw}(e),Q)_{\theta,\,C|(m,c,r,Q',\delta)} \mid S' \mid ?_m \mid S \; ; \; \mathcal{D} \; ; \; A}{(\text{call}(r\sigma),Q'\sigma)_{\delta\sigma,\,C} \mid S \; ; \; \mathcal{D} \; ; \; A} \; (\text{THROWSUCCESS}) \; \begin{array}{l} \text{if } e \notin \mathcal{V} \text{ and } \sigma = \\ mgu(c,e') \text{ for a} \\ \text{fresh variant } e' \text{ of } e \end{array}$$

$$\frac{(\text{throw}(e),Q)_{\theta,\,C|(m,c,r,Q',\delta)} \mid S' \mid ?_m \mid S \; ; \; \mathcal{D} \; ; \; A}{(\text{throw}(e),Q)_{\theta,\,C} \mid S \; ; \; \mathcal{D} \; ; \; A} \; (\text{THROWNEXT}) \; \begin{array}{l} \text{if } e \notin \mathcal{V} \text{ and} \\ mgu(c,e') = fail \\ \text{for a fresh variant} \\ e' \text{ of } e \end{array}$$

$$\frac{(\text{throw}(e),Q)_{\theta,\varepsilon} \mid S \; ; \; \mathcal{D} \; ; \; A}{\text{ERROR}} \; (\text{THROWERR}) \; \text{if } e \notin \mathcal{V} \qquad \frac{\square_{\theta,\varepsilon} \mid S \; ; \; \mathcal{D} \; ; \; A}{S \; ; \; \mathcal{D} \; ; \; A|\theta} \; (\text{SUCCESS}) \; \begin{array}{l} \text{if } S \text{ contains} \\ \text{no findall-} \\ \text{suspensions} \end{array}$$

$$\frac{\square_{\theta,\,C|(m,c,r,Q,\delta)} \mid S' \mid ?_m \mid S \; ; \; \mathcal{D} \; ; \; A}{(Q\theta)_{\delta\theta,\,C} \mid S' \mid ?_m \mid S \; ; \; \mathcal{D} \; ; \; A} \; (\text{CATCHNEXT}) \; \begin{array}{l} \text{if } S' \text{ contains no} \\ \text{findall-suspensions} \end{array}$$

$$\frac{\square_{\theta,C} \mid S' \mid \%^{r,\ell,s}_{Q',\delta',C'} \mid S \; ; \; \mathcal{D} \; ; \; A}{S' \mid \%^{r,\ell|r\theta,s}_{Q',\delta',C'} \mid S \; ; \; \mathcal{D} \; ; \; A} \; (\text{FINDNEXT}) \; \begin{array}{l} \text{if } S' \text{ contains no findall-suspensions and} \\ (C \text{ is either empty or else its last element} \\ \text{is } (m,c,r,Q,\delta) \text{ and } S' \text{ contains no } ?_m) \end{array}$$

Fig. 10. Additional Inference Rules for Prolog Programs with Error Handling

	$\text{catch}(\text{catch}(\text{findall}(X,\mathsf{p}(X),L),\mathsf{a},\text{fail}),\mathsf{b},\text{true})_{\varnothing,\varepsilon} \mid ?_0$
CATCH	$\text{call}(\text{catch}(\text{findall}(X,\mathsf{p}(X),L),\mathsf{a},\text{fail})_{\varnothing,(1,\mathsf{b},\text{true},\square,\varnothing)} \mid ?_1 \mid ?_0$
CALL	$\text{catch}(\text{findall}(X,\mathsf{p}(X),L),\mathsf{a},\text{fail})_{\varnothing,(1,\mathsf{b},\text{true},\square,\varnothing)} \mid ?_2 \mid ?_1 \mid ?_0$
CATCH	$\text{call}(\text{findall}(X,\mathsf{p}(X),L))_{\varnothing,C} \mid ?_3 \mid ?_2 \mid ?_1 \mid ?_0$
CALL	$\text{findall}(X,\mathsf{p}(X),L)_{\varnothing,C} \mid ?_4 \mid ?_3 \mid ?_2 \mid ?_1 \mid ?_0$
FINDALL	$\text{call}(\mathsf{p}(X))_{\varnothing,C} \mid \%^{X,[],L}_{\square,\varnothing,C} \mid ?_4 \mid ?_3 \mid ?_2 \mid ?_1 \mid ?_0$
CALL	$\mathsf{p}(X)_{\varnothing,C} \mid \%^{X,[],L}_{\square,\varnothing,C} \mid ?_5 \mid ?_4 \mid ?_3 \mid ?_2 \mid ?_1 \mid ?_0$
CASE	$\mathsf{p}(X)^{\mathsf{p}(\mathsf{a})}_{\varnothing,C} \mid \mathsf{p}(X)^{\mathsf{p}(Y):-\text{throw}(\mathsf{b})}_{\varnothing,C} \mid ?_6 \mid \%^{X,[],L}_{\square,\varnothing,C} \mid ?_5 \mid ?_4 \mid ?_3 \mid ?_2 \mid ?_1 \mid ?_0$
EVAL	$\square_{\{X/\mathsf{a}\},C} \mid \mathsf{p}(X)^{\mathsf{p}(Y):-\text{throw}(\mathsf{b})}_{\varnothing,C} \mid ?_6 \mid \%^{X,[],L}_{\square,\varnothing,C} \mid ?_5 \mid ?_4 \mid ?_3 \mid ?_2 \mid ?_1 \mid ?_0$
FINDNEXT	$\mathsf{p}(X)^{\mathsf{p}(Y):-\text{throw}(\mathsf{b})}_{\varnothing,C} \mid ?_6 \mid \%^{X,[\mathsf{a}],L}_{\square,\varnothing,C} \mid ?_5 \mid ?_4 \mid ?_3 \mid ?_2 \mid ?_1 \mid ?_0$
EVAL	$\text{throw}(\mathsf{b})_{\{Y/X\},C} \mid ?_6 \mid \%^{X,[\mathsf{a}],L}_{\square,\varnothing,C} \mid ?_5 \mid ?_4 \mid ?_3 \mid ?_2 \mid ?_1 \mid ?_0$
THROWNEXT	$\text{throw}(\mathsf{b})_{\{Y/X\},(1,\mathsf{b},\text{true},\square,\varnothing)} \mid ?_2 \mid ?_1 \mid ?_0$
THROWSUCCESS	$\text{call}(\text{true})_{\{Y/X\},\varepsilon} \mid ?_0$
$\vdots$	

Fig. 11. Evaluation for a Query of Nested catch- and findall-Calls

it is not necessarily correct to continue the evaluation with the goal $(Q\theta)_{\delta\theta, C}$ as in the rule (CATCHNEXT). This is because the evaluation of the catch-term may have led to a findall-call and the current "success" goal $\square_{\theta, C|(m,c,r,Q,\delta)}$ resulted from this findall-call. Then one first has to compute the remaining solutions to this findall-call and one has to keep the catch-context (m, c, r, Q, δ) since these computations may still lead to exceptions that have to be caught by this context. Thus, then we only add the computed answer substitution θ to its corresponding findall-suspension, cf. the modified (FINDNEXT) rule.

For the program with the fact p(a) and the rule p(Y) :− throw(b), an evaluation of a query with catch and findall is given in Fig. 11. Here, the clauses $\mathcal{D}$ for dynamic predicates and the list A of answer substitutions were omitted for readability. Moreover, C stands for the list $(1, b, true, \square, \varnothing) \mid (3, a, fail, \square, \varnothing)$.

7   Equivalence to the ISO Semantics

In this section, we formally define our new semantics for Prolog and show that it is equivalent to the semantics defined in the ISO standard [11,13]. All definitions and theorems refer to the *full* set of inference rules (handling full Prolog). As mentioned, all inference rules and all proofs can be found in [21].

Theorem 1 ("Mutual Exclusion" of Inference Rules). *For each state, there is at most one inference rule applicable and the result of applying this rule is unique up to renaming of variables and of fresh numbers used for markers.*

Let $s_0 \rightsquigarrow s_1$ denote that the state s_0 was transformed to the state s_1 by one of our inference rules. Any finite or infinite sequence $s_0 \rightsquigarrow s_1 \rightsquigarrow s_2 \rightsquigarrow \ldots$ is called a *derivation* of s_0. Thm. 1 implies that any state has a unique maximal derivation (which may be infinite). Now we can define our semantics for Prolog.

Definition 2 (Linear Semantics for Prolog). *Consider a Prolog program with the clauses $\mathcal{P}$ for static predicates and the clauses $\overline{\mathcal{D}}$ for dynamic predicates. Let $\mathcal{D}$ result from $\overline{\mathcal{D}}$ by labeling each clause in $\overline{\mathcal{D}}$ by a fresh natural number. Let Q be a query and let $s_Q = \langle S_Q; \mathcal{D}; \varepsilon \rangle$ be the corresponding initial state, where $S_Q = (Q[!/!_0])_{\varnothing,\varepsilon} \mid ?_0$.*

(a) We say that the execution of Q has length $\ell \in \mathbb{N} \cup \{\infty\}$ iff the maximal derivation of s_Q has length ℓ. In particular, Q is called terminating iff $\ell \neq \infty$.

(b) We say that Q leads to a program error iff the maximal derivation of s_Q is finite and ends with the state ERROR.

(c) We say that Q leads to the (finite or infinite) list of answer substitutions A iff either the maximal derivation of s_Q is finite and ends with a state $\langle \varepsilon; \mathcal{D}'; A \rangle$, or the maximal derivation of s_Q is infinite and for every finite prefix A' of A, there exists some S and $\mathcal{D}'$ with $s_Q \rightsquigarrow^* \langle S; \mathcal{D}', A' \rangle$. As usual, $\rightsquigarrow^*$ denotes the transitive and reflexive closure of $\rightsquigarrow$.

In contrast to Def. 2, the ISO standard [11,13] defines the semantics of Prolog using search trees. These search trees are constructed by a depth-first search from left to right, where of course one avoids the construction of parts of the tree that are not needed (e.g., because of cuts). In the ISO semantics, we have the following for a Prolog program $\mathcal{P}$ and a query Q:[11]

(a) The execution of Q has *length* $k \in \mathbb{N} \cup \{\infty\}$ iff k unifications are needed to construct the search tree (where the execution of a built-in predicate also counts as at least one unification step).[12] Of course, here every unification attempt is counted, no matter whether it succeeds or not. So in the program with the fact p(a), the execution of the query p(b) has length 1, since there is one (failing) unification attempt.

(b) Q leads to a *program error* iff during the construction of the search tree one reaches a goal $(\mathsf{throw}(e), Q)$ and the thrown exception is not caught.

(c) Q leads to the list of *answer substitutions* A iff Q does not lead to a program error and A is the list of answer substitutions obtained when traversing the (possibly infinite) search tree by depth-first search from left to right.

Thm. 3 (a) shows that our semantics and the ISO semantics result in the same termination behavior. Moreover, the computations according to the ISO semantics and our maximal derivations have the same length up to a constant factor. Thus, our semantics can be used for termination and complexity analysis of Prolog. Thm. 3 (b) states that our semantics and the ISO semantics lead to the same program errors and in (c), we show that the two semantics compute the same answer substitutions (up to variable renaming).[13]

Theorem 3 (Equivalence of Our Semantics and the ISO Semantics).
Consider a a Prolog program and a query Q.

(a) *Let ℓ be the length of Q's execution according to our semantics in Def. 2 and let k be the length of Q's execution according to the ISO semantics. Then we have $k \leq \ell \leq 3 \cdot k + 1$. So in particular we also obtain $\ell = \infty$ iff $k = \infty$ (i.e., the two semantics have the same termination behavior).*

(b) *Q leads to a program error according to our semantics in Def. 2 iff Q leads to a program error according to the ISO semantics.*

(c) *Q leads to a (finite or infinite) list of answer substitutions $\delta_0, \delta_1, \ldots$ according to our semantics in Def. 2 iff Q leads to a list of answer substitutions $\theta_0, \theta_1, \ldots$ according to the ISO semantics, where the two lists have the same length $n \in \mathbb{N} \cup \{\infty\}$ and for each $i < n$, there exists a variable renaming τ_i such that for all variables X in the query Q, we have $X\theta_i = X\delta_i \tau_i$.*

---

[11] See [21] for a more formal definition.

[12] In other words, even for built-in predicates p, the evaluation of an atom $p(t_1, \ldots, t_n)$ counts as at least one unification step. For example, this is needed to ensure that the execution of queries like "repeat, fail" has length ∞.

[13] Moreover, the semantics are also equivalent w.r.t. the side effects of a program (like the changes of the dynamic clauses, input and output, etc.).

To see why we do not have $\ell = k$ in Thm. 3(a), consider again the program with the fact $p(a)$ and the query $p(b)$. While the ISO semantics only needs $k = 1$ unification attempt, our semantics uses 3 steps to model the failure of this proof. Moreover, in the end we need one additional step to remove the marker $?_0$ constructed in the initial state. The evaluation is shown in Fig. 12, where we omitted the catch-contexts and the components for dynamic predicates and answer substitutions for readability. So in this example, we have $\ell = 3 \cdot k + 1 = 4$.

$$
\begin{array}{rc}
 & (p(b))_\varnothing \mid ?_0 \\ \hline
\text{CASE} & (p(b))_\varnothing^{p(a)} \mid ?_1 \mid ?_0 \\ \hline
\text{BACKTRACK} & ?_1 \mid ?_0 \\ \hline
\text{FAILURE} & ?_0 \\ \hline
\text{FAILURE} & \varepsilon
\end{array}
$$

Fig. 12. Evaluation for $p(b)$

8   Conclusion

We have presented a new operational semantics for full Prolog (as defined in the corresponding ISO standard [11,13]) including the cut, "all solution" predicates like findall, dynamic predicates, and exception handling. Our semantics is *modular* (i.e., easy to adapt to subsets of Prolog) and *linear* resp. *local* (i.e., derivations are lists instead of trees and even the cut and exceptions are local operations where the next state in a derivation only depends on the previous state).

We have proved that our semantics is equivalent to the semantics based on search trees defined in the ISO standard w.r.t. both termination behavior and computed answer substitutions. Furthermore, the number of derivation steps in our semantics is equal to the number of unifications needed for the ISO semantics (up to a constant factor). Hence, our semantics is suitable for (possibly automated) analysis of Prolog programs, for example for static analysis of termination and complexity using an abstraction of the states in our semantics as in [19,20].

In [19,20], we already successfully used a subset of our new semantics for automated termination analysis of definite logic programs with cuts. In future work, we will extend termination analysis to deal with all our inference rules in order to handle full Prolog as well as to use the new semantics for asymptotic worst-case complexity analysis. We further plan to investigate uses of our semantics for debugging and tracing applications exploiting linearity and locality.

References

1. Apt, K.R.: From Logic Programming to Prolog. Prentice Hall (1997)
2. Arbab, B., Berry, D.M.: Operational and denotational semantics of Prolog. Journal of Logic Programming 4, 309–329 (1987)
3. Börger, E., Rosenzweig, D.: A mathematical definition of full Prolog. Science of Computer Programming 24, 249–286 (1995)
4. Cerrito, S.: A linear semantics for allowed logic programs. In: LICS 1990, pp. 219–227. IEEE Press (1990)

5. Cheng, M.H.M., Horspool, R.N., Levy, M.R., van Emden, M.H.: Compositional operational semantics for Prolog programs. New Generat. Comp. 10, 315–328 (1992)
6. de Bruin, A., de Vink, E.P.: Continuation Semantics for Prolog with Cut. In: Díaz, J., Yu, Y. (eds.) CAAP 1989 and TAPSOFT 1989. LNCS, vol. 351, pp. 178–192. Springer, Heidelberg (1989)
7. de Vink, E.P.: Comparative semantics for Prolog with cut. Science of Computer Programming 13, 237–264 (1990)
8. Debray, S.K., Mishra, P.: Denotational and operational semantics for Prolog. Journal of Logic Programming 5(1), 61–91 (1988)
9. Debray, S.K., Lin, N.-W.: Cost analysis of logic programs. ACM Transactions on Programming Languages and Systems 15, 826–875 (1993)
10. Deransart, P., Ferrand, G.: An operational formal definition of Prolog: a specification method and its application. New Generation Computing 10, 121–171 (1992)
11. Deransart, P., Ed-Dbali, A., Cervoni, L.: Prolog: The Standard. Springer (1996)
12. Hirokawa, N., Moser, G.: Automated Complexity Analysis Based on the Dependency Pair Method. In: Armando, A., Baumgartner, P., Dowek, G. (eds.) IJCAR 2008. LNCS (LNAI), vol. 5195, pp. 364–379. Springer, Heidelberg (2008)
13. ISO/IEC 13211-1. Information technology - Programming languages - Prolog (1995)
14. Jeavons, J.: An alternative linear semantics for allowed logic programs. Annals of Pure and Applied Logic 84(1), 3–16 (1997)
15. Jones, N.D., Mycroft, A.: Stepwise development of operational and denotational semantics for Prolog. In: SLP 1984, pp. 281–288. IEEE Press (1984)
16. Kulaš, M., Beierle, C.: Defining standard prolog in rewriting logic. In: WRLA 2000. ENTCS, vol. 36 (2001)
17. Nicholson, T., Foo, N.: A denotational semantics for Prolog. ACM Transactions on Programming Languages and Systems 11, 650–665 (1989)
18. Noschinski, L., Emmes, F., Giesl, J.: A Dependency Pair Framework for Innermost Complexity Analysis of Term Rewrite Systems. In: Bjørner, N., Sofronie-Stokkermans, V. (eds.) CADE 2011. LNCS(LNAI), vol. 6803, pp. 422–438. Springer, Heidelberg (2011)
19. Schneider-Kamp, P., Giesl, J., Ströder, T., Serebrenik, A., Thiemann, R.: Automated termination analysis for logic programs with cut. In: ICLP 2010, Theory and Practice of Logic Programming, vol. 10(4-6), pp. 365–381 (2010)
20. Ströder, T., Schneider-Kamp, P., Giesl, J.: Dependency Triples for Improving Termination Analysis of Logic Programs with Cut. In: Alpuente, M. (ed.) LOPSTR 2010. LNCS, vol. 6564, pp. 184–199. Springer, Heidelberg (2011)
21. Ströder, T., Emmes, F., Schneider-Kamp, P., Giesl, J., Fuhs, C.: Fuhs. A linear operational semantics for termination and complexity analysis of ISO Prolog. Technical Report AIB-2011-08, RWTH Aachen (2011), http://aib.informatik.rwth-aachen.de/
22. Zankl, H., Korp, M.: Modular complexity analysis via relative complexity. In: RTA 2010. LIPIcs, vol. 6, pp. 385–400 (2010)

Author Index